The Springer Textbooks series publishes a broad portfolio of textbooks on Earth Sciences, Geography and Environmental Science. Springer textbooks provide comprehensive introductions as well as in-depth knowledge for advanced studies. A clear, reader-friendly layout and features such as end-of-chapter summaries, work examples, exercises, and glossaries help the reader to access the subject. Springer textbooks are essential for students, researchers and applied scientists.

More information about this series at https://link.springer.com/bookseries/15201

T0213856

**Springer Textbooks in Earth Sciences,
Geography and Environment**

Celia Marcos

Crystallography

Introduction to the Study of Minerals

 Springer

Celia Marcos
Área de Cristalografía y Mineralogía
Departamento de Geología
Facultad de Geología
Universidad de Oviedo
Oviedo, Spain

ISSN 2510-1307   ISSN 2510-1315   (electronic)
Springer Textbooks in Earth Sciences, Geography and Environment
ISBN 978-3-030-96785-7   ISBN 978-3-030-96783-3   (eBook)
https://doi.org/10.1007/978-3-030-96783-3

This Springer imprint is published by the registered company Springer Nature Switzerland AG
The registered company address is: Gewerbestrasse 11, 6330 Cham, Switzerland

Preface

This book is the result of my dedication to teaching crystallography to geology students.

This book is intended for those who want to enter the world of crystals and minerals. An attempt has been made to do so in a clear and concise manner.

The text is structured into three blocks: Geometric Crystallography, Crystallochemistry, and Crystallophysics. The theoretical content of each block is divided into topics, and these are presented in chapters or sections.

The Geometric Crystallography block is structured into five subjects. It studies the external morphology of crystals and their symmetry, very important for the study of minerals. It also studies the geometry and symmetry of lattices, a concept that allows studying crystalline matter from a mathematical point of view, abstracting it from its material content, essential in the understanding of this science and that of minerals. The knowledge of geometric crystallography is also essential for the identification of gems. The methods and techniques of obtaining synthetic gems are increasingly sophisticated and make their distinction more difficult. The characteristics of symmetry and growth planes are, in the case of diamond and corundum, for example, the best way to recognize their natural or synthetic origin. This section is also essential for researchers who are engaged in the structural resolution of organic and inorganic compounds. The objective of this block is the acquisition of basic knowledge about geometric crystallography. This knowledge includes the definition of crystal and mineral; properties of crystals, including periodicity, order, and translation; knowledge of crystals from a geometric point of view, including lattice, lattice elements, lattice row symbols and crystalline plane notations, symmetry, spatial groups, and point groups; and the proper use of the language of geometric crystallography.

The crystallochemistry block is structured into four themes. It studies the arrangement of atoms in a crystalline matter, that is, its structure. Subsequently, the concept of real crystal, with all its imperfections, is introduced. The objective is the acquisition of basic knowledge about crystallochemistry. This knowledge includes the definition of crystal structure, bond, types of crystals, package, coordination, position of atoms, basic structural types, the difference between ideal crystal and real crystal, and concepts of order such as disorder, defects, isomorphism, solid solution, stability and equilibrium, polymorphism, and politypism.

In the crystallophysics block, with four subjects, the physical properties of crystals and minerals are studied in more detail, including those relating to the interaction of electromagnetic waves with crystals and minerals, mainly concerning visible light and X-rays. The objective is the acquisition of basic knowledge about crystallophysics, including the definition of physical property, directional and non-directional properties, properties in crystals and minerals, representation surfaces, the interaction of electromagnetic radiation with matter, identification of minerals with a polarizing microscope, and X-ray diffraction.

Oviedo, Spain   Celia Marcos

Acknowledgements

To my parents, because they helped me with my education. Thanks to them, I have been able to study and work on subjects like crystallography.

To my teachers of Crystallography, Joaquín Solans Huguet and Dámaso Moreiras Blanco, who transmitted me a solid base of this science.

I also want to thank my students. Their feedback and questions allowed me to develop the content clearly and concisely.

I thank those of you who, not belonging to the Chrstallography community, have contacted me to solve crystallographic questions allowing me to consider different points of view.

I would especially like to thank Prof. Dr. Andreas Danilewsky for his thorough revision of this book, improving content, thanks to his comprehensive knowledge of crystals and minerals.

To Marilyn Pomeroy for editing and proofreading.

I am very grateful to Dr. *Alexis Vizcaino*, Springer editor, for his interest in bringing this book to light and for his thoughtful manner and kindness.

Finally, I would like to thank Springer for publishing this book.

Contents

Part I
Geometric Crystallography

In this part, Chaps. 1 to 5, the crystal is studied as an ideal entity, from a geometrical point of view through the concept of lattice. The geometry and symmetry of the lattices and the external morphology of the crystals are considered. From a macroscopic point of view, the crystal is considered a homogeneous and continuous, anisotropic and symmetrical medium. When the internal symmetry is studied, the crystal is considered a homogeneous and discrete medium, as well as anisotropic and symmetrical.

In this block, it is important to understand the importance of some features. First, it is important to abstract in crystallography. Considering a reference system (e.g., crystallographic coordinate system), the following two situations are equivalent: (1) move (translate) the reference system and leave the crystal static or (2) move (translate) the crystal and leave the coordinate system static. The crystal can be described from a geometrical point of view, through the lattice, not only in an unlimited way, but also in a limited way, with the concept of cell, which is the smallest space of the lattice limited by the fundamental translations, which are the smallest distances at which the crystal motif (ion, atom, molecule or groups of them) is repeated in the three-dimensional space.

Second, it is important to understand the orderly and periodic arrangement of the motif that is repeated in the crystal properties: geometric (shown in this block), chemical (shown in the crystallochemistry block), and physical (shown in the crystallophysics block). In relation to the geometric properties, it is important to understand that the external aspect of a crystal, its morphology, is a consequence of the orderly and periodic arrangement of its atomic constituents in the three-dimensional space. In this way, a parallel can be made between the external symmetry, the non-translational symmetry of the punctual groups, and the symmetry that relates the atoms; that is, the spatial symmetry that includes the translations. Thus, it is also possible to understand the parallelism between the position of a crystalline face and that of an atom of that crystal, in relation to some crystallographic coordinates. The position of a crystal face is given by Miller indices. It can be in a general position if it is only located on a monary axis of rotation (360°

rotation around the axis), or a particular (special) position if it is located on symmetry elements of a higher order than the axis of order 1 (monary). The position of an atom of that crystal is given by its coordinates in relation to those crystallographic coordinates and can also occupy a general or particular position.

Chapter 1
Introduction to Crystallography

Abstract This chapter defines the concepts of crystal and mineral, emphasizing their difference from non-crystalline materials. The most important periods, throughout history, in the development of crystallography and mineralogy and their beginnings as science are presented.

1.1 Introduction

It is known that most of the earth is solid and is formed by minerals that are crystalline; that is to say, they are constituted by elements (ions, atoms or molecules) that repeat periodically in an orderly way, in the space of three dimensions.

Minerals can be formed in various ways and in various environments. In any case, this is the result of chemical and physical processes that are verified in all geological eras and that continue to manifest themselves. There are three fundamental processes of formation: magmatic, metamorphic, and sedimentary. They are produced in a variety of ways: through solidification from a mass of melted material, consolidated as quartz, feldspar or olivine; through precipitation in a magma, like gold; hydrothermally, like fluorite, mica or barite; sublimation, from gas like halite or cinnabar; through chemical alteration from contact chemical reactions in areas of different composition, such as clays; through physical alteration from solid-state reactions with pressure or temperature increase on existing minerals, such as diamond; and biogenically from living beings such as calcite and apatite. In addition, minerals can change under certain circumstances.

Minerals are the components of rocks and ores that are distinguished by their chemical composition and physical properties.

From a genetic point of view, minerals are natural chemical combinations; that is, natural products resulting from the different physical–chemical processes that act on the earth's crust.

C. Marcos, *Crystallography*, Springer Textbooks in Earth Sciences,
Geography and Environment, https://doi.org/10.1007/978-3-030-96783-3_1

Most of these products are found in the form of minerals in solid state, possessing certain chemical and physical properties in close mutual relationship with their crystalline structure, and they are stable in certain pressure and temperature ranges. Most minerals are in a crystalline state.

1.2 Crystallography and Mineralogy

Crystallography

Crystallography is a science that deals with the study of crystalline matter, the laws that govern its formation and its geometric, chemical, and physical properties.

This science is classified into the following:

- Geometric crystallography studies the external morphology of crystals and their symmetry, and the geometry and symmetry of the lattice.
  From a macroscopic point of view, crystalline matter can be considered a homogeneous and continuous medium, anisotropic, and symmetrical. When studying internal symmetry, crystalline matter must be considered a homogeneous and discrete medium, anisotropic, and symmetrical.
- Chemical crystallography or crystallochemistry studies the arrangement of atoms; that is, their crystalline structure. The concept of the real crystal must be introduced, and its imperfections must be considered, contrary to geometric crystallography.
- Physical crystallography or crystallophysics studies the physical properties of crystals, trying to relate them to the chemical composition and structure.

In this section, it is important to consider properties derived from the interaction of X radiation with matter, since they allow us to know the arrangement of atoms in the structure, identify crystalline phases, etc.

Mineralogy

Mineralogy is the science of minerals and studies, in close relationship to each other, their chemical composition, crystal structure, physical properties, and conditions of their genesis, as well as their practical importance.

Mineralogy can be divided into the following:

- Chemical mineralogy studies the chemical properties of minerals.
- Physical mineralogy studies the physical properties of minerals, as well as the mechanical properties such as optical, electrical, magnetic, etc.
- Determinative mineralogy describes the various techniques for identifying and determining minerals.
- Descriptive mineralogy describes the crystallographic, chemical, and physical properties, as well as the associations and deposits of minerals.

– Mineralogenesis is the study of the genesis of minerals and provides data for prospection and valuation of mineral deposits.
– Applied mineralogy describes the applications of minerals in industry, prospecting and exploration, etc.

1.3 Historical Background

Mineralogy began as an applied science, dedicated to the use of mineral deposits useful to man. Along with the study of its usefulness, the descriptive aspect of the new minerals that were discovered was developed from the earliest times. This is how the first works dealing with minerals are presented. These include the texts of Aristotle,[1] Teofrasto,[2] Avicenna,[3] and Alberto Magno,[4]

During the Renaissance, works that dealt with minerals did so from the metallurgical point of view and their industrial use, such as the work *De Re Metallic* by Agricola[5] (1530) and *Pirotechnia* by Birunguccio[6] (1535).

In the first half of the eighteenth century, minerals were studied as simple chemical compounds of natural origin. In this sense, the works of Cronstedt (1758)[7] represent great progress.

The laws of Rome de l'Isle and Haüy on the characteristics of the crystalline material made it possible to improve methods for mineralogical determination.

Classical determinations are based on the most evident and observable physical properties, without the need for complicated apparatus; however, the use of the polarizing microscope allowed a great advance in mineral determination techniques.

Determination of chemical composition is very important in all mineralogical studies but, by itself, it is insufficient for identifying the different minerals since, in many of them, certain cations are interchangeable (micas, chlorites, zeolites, garnets, etc.) or different minerals correspond to compounds of identical chemical composition (diamond and graphite, calcite and aragonite, etc.).

Around 1735, Linné[8] divided crystals into classes, depending on their external forms. Wooden models were built and drawings of crystals were published. Around 1780, a great step forward was taken in crystallographic science when A. Carangeot built an instrument, a contact goniometer, to measure the angles between the faces of crystals. Stensen (1669) presented the constancy of the dihedral angles of the

---

[1] Aristotle (315 before JC), Book of the stones.

[2] Teofrasto (77 before JC), Naturalis historia.

[3] Avicenna, Treatise on the stones (in which a mineral classification is outlined).

[4] Alberto Magno (1262), In De Mineralibus et Rebus Metallius.

[5] Agricola (1530), De Mineralibus et Rebus Metallius.

[6] Birunguccio (1535), Pirotechnia.

[7] Cronstedt [1].

[8] Linné (1707–1778), Nature system.

faces of quartz crystals, on the basis of Carangeot's measurements, and later Rome de l'Isle[9] (1783) generalized the discoveries of Stensen. Haüy (1784)[10] showed that crystals are formed by stacking identical, very small blocks that he called integral molecules. This concept would be equivalent to the unit cell concept in modern crystallography. Later, Haüy (1801)[11] developed the theory of rational indices for crystal faces.

Around 1800, there was a gradual change in the attitude toward crystals and minerals,[12] as interest in collecting gave way to scientists investigating them and exhibiting them in museums. At the same time, the first scientific societies and journals dedicated to minerals were born.

Chemists like M.H. Klaproth and C.F. Bucholz in Germany, L.N. Vauquelin in France, and J.J. Berzelius in Sweden had perfected methods for analyzing minerals and rocks, which led to the discovery of many new elements. The discovery of the elements and the possibilities of chemical analysis gave rise to one of the great controversies in the world of crystallography—what affects the polymorphism of calcium carbonate. Another problem was that of isomorphism. The configuration of the processes that explain these phenomena has been of great importance in crystallography and mineralogy.

Important advances in the domain of crystallography are due to Russian science, specifically Federov[13] (1890). Another important contribution of Federov[14] (1891) to science relates to the microscopic study of minerals. With the invention of a polarization device by Nicol[15] in 1828, the usefulness of optical microscopy for the identification of minerals was greatly enhanced.

At the end of the nineteenth century, Fedorov[16] and Schoenflies[17] worked almost simultaneously, but independently, on the order and internal symmetry within crystals.

The discoveries of the physicist Laue,[18] in 1912, regarding the diffraction of X-rays when passing through a crystal, and subsequent investigations in this field carried out by the Russian physicist Wulff,[19] the Braggs[20] (father and son), Pauling, and others made it possible to clearly verify the close relationship between the crystalline structure of minerals, their chemical composition, and physical

---

[9] de Romé Delisle [2].

[10] Haüy [3].

[11] Haüy [4].

[12] Wilson [5].

[13] Federov (1890), Symmetry of the regular systems of figures.

[14] Fedorow [6].

[15] Nicol [7].

[16] Fedorov [8].

[17] Schoenflies, A. (1891) Krystallsysteme und Krystallstructur.

[18] Eckert [9].

[19] Wulff [10].

[20] Bragg [11, 12].

properties. Thanks to these advances, crystallochemistry was born, a science that studies the laws of the spatial arrangement of atoms or ions in crystals and the relationship between the crystalline structure of minerals and their chemical and physical properties.

Theory of the phases and of the equilibria of the physical–chemical systems in the field of physical chemistry is also very important. In this field, much is due to Gibbs,[21] author of the theory of phases.

1.4 Crystalline State

Crystalline state is the state of thermodynamic equilibrium of a solid that, under certain thermodynamic conditions (pressure—P and temperature—T) and with a determined composition, corresponds to a determined crystalline structure.

The principal property of solids in a crystalline state is periodicity, from which other macroscopic characteristics are derived: homogeneity, anisotropy, and symmetry.

– Homogeneity

Macroscopically, it means invariance of a property F measured at one point x, relative to its measure at another point x + x′; that is,

$$F(x) = F(x + x') \tag{1.1}$$

From this condition at macroscopic level, the constancy of the chemical composition and phase state through the entire volume of the substance in the crystalline state is obtained.

The concept of homogeneity makes it possible to consider a crystalline substance as a continuous medium.

This concept is very important in crystallography since phenomenological descriptions of many physical properties of crystals can be given without reference to their discrete atomic structure. From a macroscopic point of view, the distances in the crystal are considerably larger than the interplanar spacing (distance between two parallel crystalline planes) of the crystal at the microscopic level and with volumes that far exceed that of the unit cell (space limited by the smallest distances at which the motif is repeated in all three dimensions of space).

– Anisotropy

There are certain properties of crystals that are independent of the direction in which they are measured; they are said to be scalar properties, such as specific weight, heat capacity, etc.

---

[21] Gibbs [13].

There are other properties that depend on the direction in which they are measured; some are said to be vector properties and others are tensorial, such as thermal conductivity, dielectric constant, refractive index, etc.

If the description of a property is independent of any orientation, the substance is said to be *isotropic* for that property.

If a property is orientation-dependent, the substance is said to be *anisotropic* for that property.

In any case, a crystalline substance will always be anisotropic for some properties, such as the different arrangement of atoms in different directions (structural anisotropy).

– Symmetry

Symmetry is the property that makes an object not distinguishable from its original position after transforming it.

Taking these characteristics into account at a macroscopic level, a crystalline substance can be defined as a material that is homogeneous, continuous, anisotropic, and symmetrical.

However, as will be seen later, a crystalline substance is not a static entity: the atoms vibrate and, when the temperature increases, they vibrate more. This affects their physical properties.

Crystals show local defects and variations in composition and also deviation of the structure ideal.

These imperfections are not considered when treating the crystal from a macroscopic point of view.

There will be substances whose properties are not very sensitive to structural defects and can be described using an ideal crystal model; it will be necessary to consider their real structure, since they will present properties that will depend to a greater or lesser extent on structural defects.

1.5 Crystal, Monocrystal, and Crystalline Aggregate

Crystal was first defined as a solid that, under certain formation conditions, appears as a polyhedron; that is, limited by crystalline faces (Fig. 1.1).

Crystal is actually defined as a solid material whose constituents (such as atoms, molecules, or ions) are arranged in a highly ordered microscopic structure and has a well-defined non-diffuse X-ray diffraction pattern (see some examples in Appendix I of Chap. 17).

A monocrystal is defined as a unique crystal; that is to say, the periodicity is maintained in it.

A crystalline aggregate is defined as a group of crystallites (small crystals) of the same species that grow together. They can appear in various forms:

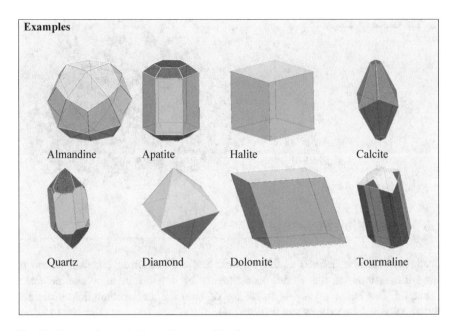

Examples

Almandine   Apatite   Halite   Calcite

Quartz   Diamond   Dolomite   Tourmaline

Fig. 1.1 Draws of crystals limited by crystalline faces

– Druses are aggregates formed by crystals that grow inside a convex or more or less flat surface (Fig. 1.2).
– Geodes are aggregates similar to druses but the crystals grow on a concave surface (in a cavity) (Fig. 1.3).
– Irregular aggregates are aggregates formed by crystals growing irregularly, giving curious or irregular shapes.
– Regular aggregates are aggregates formed by crystals growing regularly, in parallel, one inside the other, etc. Many of them constitute twins.

Fig. 1.2 Drusa of calcite (photo courtesy of Luis Miguel Rodríguez Terente, Geology Museum Conservator of Geology Department of Oviedo University, Spain)

Fig. 1.3 Geode of amethyst
quartz

1.6 Crystalline Structure

Crystal structure is defined as the periodic and ordered arrangement in space of
three dimensions of the atomic constituents of a solid in a crystalline state.
Example: The halite crystal is made up of chlorine and sodium ions arranged
periodically and orderly in space (Fig. 1.4).

Fig. 1.4 Crystalline structure
of halite

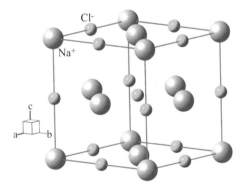

References

1. Cronstedt AF (1758) Försök til Mineralogie, eller Mineral-Rikets Upställning. Stockholm, 251 p. Translated into Danish in 1770 and German in 1780
2. de Romé Delisle JBL (1783) Cristallographie ou Description des Formes Propres à tous le Corps du Regne Minéral…, 4 vols, Paris
3. Haüy RJ (1784) Essai d´une Théorie sur la Structure des Crystaux. Gogué & Née, Paris, 236 p
4. Haüy RJ (1801) Traité de Minéralogie, vols. 1–5. Conseil des Mines, Paris
5. Wilson WE (1994) The history of mineral collecting 1530–1799. Mineral Rec 25(6):264
6. Fedorow EV (1891) Eine neue Methode der optischen Untersuchung von Krystallplatten in parallellem Lichte. Mineral. Petrogr. Mitth. 12, 505–509. On the Universal stage, more in Zeitschr. Kristallogr. 21, 574–714 (1893)
7. Nicol W (1829) On a method of so far increasing the divergency of the two rays in calcareous-spar, that only one image may be seen at a time. Edinb New Philos J 6:83–84
8. Fedorov ES (1890) Protokol'naia Zapis'. *Zap. Min. Obshch.* (Official report. *Trans. Mineral. Soc.*), vol 26, p 454 (1890)
9. Eckert M (2012) Max von Laue and the discovery of X-ray diffraction in 1912. Ann Phys (Berlin) 524(5):A83–A85
10. Wulff G (1901) XXV. Zur Frage der Geschwindigkeit des Wachsthums und der Auflösung der Krystallflächen. Zeitschrift fur Krystallographie und Mineralogie 34(5/6):449–530
11. Bragg WH (1969) The intensity of Rejlexion of X-Rays by crystal. Acta Cryst A 25:3
12. Bragg L (1969) The early history of intensity measurements. Acta Cryst A 25:1
13. Gibbs JW (1874) On the equilibrium of heterogeneous substances. transactions of the connecticut academy of arts and sciences

Chapter 2
Periodicity, Crystalline Lattices, Symbols, and Notations

Abstract This chapter builds on the fact (explained in the previous chapter) that the material elements (ions, atoms, molecules or groups of them) that constitute crystalline matter, the motif, are repeated in an orderly manner in the three-dimensional space. When a motif is systematically repeated, the result is a periodic pattern. The concepts of fundamental translation and crystalline lattice are introduced, a concept that allows the study of the crystal from a mathematical point of view, facilitating its study from the point of view of symmetry, and indicating the types of plane and three-dimensional lattices (Bravais lattices). The geometrical elements of the lattice (nodes, rows, and planes) and their identification using vectors, symbols, and Miller indices are explained, from vectors and coordinates in relation to the crystallographic coordinates a, b, and c (or cell parameters, lattice parameters, and fundamental translations). The idea that the lattice has an infinite extension and the cell is the smallest space of the lattice limited by the fundamental translations is emphasized, which allows us to describe the symmetry or the atomic content of a crystal. Bearing in mind that a direction can be common to a beam of planes, the concepts of zone, zone axis, and tautozonality, defined by Weiss at the beginning of the nineteenth century, are introduced to facilitate the morphological study of crystals. Given the existence of periodicity, it is the same to consider the zone axis as an edge common to a series of planes, or as an external line parallel to them, since all lines equivalent to a given line are parallel to it. The concepts of interplanar spacing of a family of planes and reticular density are introduced, along with their relationship with the cell parameters and Miller indices of those planes. Interplanar spacing is very important in relation to the crystal structure and is obtained experimentally (for example, from X-ray diffraction). To finish the chapter, the reciprocal lattice is discussed, which is of special importance in the interpretation of the diffraction of X-rays, electrons, and neutrons by crystals.

2.1   Crystal Lattice

A crystal lattice is a three-dimensional representation of nodes. Each lattice node represents the motif (ion, atom, molecule or groups of them) that is repeated periodically in the crystal structure. In Fig. 2.1, the nodes (black circles) represent the Na^+ ions of the NaCl structure; in this case, they represent the motif that is repeated.

– **One-dimensional lattice**

A one-dimensional lattice is an arrangement of nodes in one direction (Fig. 2.2).

– **Two-dimensional lattice**

A two-dimensional lattice is an arrangement of nodes in two directions (Fig. 2.3).

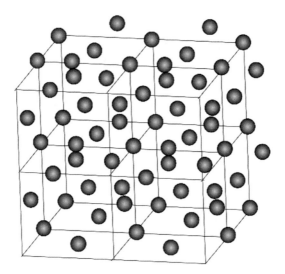

Fig. 2.1 Three-dimensional lattice

Fig. 2.2 One-dimensional lattice, with **t** = translation vector

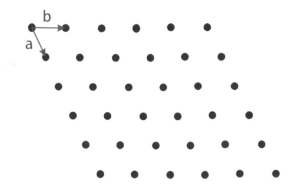

Fig. 2.3 Two-dimensional lattice, with **a** and **b** = translation vectors in a- and b- direction, respectively

2.2 Translation

Translation is the distance at which the motif is repeated parallel and identically along a given direction of the crystal structure (Fig. 2.4).

Because this monotonous repetition of the motif constitutes the fundamental characteristic of a crystal, the crystalline medium can be abstracted from its material content and can only be considered based on the present translations.

The *translation* is a *transformation*; that is, an *operation of symmetry*. It is the simplest symmetry operation inherent in the crystal structure. Translation is represented by a vector called *translation vector*. It is defined by sense, direction and module.

Fundamental or *unit translations* are the smallest translations in the three directions of space. Its modules are represented by a, b, and c, and they are assigned

Fig. 2.4 Translations in a plane lattice, with **a** and **b** = translation vectors in a- and b- direction, respectively

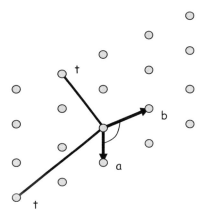

the value unit. This trio of translations constitutes the coordinate system of the lattice. They are called *lattice constants* because their values are fixed for a crystal. They are expressed by the aforementioned modules—*a, b,* and *c*—and by the angles between them α, β, and γ.

Therefore, lattice is the infinite arrangement that results from the translational symmetry in any direction.

2.3   Plane Lattices

There are five plane lattices, four primitive lattices, and one centered lattice. The primitive lattices are oblique, square, rectangular, and hexagonal. The centered lattice is the orthorhombic lattice, which is equivalent to the centered rectangular lattice. The relationships between the fundamental translations and the angles between them are shown in Table 2.1.

2.4   Space Lattices

There are 14 space lattices, and they are called Bravais lattices. They are obtained by stacking plane lattices. These lattices can be primitive and centered.

Primitive lattices are symbolized by P. Multiple lattices can be:

– Base-centered (A, B or C)
– Face-centered (F)
– Interior-centered (I)

Space lattices are named according to the relationships between the fundamental translations and the angles between them (Table 2.2).

2.5   Lattice Origin

The lattice origin is a point on the lattice, chosen as the starting point in the lattice description.

It can be chosen at any position in the crystal structure and does not need to coincide with any atom or ion.

Table 2.1 Plane lattices and lattice parameters

Lattice		Lattice parameters	
Oblique		$a \neq b$	$\gamma \neq 90°$
Square		$a = b$	$\gamma = 90°$
Rectangular or orthorhombic primitive		$a \neq b$	$\gamma = 90°$
Hexagonal		$a = b$	$\gamma = 60°$ or $120°$
Centered orthorhombic/ rectangular		$a = b$	$\gamma \neq 90°$, $60°$ or $120°$ and cos $(\gamma) = b/2a$

Table 2.2 Space lattices and lattice parameters

Lattice	Type	Lattice parameters	
Triclinic	P	$a \neq b \neq c$	$\alpha \neq \beta \neq \gamma \neq 90°$
Monoclinic	P, A (B,C)	$a \neq b \neq c$	$\alpha = \gamma = 90° \neq \beta$
Orthorhombic	P, I, F, A (B,C)	$a \neq b \neq c$	$\alpha = \beta = \gamma = 90°$
Tetragonal	P, I	$a = b \neq c$	$\alpha = \beta = \gamma = 90°$
Hexagonal	P	$a = b \neq c$	$\alpha = \beta = 90° \; \gamma = 120°$
Trigonal	P	$a = b \neq c$	$\alpha = \beta = 90° \; \gamma = 120°$
Rhombohedral	R	$a = b = c$	$\alpha = \beta = \gamma \neq 90°$ (or oblique)
Cubic	P, I, F	$a = b = c$	$\alpha = \beta = \gamma = 90°$

2.6   Elemental Cell

The lattice can be described by the unit cell in two and three dimensions, respectively, as the translational unit; the cell may be primitive or centered.

In two dimensions, the elemental cell is a parallelogram. It is limited by the fundamental translations in a lattice and constitutes the smallest characteristic part of the crystal.

In three dimensions, the elemental cell is a parallelepiped. It is limited by the fundamental translations in a lattice and constitutes the smallest characteristic part of the crystal.

The primitive elemental cell is the cell containing one lattice node.

The multiple elemental cell is the cell containing more than one node.

Examples of these in two dimensions can be seen in Fig. 2.5.

The 14 elemental cells in three dimensions are named as the corresponding Bravais lattices (Table 2.3).

Fig. 2.5 Cells A and B are primitive and cell C is multiple

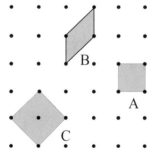

Table 2.3 Elemental cells in three dimensions, corresponding to the Bravais lattices

Cell	Type				Cell parameters
Triclinic	Primitive P				$a \neq b \neq c$ $\alpha \neq \beta \neq \gamma \neq 90°$
Monoclinic	Primitiva P			Centered-base A (B,C)	$a \neq b \neq c$ $\alpha = \gamma = 90° \neq \beta$
Orthorhombic	Primitive P	Centered-interior I	Centered-face F	Centered-base A (B,C)	$a \neq b \neq c$ $\alpha = \beta = \gamma = 90°$

(continued)

Table 2.3 (continued)

Cell	Type		Cell parameters
Tetragonal	Primitiva P 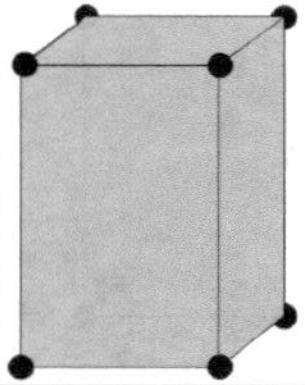	Centered-interior I 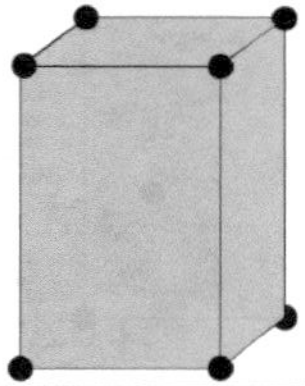	$a = b \neq c$ $\alpha = \beta = \gamma = 90°$
Hexagonal	Primitive P 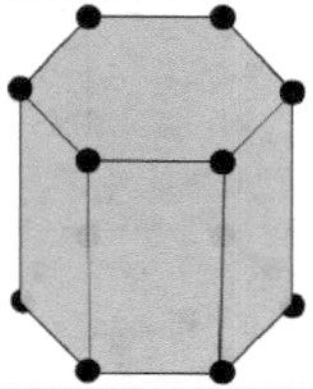		$a = b \neq c$ $\alpha = \beta = 90°\ \gamma = 120°$
Trigonal/Rombohedral	Primitive P with the rhombohedral cell inside 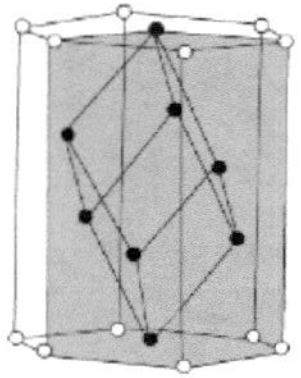		$a = b \neq c$ $\alpha = \beta = 90°\ \gamma = 120°$

(continued)

Table 2.3 (continued)

Cell	Type			Cell parameters
Cubic	Primitive P	Centered-interior I	Centered-face F	$a \neq b \neq c$ $\alpha = \beta = \gamma = 90°$

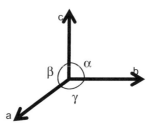

Fig. 2.6 Fundamental translations and the angles between them

2.7   Unit Cell

The unit cell is a primitive or multiple cell, selected according to certain requirements concerning metric and symmetry.

2.8   Cell Parameters

The metric is defined by the cell parameters, the fundamental translations a, b, c,
and the angles between them α, β, and γ (Fig. 2.6).

2.9   Cell Volume

The cell volume is given by the expression (2.1):

$$V = abc\sqrt{1 - \cos^2 \alpha - \cos^2 \beta - \cos^2 \gamma + 2 \cos \alpha \cos \beta \cos \gamma} \qquad (2.1)$$

2.10   Properties of Crystal Lattice

The lattice is characterized by the same geometrical properties as crystal. These
properties are:

– Homogeneity

In the lattice, all nodes are equivalent, but the distance between a node and its
neighbors is not constant and depends on the direction taken to measure said
distance, as shown in Fig. 2.7.

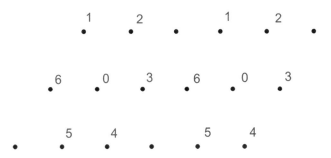

Fig. 2.7 Homogeneity of lattice. The arrangement of the blue nodes around the central blue node 0 is the same as that of the red nodes around the central red node 0

OA<OB>OC

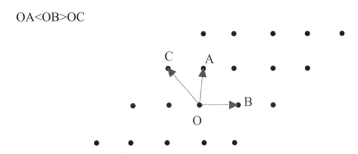

Fig. 2.8 Anisotropy of lattice

– Anisotropy

In the lattice, all nodes are equivalent, but the distance between a node and its neighbors is not constant and depends on the direction taken to measure said distance.

Example: In Fig. 2.8, the distance from the origin O to any other node around it (A, B or C) is not the same.

– Symmetry

Nodes O, 1, 2, etc., are equivalent by the application of translation b (Fig. 2.9).

Fig. 2.9 Translation as trivial symmetry

2.11   The Crystal as Interpenetrated Parallel Lattices

The crystal structure may be made up of an infinite number of parallel lattices of constant dimensions, interpenetrating each other.

In NaCl (Fig. 2.10a), two sub-lattices can be considered: One is formed from the Cl$^-$ ions, taking this ion as a repetition motif (Fig. 2.10b). The other is formed from the Na$^+$ ions, taking this ion as a repetition motif (Fig. 2.10c). Both lattices are identical in dimensions and are parallel, but one of them is displaced one half of the translation, in the three dimensions of space, with respect to the other (Fig. 2.10d and e). However, the crystal structure is defined by a lattice, which may consist of

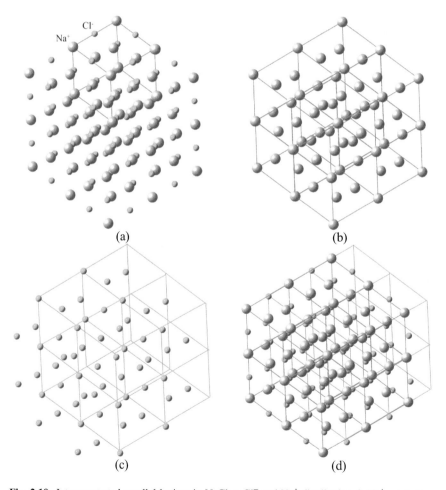

Fig. 2.10 Interpenetrated parallel lattices in NaCl: **a** Cl$^-$ and Na$^+$ distribution. **b** Na$^+$ sub-lattice. **c** Cl$^-$ sub-lattice. **d** Cl$^-$ and Na$^+$ sub-lattices interpenetrated and displaced one half forming the NaCl structure

various sub-lattices with atoms/ions or molecules on or between the nodes. In the example of NaCl, either the chlorine ions or the sodium ions lattice would describe this.

2.12 Lattice Elements

The elements of the lattice are nodes, rows, and planes.

2.12.1 Node

The node is an equivalent point by the translation. The node of the crystal lattice replaces the motif that is repeated in the crystal structure.

There are points in the crystal structure, in addition to the nodes. The points are not equivalent and are located between the nodes.

Considering the lattice definition, atoms or building units in the crystal structure can occupy the position of a node or a point or in between.

The position of a node can be specified by a translation t

$$t = ma + nb + pc \qquad (2.2)$$

where

m, n, and p are the coordinates of a node that is taken as the origin of the coordinate system defined by a, b, c.

The position of any point can be specified by the vector $\mathbf{r}$ which is given by the expression (2.3).

$$r = r_i + ma + nb + pc \qquad (2.3)$$

where

$\mathbf{r_i}$ is the distance between the considered point and its counterpart.

The vector $\mathbf{r_i}$ is given by the expression (2.4)

$$r_i = x_i a + y_i b + z_i c \qquad (2.4)$$

where

x_i, y_i, and z_i are the coordinates of the point defined by $\mathbf{r_i}$ and are three positive numbers less than unity.

All the points defined by $\mathbf{r_i}$ occupy a space limited by the fundamental translations; that is, it constitutes the elementary cell. This space fills the entire crystalline space by applying the fundamental translations.

Example

In Fig. E1, the position vector of the node N is given by $\mathbf{r} = ON$:

$$r = ma + nb + pc = 2a + 3b + oc$$

The coordinates m, n, q of node N are 2, 3, 0
The position vector of point A is given by $\mathbf{r}_i = OA$:

$$r_i = x_i a + y_i b + z_i c = 0.74a + 0.72b + 0c$$

The coordinates x_i, y_i, z_i of point A are 0.74, 0.72, 0
The position vector of point B, homologous of A, is given by $\mathbf{r} = OB$:

$$r = r_i + ma + nb + pc$$

thus

$$r = r_i + 2a + 3b + 0c$$

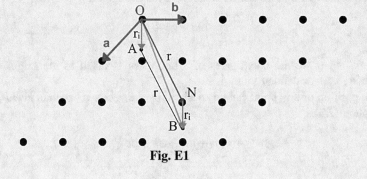

Fig. E1

2.12.2   Lattice Row

The lattice row is an arrangement of nodes along one direction.
   A lattice direction is, therefore a direction that contains nodes.
   Each pair of lattice nodes defines a lattice row.
   The lattice row symbol defines a lattice direction and is formed by three integer numbers, positive or negative, enclosed in square brackets [uvw].

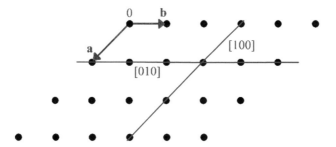

Fig. 2.11 Fundamental reticular rows

When a negative number appears, a line is placed above it, for example $[1\overline{2}0]$, and read as, "one, bar two, zero", and the bar two is mathematically handled, like -2.

If two-digit numbers appear, they are separated by commas, for example [10,2,2].

Those numbers are the direct coordinates u, v, and w of a node of the lattice, contiguous to another node taken as the origin.

The rows parallel to the fundamental translations *a*, *b*, and *c*, respectively, are called *fundamental reticular rows,* and their symbols are [100], [010], and [001] (Fig. 2.11). They define the edges of the lattice elemental cell. They are parallel to the fundamental translations *a*, *b,* and *c*, respectively.

The set of symmetry-equivalent directions is symbolized by <uvw>.

Example

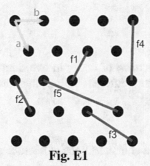

Fig. E1

Remarks:

- In this example, two solution equivalents must be considered, since the lattice row sense of Fig. E1 has not been indicated, and any node of the lattice, due to the property of homogeneity, can be chosen as the origin.
- The lattice row passes through at least two points. One of them is taken as the coordinate origin. The other is specified by the coordinates in relation to the node taken as the origin.

Solution 1

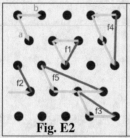

row	coordinates		symbol
	a	*b*	
f1	$1 \cdot a = 1 \cdot 1 = 1$	$(-1 \cdot b) = (-1 \cdot 1) = -1$	$[1\bar{1}]$
f2	$1 \cdot a = 1 \cdot 1 = 1$	$0 \cdot b = 0 \cdot 1 = 0$	$[10]$
f3	$1 \cdot a = 1 \cdot 1 = 1$	$1 \cdot b = 1 \cdot 1 = 1$	$[11]$
f4	$2 \cdot a = 2 \cdot 1 = 2$	$(-1 \cdot b) = (-1 \cdot 1) = -1$	$[2\bar{1}]$
f5	$1 \cdot a = 1 \cdot 1 = 1$	$2 \cdot b = 2 \cdot 1 = 2$	$[12]$

Fig. E2

Solution 2

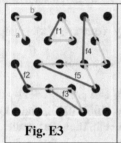

row	*a* coordinate	*b* coordinate	symbol
f1	$(-1a) = (-1 \cdot 1) = -1$	$1b = 1 \cdot 1 = 1$	$[\bar{1}1]$
f2	$-1 \cdot a = -1 \cdot 1 = -1$	$(0 \cdot b) = 0 \cdot 1 = 0$	$[\bar{1}0]$
f3	$(-1a) = (-1 \cdot 1) = -1$	$(-1b) = (-1 \cdot 1) = -1$	$[\bar{1}\bar{1}]$
f4	$(-2 \cdot a) = (-2 \cdot 1) = -2$	$1 \cdot b = 1 \cdot 1 = 1$	$[\bar{2}1]$
f5	$(-1a) = (-1 \cdot 1) = -1$	$(-2 \cdot b) = (-2 \cdot 1) = -2$	$[\bar{1}\bar{2}]$

Fig. E3

2.12.3   Lattice Plane

A lattice plane is an arrangement of nodes in two directions, forming a net. Each trio of nodes arranged in three different lattice directions defines a reticular plane.
Lattice Plane Symbol

–   *Weiss indices*

Weiss parameters are the direct intersections of a lattice plane with the fundamental translations.

–   *Miller indices*

Miller indices are a series of three positive or negative integers in parentheses, (*hkl*), which refer to the reciprocal intersections of a plane with the fundamental translations; *h, k, l* are integer numbers and coprime.

These indices indicate the position of a plane in the lattice; that is, the inclination with respect to the fundamental translations and its orientation and distance with respect to the origin of coordinates (fundamental translations of the lattice or the cell).

When a negative number appears, a line is placed above it; for example, $(1\bar{2}0)$ and it is read one, minus two, zero.

If two-digit numbers appear, they are separated by commas; for example (10, 2 1).

The plane closest to the origin represents a family of lattice planes and passes through three nodes, one in each lattice fundamental row, and whose coordinates are as follows:

$$A = Ha \qquad B = Kb \qquad C = Lc$$

where

H, K, and L represent the intersections of the plane with the fundamental rows; that is, the *Weiss indices*

a, b, and c represent the fundamental translation or coordinate axis.

The number of planes existing between the plane with coordinates A, B, and C and the lattice origin is given by the product $H \cdot K \cdot L = N$.

The ratios $N/H = h$, $N/K = k$, and $N/L = l$, are integers and are the Miller indices. They are the inverse of the Weiss parameters.

Example
Figure E1 shows the plane AB, parallel to the crystallographic axis c,

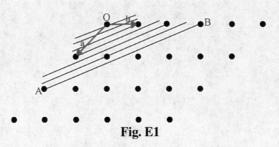

Fig. E1

A and B being their coordinates, so

$$A = Ha = 2 \cdot 1 = 2$$

$$B = Kb = 3 \cdot 1 = 3$$

$$C = Lc = \infty \cdot 1 = \infty$$

$$N = H \cdot K = 2 \cdot 3 = 6$$

$$h = N/H = 6/2 = 3$$

$$k = N/K = 6/3 = 2$$

$$l = N/L = 6/\infty = 0$$

Thus, its Miller indices are (320).

The crystalline planes that intersect the three coordinate axes of the lattice are symbolized by (*hkl*).

The planes that intersect two axes and are parallel to the third are symbolized by (*hk0*), (*h0l*), and (*0kl*), depending on whether they are parallel to axis *c*, *b* or *a*, respectively.

Planes intersecting an axis and parallel to the other two are symbolized as (*h00*), (*0k0*), and (*00l*).

The simplest symbol planes are (*100*), (*010*), and (*001*) and are called fundamental planes. These fundamental planes define the faces of the elemental cell of the lattice.

Other planes with simple symbols are (*110*), (*101*), and (*011*).

Example

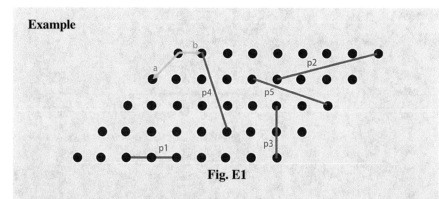

Fig. E1

Remarks:

In this example, the crystallographic coordinate *c* is perpendicular to the paper plane and the planes p are parallel to this axis; that is, they are its traces.

The trace of the lattice plane passes through at least two points. Each trace can be parallel to one of the coordinates *a* or *b*, respectively, or it can intersect both. When said trace is parallel to one of the two coordinates, it means that its intersection is infinite.

It must be considered whether the intersections of the coordinate axis a, b, c with the plane trace are positive or negative.

It must be considered, that h, k, l in Miller indices are coprime; therefore, the greatest common devisor allowed is 1.

N represents the number of planes between a rational intersection and the origin of coordinates.

In this example, two solution equivalents must be considered, since any node of the lattice, due to the property of homogeneity, can be chosen as the origin.

Solution 1

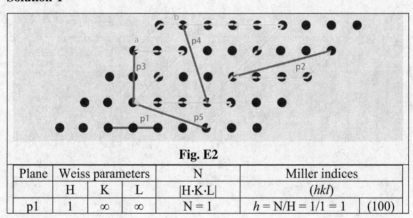

Fig. E2

Plane	Weiss parameters			N	Miller indices	
	H	K	L	|H·K·L|	(hkl)	
p1	1	∞	∞	N = 1	h = N/H = 1/1 = 1	(100)

					$k = N/H = 1/\infty = 0$ $l = N/H = 1/\infty = 0$	
p2	-1	-3	∞	$N = 1 \cdot 3 = 3$	$h = N/H = 3/-1 = -3$ $k = N/H = 3/-3 = -1$ $l = N/H = 3/\infty = 0$	$(\overline{3}\overline{1}0)$
p3	2	-2	∞	$N = 2 \cdot 2 = 4$	$h = N/H = 4/2 = 2$ $k = N/H = 4/-2 = -2$ $l = N/H = 4/\infty = 0$	$(2\overline{2}0)$ $(1\overline{1}0)$
p4	3	-4	∞	$N = 3 \cdot 4 = 12$	$h = N/H = 12/3 = 4$ $k = N/H = 12/-4 = -3$ $l = N/H = 12/\infty = 0$	$(4\overline{3}0)$
p5	1	-4	∞	$N = 1 \cdot 4 = 4$	$h = N/H = 4/1 = 4$ $k = N/H = 4/-4 = -1$ $l = N/H = 4/\infty = 0$	$(4\overline{1}0)$

Plane	Plane coordinates		
	$A = H \cdot \boldsymbol{a}$	$B = K \cdot \boldsymbol{b}$	$C = L \cdot \boldsymbol{c}$
p1	$A = 1 \cdot 1 = 1$	$B = \infty \cdot 1 = \infty$	$C = \infty \cdot 1 = \infty$
p2	$A = 1 \cdot 1 = -1$	$B = 3 \cdot 1 = 3$	$C = \infty \cdot 1 = \infty$
p3	$A = 2 \cdot 1 = 2$	$B = 2 \cdot 1 = 2$	$C = \infty \cdot 1 = \infty$
p4	$A = 3 \cdot 1 = 3$	$B = 4 \cdot 1 = 4$	$C = \infty \cdot 1 = \infty$
p5	$A = 1 \cdot 1 = 1$	$B = 4 \cdot 1 = 4$	$C = \infty \cdot 1 = \infty$

Solution 2

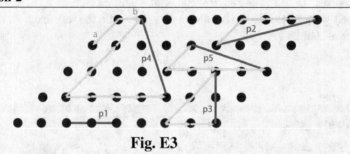

Fig. E3

Plane	Weiss parameters			N	Miller indices	
	H	K	L	|H·K·L|	(hkl)	
p1	-1	∞	∞	N = 1	$h = N/H = 1/-1 = -1$ $k = N/H = 1/\infty = 0$ $l = N/H = 1/\infty = 0$	$(\bar{1}00)$
p2	1	3	∞	N = 1 · 3 = 3	$h = N/H = 3/1 = 3$ $k = N/H = 3/3 = 1$ $l = N/H = 3/\infty = 0$	(310)
p3	-2	2	∞	N = 2 · 2 = 4	$h = N/H = 4/-2 = -2$ $k = N/H = 4/2 = 2$ $l = N/H = 4/\infty = 0$	$(\bar{2}20)$ $(\bar{1}10)$
p4	-3	4	∞	N = 3 · 4 = 12	$h = N/H = 12/-3 = -4$ $k = N/H = 12/4 = 3$ $l = N/H = 12/\infty = 0$	$(\bar{4}30)$
p5	-1	4	∞	N = 1 · 4 = 4	$h = N/H = 4/-1 = -4$ $k = N/H = 4/4 = 1$ $l = N/H = 4/\infty = 0$	$(\bar{4}10)$

Plane	Plane coordinates		
	A = H·**a**	B = K·**b**	C = L·**c**
p1	A = (-1)·1 = -1	B = ∞·1 = ∞	C = ∞·1 = ∞
p2	A = 1·1 = -1	B = 3·1 = 3	C = ∞·1 = ∞
p3	A = (-2)·1 = -2	B = (-2)·1 = -2	C = ∞·1 = ∞
p4	A = (-3)·1 = -3	B = (-4)·1 = -4	C = ∞·1 = ∞
p5	A = (-1)·1 = -1	B = (-4)·1 = -4	C = ∞·1 = ∞

In the case of a trigonal or hexagonal lattice, four axes are used, a_1, a_2, a_3, c, and four indices, $(h\ k\ \bar{i}\ l)$ called Bravais–Miller indices, where h, k, $\bar{i}$, l are again inversely proportional to the intercepts of a plane with the four axes. The indices h, k, $\bar{i}$ are related by

$$h + k = \bar{i} \tag{2.5}$$

Example

1. The Bravais–Miller indices of a plane that intersects with the hexagonal axis $a = 2$, $b = 3$, $c = \infty$, are obtained as follows:

   (1°) hkl are obtained:

   $$N = H \cdot K \cdot L = 2 \cdot 3 \cdot 1 = 6$$

   $$h = N/H = 6/2 = 3$$

   $$k = N/K = 6/3 = 2$$

   $$l = N/L = 6/\infty = 0$$

   (2°) $\bar{i}$ is obtained applying the expression (2.5)

   $$\bar{i} = -(3+2) = -5$$

   So, the Bravais–Miller índices are $(32\ \bar{5}\ 0)$.
   Because i is linearly dependent on h and k, it can be written as $(32.\bar{5}\ 0)$ or (320)

Rational Index Law

Haüy (1784, 1801)[1] long ago deduced the rational indices law which states that the intercepts of the natural faces of a crystal with the coordinates axes (or unit-cell axes)—a, b, c—are inversely proportional to Miller indices (hkl) (prime integers).

A set of crystal faces that are related to each other by symmetry is a crystal form. A crystal form (which could imply many faces) is designated by the Miller index or Miller–Bravais index notation, enclosing the indices in braces, i.e., $\{101\}$ or $\{11\bar{2}1\}$.

---

[1] Haüy [1, 2].

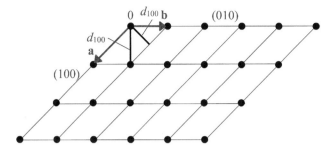

Fig. 2.12 Lattice spacing

2.13 Lattice Spacing

Lattice spacing or interplanar spacing is the distance between the planes of a family of lattice planes (Fig. 2.12).

It is measured perpendicular to the planes of the plane family.

The spacing of a plane family (hkl) is symbolized by d_{hkl} and is characteristic of each family.

The lattice spacing is related to the lattice constants (fundamental translations or coordinate systems) and to the Miller indices, using the so-called quadratic form given by expression (2.6).

$$\frac{1}{d_{hkl}^2} = \frac{1}{V^2} \left[\begin{array}{c} h^2 b^2 c^2 \sin^2\alpha + k^2 a^2 c^2 \sin^2\beta + l^2 a^2 b^2 \sin^2\gamma + \\ 2hkabc^2(\cos\alpha\ \cos\beta - \cos\gamma) + \\ 2hlab^2c(\cos\alpha\ \cos\gamma - \cos\beta) + \\ 2kla^2bc(\cos\beta\ \cos\gamma - \cos\alpha) \end{array} \right] \tag{2.6}$$

With volume V given in Eq. (2.1).

Example

1. For cubic cells, the expression (2.6) reduces to

$$d_{hkl} = a/\sqrt{h^2 + k^2 + l^2}$$

and the lattice spacing of a family of planes (121) of a cubic cell with $a = 2$ Å is

$$d_{hkl} = 2/\sqrt{1^2 + 2^2 + 2^2} = 0.8197\,\text{Å}$$

2. For orthorhombic cells, the expression (2.6) is

$$d_{hkl} = 1/\left(h^2/a^2\right) + \left(k^2/b^2\right) + \left(l^2/c^2\right)$$

and the lattice spacing of a family of planes (121) of an orthorhombic cell with a = 2 Å, b = 3 Å 3 Å, and c = 1 Å is

$$d_{hkl} = \sqrt{\left(1^2/2^2\right) + \left(2^2/3^2\right) + \left(1^2/1^2\right)} = 8\cdot 10^{-4}\,\text{Å}$$

2.14   Tautozonal Planes

Tautozonal planes are a set of non-parallel planes that are characterized by having a common crystallographic direction, called a zone axis, and whose symbol is [uvw].

Figure 2.13 shows some families of planes parallel to the c axis (perpendicular to the writing plane), with the same symbol and equal interplanar distance.

The condition for a plane (*hkl*) to be parallel to an axis [uvw] is given by *Weiss zone law*

$$hu + kv + lw = 0 \qquad\qquad (2.7)$$

Fig. 2.13 Tautozonal planes. Families of planes parallel to the *c* axis (zone axis)

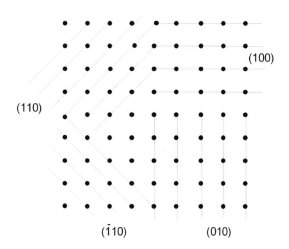

If two planes (hkl) and $(h'k'l')$ belong to the same zone, they are parallel to the same crystallographic direction [uvw] and must be fulfilled

$$hu + kv + lw = 0$$
$$h'u + k'v + l'w = 0 \tag{2.8}$$

Solving both equations, the symbol of the zone axis is obtained

$$u = (kl' - k'l)$$
$$v = (lh' - l'h) \tag{2.9}$$
$$w = (hk' - k'h)$$

Example

1. The plane (100) is parallel to the axis [001]

$$1 \cdot 0 + 0 \cdot 0 + 0 \cdot 1 = 0$$

2. The plane (100) is not parallel to the axis [210]

$$1 \cdot 2 + 0 \cdot 1 + 0 \cdot 0 = 2 \neq 0$$

3. To know if planes (110) and (320) belong to the same zone [001], expression (2.7) is applied

$$1 \cdot 0 + 1 \cdot 0 + 0 \cdot 1 = 0$$
$$3 \cdot 0 + 2 \cdot 0 + 0 \cdot 1 = 0$$

Both planes belong to the zone with zone axis [001]
To know the symbol of the zone axis to which the planes (211) and (321) belong, expression (2.8) is applied.

$$u = 1 \cdot 1 - 2 \cdot 1 = -1$$
$$v = 1 \cdot 3 - 1 \cdot 2 = 1 \qquad \text{The zone symbol is } [\bar{1}10]$$
$$w = 2 \cdot 2 - 2 \cdot 2 = 0$$

2.15 Reticular Density

– Spatial lattice density or reticular density

Spatial lattice density is the number of nodes per volume unit. It is the inverse of the cell volume, V_{hkl}

$$\rho_{hkl} = 1/V_{hkl} \tag{2.10}$$

– Planar lattice density

Planar lattice density is the number of nodes per unit area. It is the inverse of the plan cell area, S_{hkl}

$$\rho_{hkl} = 1/S_{hkl} = d_{hkl}/V_{hkl} \tag{2.11}$$

$$S_{hkl} = V_{hkl}/d_{hkl} \tag{2.12}$$

The planes with the highest node density are those with the simplest Miller indices and are more distant from the origin, which may constitute crystalline faces.

– Linear lattice density

The linear lattice density is the number of nodes per unit length. It is the inverse of the module of the translation vector.

The rows with the simplest Miller indices are those with the highest node density. They are: [100], [010], and [001].

Example

In Fig. E1, the plane family p1, p2, and p3 have different Miller indices—p1 (110), p2 $(\bar{1}20)$, p3 $(\bar{1}30)$—interplanar spacing and reticular density.

The planes are represented by their traces.

The reticular density can be considered in this case as the number of nodes per length unit. For this example, the green line has been taken as the unit of length. Then

$$\rho_{110} = 3 \text{ nodes}$$
$$\rho_{\bar{1}20} = 2 \text{ nodes}$$
$$\rho_{\bar{1}30} = 1 \text{ node}$$

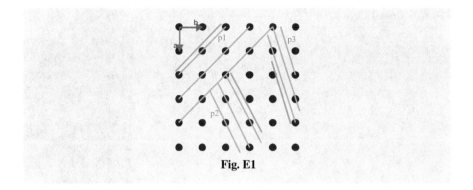

Fig. E1

2.16 Crystal Face

A crystal face is a manifestation in the crystal of planes that are characterized by having the highest plane lattice density, and their Miller indices are simple.

Example: In Fig. 2.14, the different numbered plane surfaces limiting the pyrite crystal are crystal faces.

2.17 Crystal Edge

A crystalline edge is a manifestation in the crystal of a lattice row characterized by having the highest linear lattice density and having the simplest symbol.

In the gypsum crystal image (Fig. 2.15), some edges have been highlighted with a red line.

Fig. 2.14 Crystal faces

Fig. 2.15 Crystal edges

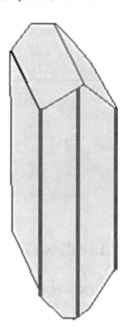

2.18   Reciprocal Lattice

The reciprocal lattice is, as its name indicates, the reciprocal of the crystal lattice or direct lattice.

This mathematical construction is helpful, e.g., for X-ray structure analysis.

It has an origin.

Lattice parameters are symbolized with an asterisk*

$$
\begin{aligned}
a^* &= 1/d_{100} \\
b^* &= 1/d_{010} \\
c^* &= 1/d_{001}
\end{aligned}
\tag{2.13}
$$

Axis a^*, denoted as [100]* is perpendicular to (100) plane of the direct lattice.
Axis b^*, denoted as [010]* is perpendicular to (010) plane of the direct lattice.
Axis c^*, denoted as [001]* is perpendicular to (001) plane of the direct lattice.

The reciprocal lattice has an elementary cell defined by the reciprocal fundamental translations, previously defined.

Its volume is the inverse of the volume of the elementary cell of the direct lattice.

If α, β, and γ are the fundamental angles of the direct lattice and A, B, and C are the dihedral angles between the fundamental planes of the direct lattice, the corresponding fundamental angles of the reciprocal lattice are given by

$$\cos \alpha^* = \frac{\cos \beta \cos \gamma - \cos \alpha}{\sin \beta \sin \gamma}$$
$$\cos \beta^* = \frac{\cos \beta \gamma \cos \alpha - \cos \beta}{\sin \alpha \sin \gamma} \qquad (2.14)$$
$$\cos \gamma^* = \frac{\cos \alpha \cos \beta - \cos \gamma}{\sin \alpha \sin \beta}$$

The relationships between the translations of the direct lattice and the reciprocal lattice are

$$a^* = 1/V \, \boldsymbol{b} \cdot \boldsymbol{c} = (1/V)\boldsymbol{bc} \sin \alpha$$
$$\boldsymbol{b}^* = (1/V)\boldsymbol{ca} \sin \beta \qquad (2.15)$$
$$\boldsymbol{c}^* = (1/V)\boldsymbol{ab} \sin \gamma$$

Example (2D)
Obtaining the Reciprocal Lattice

The reciprocal lattice is obtained as follows (Fig. E1):

- Any point of the direct lattice can be chosen as the origin of the new lattice.
- Normals to the fundamental planes—(100), (010), and (001)—of the direct lattice; that is, the interplanar spacings d_{100}, d_{010}, and d_{001} are drawn, and their inverses are taken as coordinate axes of the reciprocal lattice.
- Lines are drawn on the coordinate axes, and reciprocal translations are marked on them.
- Lines intersecting with the translations b^* are drawn parallel to the lines containing a^*.
- Lines intersecting with the translations a^* are drawn parallel to the lines containing b^*.
- The nodes of the reciprocal lattice are drawn on the intersections of the translations a^* and b^*.

The reciprocal lattice node symbol is obtained as follows:

Node 01: It is at the intersection of row 0 parallel to translation a^* and row 1 parallel to translation b^*; therefore, it is 01.
Node 22: It is at the intersection of row 2 parallel to translation a^* and row 2 parallel to translation b^*; therefore, it is 22.

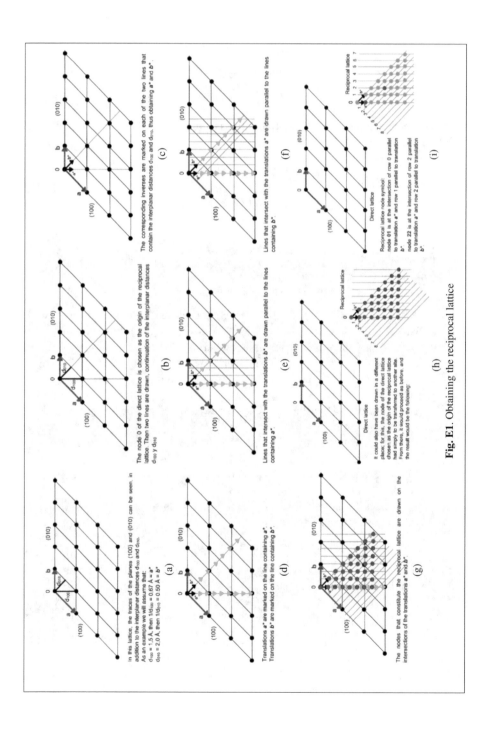

Fig. E1. Obtaining the reciprocal lattice

2.19 Relations Between the Direct and Reciprocal Lattices

– Each direct lattice corresponds to a single reciprocal lattice.
– Every reciprocal lattice has an origin and the same symmetry + inversion symmetry as the direct lattice from which it comes.
– In orthogonal crystalline systems and in two dimensions, a reciprocal lattice derives from the primitive direct lattice by a 90° turn around the node taken as the origin.
– If the direct lattice is face centered, its reciprocal will be interior centered.
– If the direct lattice is centered inside, its reciprocal will be face centered.
– If the direct lattice is bases centered, its reciprocal is also.
– If the direct lattice is primitive, its reciprocal is also primitive.
– The direct lattice is homogeneous, but the reciprocal lattice is not. In the direct lattice all the nodes are equivalent and the origin can be taken at any node. In the reciprocal lattice there is an origin, and the nodes are not interchangeable.
– The reciprocal lattice is made up of lattice rows, which are perpendicular to families of planes of the direct lattice, and families of lattice planes that are perpendicular to lattice rows of the direct lattice.
– The planes of the same family in the reciprocal lattice are not equivalent to each other, contrary to what happens in the direct lattice.
– There are 14 types of direct lattices (Bravais networks) and 14 types of reciprocal lattices.

Exercises

1. Select a repeat motif and two translations, the smaller ones in two directions, from Fig. E1a–d.

   (a) Draw the corresponding plane lattice.
   (b) Indicate the type of plane lattice in each case.
   (c) Select a cell and indicate the following:

      (1) If it is a primitive or multiple.
      (2) The cell content, depending on the motif that is repeated.
      (3) The values of a and b parameters (cm), γ (°), and the area (cm^2).

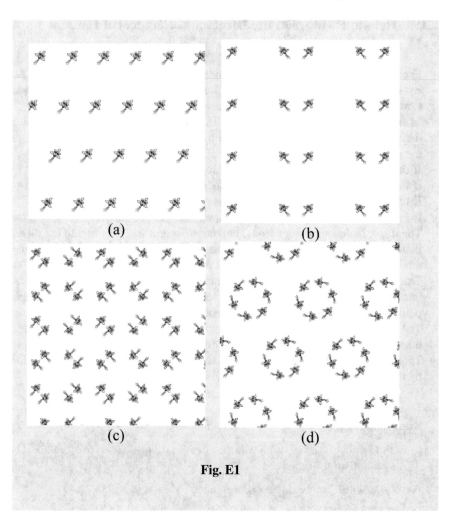

(a)

(b)

(c)

(d)

Fig. E1

2. Obtain the row symbol from Fig. E2a–f.

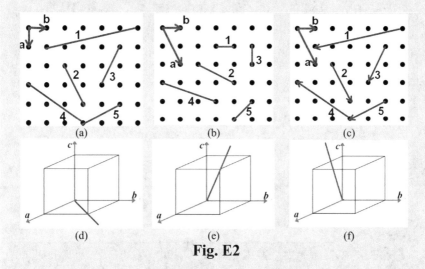

Fig. E2

3. Obtain the Miller indices of the planes of Fig. E3a–f.

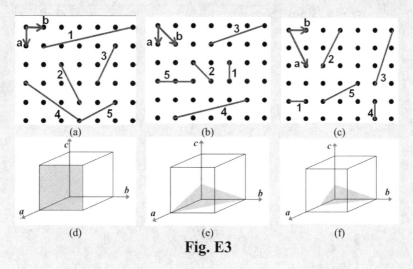

Fig. E3

Questions

1. What is the property represented in the lattice of Fig. Q1, in which it can be seen that the distances between nodes keep the relation OA < OB < OC?

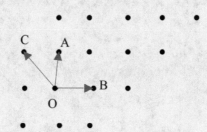

Response: [_____]

2. What is the name of the property that, from the macroscopic point of view, means invariance of a physical property F measured at a point x, in relation to its measure at another point $x + x'$; that is, $F(x) = F(x + x')$?

Response: [_____]

3. To describe a material in a crystalline state (mineral, for example), it is only necessary to use one of the infinite lattices that can represent it, all of them characterized by being interpenetrated with each other, although they are not parallel.
   - ○ True
   - ○ False

4. Symmetry means periodicity.
   - ○ True
   - ○ False

5. The cell content of Fig. Q2 is one node

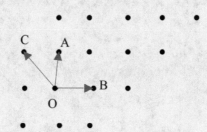

Fig. Q2

   - ○ True
   - ○ False

6. A fundamental translation is equivalent to a lattice constant
   - ○ True
   - ○ False

7. How is the fundamental translation that is related to the crystallographic direction symbolized as [010]?

   Response:

8. What should the Miller indices be?

   (a) integers
   (b) integers or fractional

   Response:

9. What are planes that have a common direction and are not parallel?

   Response:

10. The Miller indices of the plane that cuts the fundamental translation b are zeros? Yes/No

    Response:

11. What is the form in which a solid appears in a crystalline state under certain formation conditions?

    Response:

12. The coordinates of the atoms in a crystal structure can only be integers.
    ○ True
    ○ False

13. In the tetragonal lattices, α, β, and γ are equal to 90°
    ○ True
    ○ False

14. Translation is the simplest operation of symmetry and is inherent in the crystal structure.
    ○ True
    ○ False

15. Weiss parameters are three positive or negative integers in parentheses.
    ○ True
    ○ False

References

1. Haüy RJ (1784) Essai d´une Théorie sur la Structure des Crystaux. Gogué & Née, Paris, 236 p
2. Haüy RJ (1801) Traité de Minéralogie, vols 1–5. Conseil des Mines, Paris

Chapter 3
Symmetry and Lattices

Abstract This chapter explains symmetry, one of the properties of crystal, as a consequence of the periodicity of its atomic constituents in the three-dimensional space. The operations and operators of symmetry in two and three dimensions are defined. The proper and improper rotations and the symmetry with translation (reflection—translation and rotation—translation) and their symbols are described.

3.1 Concept of Symmetry

Symmetry is a property that makes objects appear indistinguishable after they have been subjected to transformation in space.

These transformations are symmetry operations, which are performed by symmetry operators or elements.

Mathematically, symmetry corresponds to a set of linear transformations that make some directions equivalent to others.

The definition of *equivalence*, from a mathematical point of view, includes the conditions of:

- identity
  $a = a$
- reflexivity
  if $a = b$, then $b = a$
- transitivity
  if $a = b$ and $b = c$, then $a = c$.

© The Author(s), under exclusive license to Springer Nature Switzerland AG 2022   49
C. Marcos, *Crystallography*, Springer Textbooks in Earth Sciences,
Geography and Environment, https://doi.org/10.1007/978-3-030-96783-3_3

3.2 Symmetry Contained in the Lattices

The crystal is symmetrical because it is periodic. Its symmetry can be deduced as a consequence of the theory of crystalline lattices.

Translation is the trivial symmetry of lattices. It is the shortest distance between two contiguous nodes in each of the three dimensions of space.

Centering is a symmetry operation of the lattice. It results from adding new nodes in the center of each parallelogram generator of the plan lattice. It is only considered possible when the resulting lattice is morphologically different from the original one.

In plane lattices, only rectangular (Fig. 3.1) or orthorhombic (Fig. 3.2) lattices can be centered.

Relations Between Lattice Elements and Symmetry Operators

– The number and type of operators that appear in a lattice depend on the lattice metric.
– The principle of reticular homogeneity means that every element of symmetry that passes through a node is repeated in parallel and indefinitely at each node of the lattice.

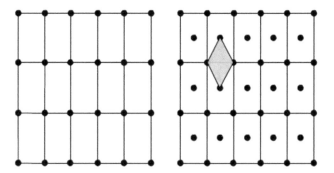

Fig. 3.1 Centered rectangular lattice

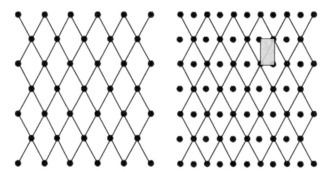

Fig. 3.2 Centered orthorhombic lattice

- Every node in the lattice is a center of symmetry.
- Every axis of symmetry is a reticular row.
- Every plane of symmetry is a reticular plane.
- Perpendicular to every axis of symmetry there is a family of reticular planes.
- Every reticular plane that is a plane of symmetry has a family of reticular rows perpendicular to it, and each reticular row of this family is an axis of symmetry.
- Every reticular row that is an axis of even order (2, 4 or 6) has a reticular plane perpendicular to it, which is a plane of symmetry.
- A reticular row that is a symmetry axis of order 4 or 6 has 4 or 6 families of rows perpendicular to it that are binary axes, and 4 or 6 families of symmetry planes that contain that row.
- The intersection of an even-ordered axis on a plane of symmetry that is perpendicular to that axis is a center of symmetry.

3.3 Symmetry Operation

Symmetry operation is a transformation that, when applied to an object, leads to a configuration indistinguishable from the original. In Table 3.1, the symmetry operations in two and three dimensions are presented.

3.4 Element of Symmetry

Symmetry element is an operator that allows the symmetry operation. There are several types of symmetry elements (Table 3.2).

Table 3.1 Symmetry operations in two and three dimensions

Symmetry operations	
In two dimensions	In three dimensions
Translation	Translation
Proper	
Rotation	Rotation
Improper	
	Rotation-inversion
With associated translation	
Reflexion-translation (gliding)	Reflexion-translation (gliding) Rotation-translation

Table 3.2 Symmetry elements in two and three dimensions	*Symmetry elements*	
	In two dimensions	In three dimensions
	– Translation vector	– Translation vector
	– Rotation point	– Rotation axis
	– Center	– Center
	– Reflexion line	– Reflexion plane
	– Gliding line	– Rotation-inversion axis
		– Rotating-reflexion axis
		– Gliding plane
		– Screw axis

3.5 Translation

Translation is the trivial symmetry of lattices. It is the shortest distance between two contiguous nodes in each of the three dimensions of space (Figs. 3.3 and 3.4).

Fig. 3.3 Translational symmetry

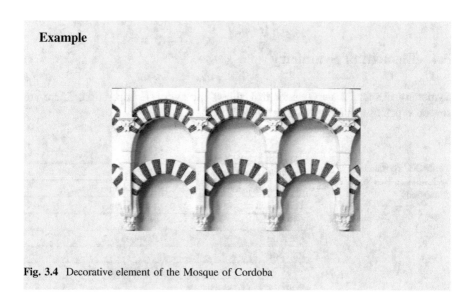

Example

Fig. 3.4 Decorative element of the Mosque of Cordoba

3.6 Symmetry Proper Operations

3.6.1 Rotations

- **In three dimensions**

  Rotation is a symmetry operation consisting of a 360°/n rotation around a symmetry axis or rotation axis (which is its corresponding symmetry element). The order of that axis is **n**, which can be **1, 2**, **3**, **4,** and **6**.

  Various notations are used to symbolize the axes of rotation, although the most commonly used are Hermann–Mauguin or international notation and Schoenflies (Table 3.3).

  In the plane, the operators of the rotation operation are *rotation points* and the order, *n,* can be **1, 2, 3, 4,** and **6**.

  This operation is called a *proper operation* and the operators are called *proper axes*.

  A rotation axis of order *n* generates a total of n operations.

  A rotation axis that involves 360°/*n* rotations also involves $m \cdot (360°/n)$ rotations, where *m* can be 1, 2, 3, 4, 5 and 6 (Table 3.4).

- **In two dimensions**

  Rotation in the plane is a symmetry operation that consists of a 360°/*n* rotation around a symmetry point or rotation point (which is its corresponding symmetry element).

  The order of that point is *n*, which can be 1, 2, 3, 4, and 6.

Table 3.3 Notations used to symbolize the rotation axes

Hermann–Mauguin	Schoenflies	Rotation axis	degrees (°)
1	C_1	Monary (identity)	360
2	C_2	Binary	180
3	C_3	Ternary	120
4	C_4	Quaternary	90
6	C_6	Senario	60

Table 3.4 Rotations involved by the rotation axes of order 1, 2, 3, 4, and 6

Rotation axis	Rotations					
1	360°/1	$1 \cdot (360°/1)$				
2	360°/2	$1 \cdot (360°/2)$	$2 \cdot (360°/2)$			
3	360°/3	$1 \cdot (360°/3)$	$2 \cdot (360°/3)$	$3 \cdot (360°/3)$		
4	360°/4	$1 \cdot (360°/4)$	$2 \cdot (360°/4)$	$3 \cdot (360°/4)$	$4 \cdot (360°/4)$	
6	360°/6	$1 \cdot (360°/6)$	$2 \cdot (360°/6)$	$3 \cdot (360°/6)$	$4 \cdot (360°/6)$	$5 \cdot (360°/6)$	$6 \cdot (360°/6)$

Table 3.5 shows the Hermann–Mauguin notation for the rotation points.

Table 3.5 Notations used to symbolize the rotation points

Hermann–Mauguin	Rotation axis	Degrees (°)
1	Monarium (identity)	360
2	Binary	180
3	Ternary	120
4	Quaternary	90
6	Senario	60

Examples

Monary rotation: Identity: 1 (Fig. 3.5).

Fig. 3.5 360° rotation

Binary rotation: 2 (Fig. 3.6).

Fig. 3.6 180° rotation

Ternary rotation: 3 (Fig. 3.7).

Fig. 3.7 120° rotation

Quaternary rotation: 4 (Fig. 3.8).

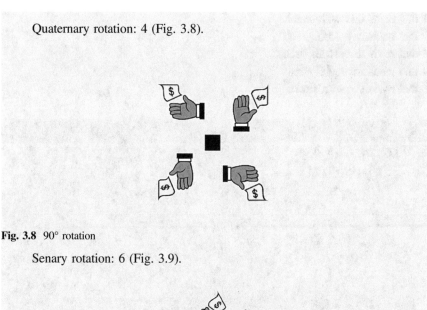

Fig. 3.8 90° rotation

Senary rotation: 6 (Fig. 3.9).

Fig. 3.9 60° rotation

3.7 Symmetry Proper Operations

3.7.1 Rotation-Inversion

Rotation-inversion is a symmetry operation that consists of a 360°/n rotation and an inversion around a symmetry element called the inversion axis of rotation. The order of that axis is **n**, and it can be 1, 2, 3, 4, and 6.

The symbol of these axes for the most used notation, Hermann–Mauguin or international, is the following:

$\bar{1}$ that reads one with streak
$\bar{2}$ that reads two with streak
$\bar{3}$ that reads three with streak
$\bar{4}$ that reads four with streak
$\bar{6}$ that reads six with streak

Examples
  Monary rotation: $\bar{1}$
  360° rotation and simultaneous inversion
  Binary rotation: $\bar{2} \equiv m$ (Fig. 3.10).

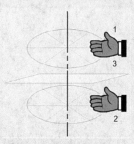

Fig. 3.10 180° rotation and simultaneous inversion

  Ternary rotation: $\bar{3}$ (Fig. 3.11)

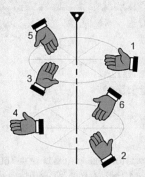

Fig. 3.11 120° rotation and simultaneous inversion

Quaternary rotation: $\bar{4}$ (Fig. 3.12)

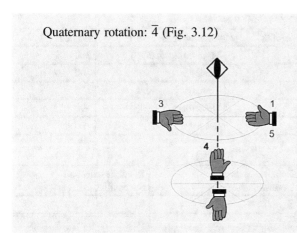

Fig. 3.12 90° rotation and simultaneous inversion

Senary rotation: $\bar{6}$ (Fig. 3.13)

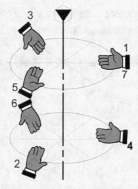

Fig. 3.13 60° rotation and simultaneous inversion

Table 3.6 shows the Hermann–Mauguin notation of these axes, as well as the Schoenflies notation, its denomination, and the number of degrees of rotation of the axes.

The equivalence with the rotation-reflection operation, which does not exist in the Hermann–Mauguin notation, is shown in Table 3.7.

Axis $\bar{1}$ is the center of symmetry.

Table 3.6 Summary of symmetry improper operations

Hermann–Mauguin	Schoenflies	Rotation axis	Degrees (°)
$\bar{1}$	C_1	Monarium (identity)	360
$\bar{2}$	C_2	Binary	180
$\bar{3}$	C_3	Ternary	120
$\bar{4}$	C_4	Quaternary	90
$\bar{6}$	C_6	Senary	60

Table 3.7 Equivalence between the rotation-reflection and rotation-inversion operations

Rotation-inversion Hermann–Mauguin	Rotation-reflection Schoenflies
$\bar{1}$	S_2
$\bar{2}$	σ
$\bar{3}$	S_6
$\bar{4}$	S_4
$\bar{6}$	S_3

The operation that makes an object with initial x, y, z coordinates be transformed by the center of symmetry into another with $-x, -y, -z$ coordinates is called inversion (Fig. 3.14).

Example

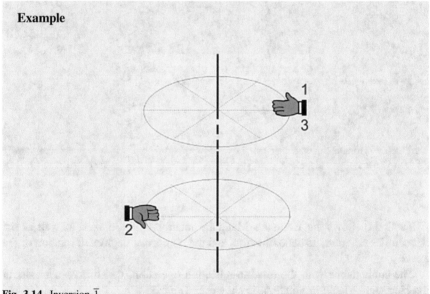

Fig. 3.14 Inversion $\bar{1}$

The axis is $\bar{2}$ equivalent to the plane of symmetry, m.

Reflection is the operation of symmetry that makes any motif or object that appears on one side of the element of symmetry, called a reflection plane, appear on the other side of the same element and at the same distance.

In three dimensions, the element of symmetry is the reflection plane.

The symbol of the reflection plane according to Hermann–Mauguin notation is m (Fig. 3.15).

Example

Fig. 3.15 Reflection plane

In the Schoenflies notation, the symbol is s,

s_h horizontal plane perpendicular to the main axis of rotation (which has the highest order).

s_n vertical plane including the rotation axis.

s_d diagonal plane including the main rotation axis and divides in two the angle between two C_2 axes, which are normal to the main rotation axis.

In two dimensions, the element of symmetry is the reflection line. The symbol of the reflection line according to Hermann–Mauguin notation is m (Fig. 3.16).

Example

Fig. 3.16 Vertical and horizontal symmetry lines

Table 3.8 Summary of symmetry operations

Symmetry operation	Rotations	Rotation axes	Notes
Proper rotations	$360°/n$	1, 2, 3, 4 and 6	1 is the *identity*
Improper rotations	$360°/n$ and simultaneous inversion	$\bar{1}, \bar{2}, \bar{3}, \bar{4}$ and $\bar{6}$	$\bar{1}$ Equivalent to a *symmetry center* $\bar{2}$ Equivalent to a *symmetry plane* $\bar{3}$ Equivalent to *ternary axis + center of symmetry* $\bar{6}$ Equivalent to a *ternary axis perpendicular to a symmetry plane*

The axis is $\bar{3}$ equivalent to axis 3 and the symmetry center.

The axis is $\bar{6}$ equivalent to axis 3 and a plane m perpendicular to it ($3/m$).

In Table 3.8, a summary of symmetry operations is presented.

3.7.2   Reflection-Translation (Glide)

– **In three dimensions**

Gliding is a symmetrical operation consisting of a reflection and a translation.

The symmetry operator is the glide plane. The translation must be contained in the gliding plane. The travel distance must be half the unit translation in that direction (Fig. 3.17).

Example
Gliding plane

Fig. 3.17 Gliding plane

In Hermann–Mauguin notation, the following gliding planes are distinguished: (1) axial, (2) diagonal, and (3) diamond.

(1) **Axial gliding plane**

Axial gliding plane is a plane whose gliding component is parallel to a crystallographic axis. Its length is half the period of the translation along this axis. It is symbolized as *a*, *b* or *c,* depending on whether the glide is along the crystallographic axes *a*, *b* or *c,* respectively.

Examples

Effect of the axial glide plane, depending on the type and in relation to different crystalline planes of an orthorhombic cell (Fig. 3.18).

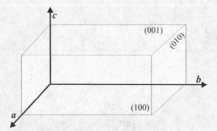

Fig. 3.18 Orthorhombic cell

1. Axial *a* perpendicular plane to crystalline plane (001) (Fig. 3.19a) and (010) (Fig. 3.19b), respectively, of an orthorhombic cell.

(a)   (b)

Fig. 3.19 Axial *a* Perpendicular Plane to Crystalline Plane (001) (**a**) and (010) (**b**)

2. Axial plane **b** perpendicular to the crystalline plane (001) (Fig. 3.20a and b) and (100) (Fig. 3.20c), respectively, of an orthorhombic cell.

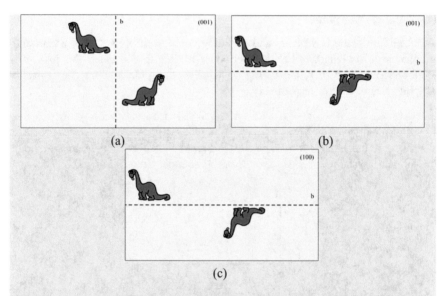

Fig. 3.20 (a) and (b) Axial **b** Perpendicular Plane to the Crystalline Plane (001); (c) Axial **b** Perpendicular Plane to Crystalline Plane (100)

3. Axial plane **c** perpendicular to the crystalline plane (100) (Fig. 3.21a) and (010) (Fig. 3.21b), respectively, of an orthorhombic cell.

Fig. 3.21 Axial Plane **c** Perpendicular to the Crystalline Plane (100) (a) and (010) (b)

(2) **Diagonal gliding plane**

Diagonal gliding plane is a plane whose gliding component is:

(a+b)/2
(a+c)/2
(b+c)/2

Its symbol is *n*.

Examples

Effect of the axial glide plane, depending on the type and in relation to different crystalline planes of an orthorhombic cell (Fig. 3.22).

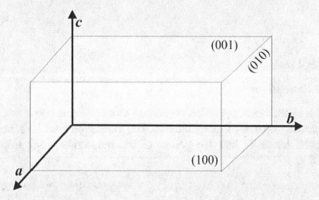

Fig. 3.22 Orthorhombic cell

1. Diagonal gliding plane n perpendicular to the crystalline plane (001), gliding (b+c)/2b (Fig. 3.23).

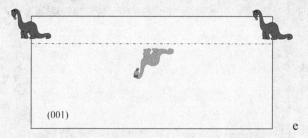

Fig. 3.23 Diagonal gliding plane n perpendicular to the crystalline plane (001)

2. Diagonal gliding plane n perpendicular to the crystalline plane (100), gliding (c + a)/2 (Fig. 3.24).

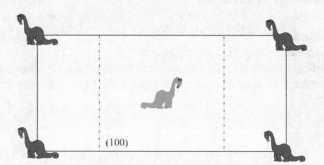

Fig. 3.24 Diagonal gliding plane n perpendicular to the crystalline plane (100)

(3) **Diamond gliding plane**

Diamond gliding planes are planes whose gliding component is:

(a+b)/4
(a+c)/4
(b+c)/4

Its symbol is *d*.

– **In two dimensions**

Gliding is a symmetrical operation consisting of a reflection and a translation.

The symmetry operator is the glide line. The translation must be contained in the glide line. Its length is half the period of the translation along this axis. It is symbolized as *g*.

Example
Gliding line (Fig. 3.25).

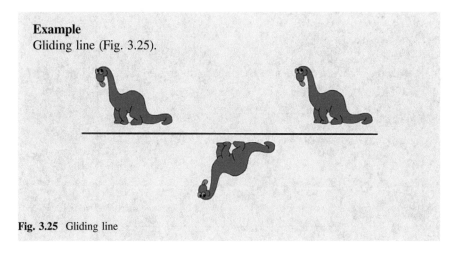

Fig. 3.25 Gliding line

3.7.3   Rotation-Translation

Rotation-translation is the operation involving rotation of order 2, 3, 4 or 6 and constant translation along the axis of rotation.

The rotation is taken counterclockwise and the translation is taken upwards.

The operator that allows the operation to be carried out is the screw axis.

Table 3.9 Screw axes

Order of the axis	Screw axes
2	2_1
3	$3_1\ 3_2$
4	$4_1\ 4_2\ 4_3$
6	$6_1\ 6_2\ 6_3\ 6_4\ 6_5$

The number of existing screw axes is $n - 1$, where n is the order of the axis. Thus, the existing screw axes are those shown in Table 3.9.

Enantiomorphic screw axes (each is the mirror image of the other)

3_1 and 3_2
4_1 and 4_3
6_1 and 6_5
6_2 and 6_4

Table 3.10 shows the symbol of the different screw axes in the Hermann–Mauguin notation, the degrees of rotation, and the translation involved (Fig. 3.26).

Table 3.10 Rotations and translations of the screw axes

Axis order	Symbol	Rotation (°)	Translation (UVW)
Binary	2_1	+180	1/2
Ternary	3_1	+120	1/3
	3_2		2/3
Quaternary	4_1	+90	1/4
	4_2		1/2
	4_3		3/4
Senary	6_1	+60	1/6
	6_2		1/3
	6_3		1/2
	6_4		2/3
	6_5		5/6

Examples

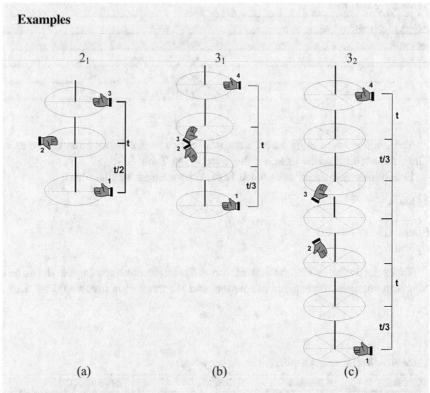

(a)   (b)   (c)

Fig. 3.26 Screw axes examples: 2_1 **(a)**, 3_1 **(b)**, 3_2 **(c)**

Exercises
Non-translational Symmetry

1. Draw all the symmetrical reticular rows to row [13], indicating their symbol, as a consequence of the action of the quaternary rotation point passing through the node taken as origin, in the net of Fig. E1.

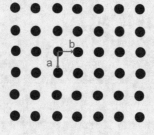

Fig. E1

2. Draw all the elements of symmetry contained in the two-dimensional lattices of Fig. E2a and b.

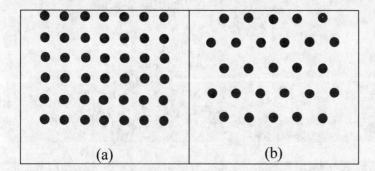

(a)   (b)

Fig. E2

3. Draw all the grid rows symmetrical to F1 row of the lattice in Fig. E3, indicating their symbol, as a consequence of the non-translational symmetry contained in it.

Fig. E3

4. Draw the elements of symmetry (proper, improper, and with associated translation) in the two-dimensional periodic models in Fig. E4.

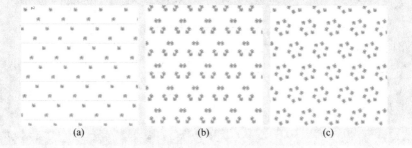

(a)   (b)   (c)

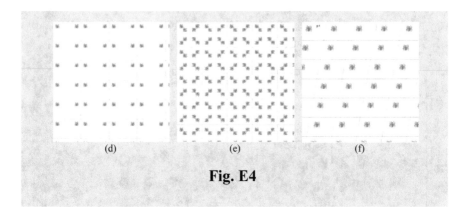

(d)   (e)   (f)

Fig. E4

Questions

1. If a rotation axis of order n has binary axes perpendicular to it:
   - ○ a. there shall be binary axes in a number independent of the order of the axis
   - ○ b. there shall be an even number of binary axes
   - ○ c. there will be no binary axes
   - ○ d. there shall be as many binary axes as the order of the axis
2. A binary axis implies,
   - ○ a. 180° rotation
   - ○ b. 180° and 360° rotations
   - ○ c. 120° and 240° rotations
   - ○ d. 120°, 240° and 360° rotations
3. A screw axis 6_3 parallel to the crystallographic axis c implies:
   - ○ a. 60° rotation and 1/2 translation along the direction of the crystallographic axis b
   - ○ b. 60° rotation and 3/4 translation along the direction of the crystallographic axis b
   - ○ c. 60° rotation and 1/3 translation along the direction of the crystallographic axis $\underline{c}$
   - ○ d. 60° rotation and 1/2 translation along the crystallographic axis c
4. One element of symmetry that allows the operation of reflection is:
   - ○ a. a reflexion plane with translation associated with the crystallographic axis c
   - ○ b. a reflexion plane with associated translation in the direction of the crystallographic axis b
   - ○ c. a reflexion plane without associated translation
   - ○ d. a plane of reflexion with associated translation in the direction of the crystallographic axis a

5. The symbol *m* means:
   ○ a. translation and gliding plane
   ○ b. reflexion plane
   ○ c. translation and axis of rotation
   ○ d. translation and reflexion plane
6. What is the name of the symmetry operation involving 90° rotations and simultaneous inversion?
   Answer: [   ]
7. How many degrees of rotation does the rotation axis of order 3 involve?
   Answer: [   ]
8. What is the name of the symmetry operation whose operator is the center of symmetry?
   Answer: [   ]
9. Write the Hermann–Mauguin symbol of an element of symmetry involving 120° rotation and 2/3 translation
   Answer: [   ]
10. Write the name of the element of symmetry in the plane that implies reflection and translation.
    Answer: [   ]
11. Write the name(s) of the symmetry operation(s) shown in Fig. Q1.

Fig. Q1

    Answer: [   ]
12. What is the name of the vector by which the position of a node in the lattice can be defined?
    Answer: [   ]
13. The line of symmetry is the element that allows the operation of reflection in the space of three dimensions
    ○ True
    ○ False
14. Crystallographic symmetry axes are those of order 1, 2, 3, 4, 5, and 6.
    ○ True
    ○ False

15. In a gliding plane, the glide can be along:
- a. any direction other than a crystallographic axis
- b. the direction of a crystallographic axis
- c. the direction of the three crystallographic axes simultaneously
- d. the direction of two crystallographic axes simultaneously.

Chapter 4
Point Symmetry

Abstract The concept of point group is defined, and the wo-dimensional point groups and three-dimensional groups are presented as the groups of non-translational symmetry, both of the lattices and of the crystals. The symbol is described using the international notation of Hermann–Mauguin, showing also that of Schoenflies. The directions of symmetry and their relation to the crystallographic directions for each type of lattice, both two and three-dimensional, are presented. The equivalence between the point groups and the crystalline classes is shown, since in many texts of mineralogy they appear extensively described. The concept of crystalline form is defined, the types that exist, and how they can be named using a symbol or a name that alludes to a polyhedral form. The concept of the crystalline system is defined as a set of point groups compatible with certain types of Bravais lattices. From the study of the faces of a crystal, its geometrical and symmetrical relations were deduced by the laws of classical crystallography, and they are presented. The representation of the symmetry and faces of a crystal through a projection is described. It starts with the three-dimensional projection on the pole sphere and then the stereographic projection in two dimensions. To do this, the concept of a normal bundle to the faces is used.

4.1 Introduction

Crystalline morphology provided experimental data for the development of mathematical crystallography until the discovery of X-ray diffraction by crystals in 1912. There are three fundamental laws of crystallography:

(1) Dihedral angles law.[1]

The dihedral angles that form the equivalent faces of various crystals of a substance are equal and characteristic of that substance, whatever the shape of the crystal (see Fig. 4.1).

---

[1] de Romé Delisle [1].

© The Author(s), under exclusive license to Springer Nature Switzerland AG 2022
C. Marcos, *Crystallography*, Springer Textbooks in Earth Sciences, Geography and Environment, https://doi.org/10.1007/978-3-030-96783-3_4

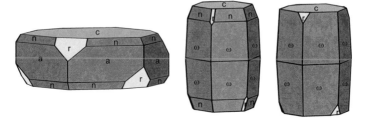

Fig. 4.1 Corundum showing different habits in which the angle between the *r* and *n* faces of the figures in the center and on the right are the same. The angle between the *r* and *n* faces of the right and central crystals is also the same

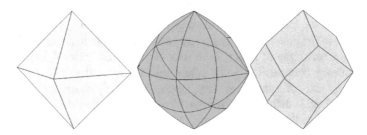

Fig. 4.2 Crystals of the same crystalline species with cubic symmetry showing different habits

(2) Symmetry law.[2]

All the crystals of the same substance have the same symmetry, whatever their faces (Fig. 4.2).

(3) Law of the rationality of the indices.[3]

The intersecting edges of three sides of a crystal allow definition of a coordinate axis system. The distance a fourth face cuts to each axis is considered the unit of measurement on this axis. All the other faces of the crystal cut to these axes at distances whose ratio to the lengths defined as units are rational and generally simple numbers (Fig. 4.3).

4.2   Point Group Definition

A point group is defined as the set of symmetry operations existing in a crystal lattice. It has all the characteristics of a mathematical group. There is a point in space that is equivalent to itself, which is usually taken as the origin of coordinates.

---

[2] Haüy [2].

[3] Haüy [3].

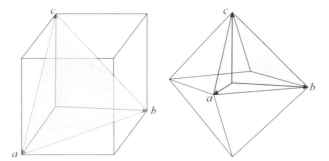

Fig. 4.3 Crystalline face (111) intersecting the fundamental translations *a, b,* and *c* to the unit distance

Order of the group is the number of elements that constitute it. If the group has n elements, the group is of order n.

A subgroup is defined as the set of elements of a group that by themselves meet the group conditions.

Box 4.1. Characteristics of a mathematical group

Any combination of two or more elements (or operations) must be equivalent to an element that also belongs to the group.

The combination is a multiplication; that is, the successive execution of symmetrical operations.

This operation can be expressed as follows:

$$AB = C$$

Where

A, B, and C are elements of the group, which we consider to be finite since the number of elements of symmetry is finite.

In the point group 2/*m*, 2 and *m* are elements that belong to the group and their combination is equivalent to another element $\bar{1}$, which also belongs to the group (Fig. 4.4).

In Fig. 4.4a, point 2 is obtained by applying the reflection to point 1 and then axis 2. The combination of the reflection (applied to point 1) and then the binary rotation (applied to the reflected point 1) gives rise to point 2. Therefore, the line of symmetry, the binary rotation, and the point of rotation-monetary inversion belong to the group.

In Fig. 4.4b, point 2 is obtained by applying the center of symmetry to point 1. The combination of the plane m and the axis 2 on point 1 is equivalent to the action of the center of symmetry on point 1.

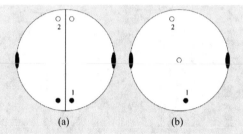

Fig. 4.4 Point group 2/m

In the group, there must be an element that can be combined with all the other elements of the group, leaving them all unaltered. This is the monary axis or the identity. This property can be expressed as follows:

$$AE = A$$

Where:
A is any element of the group
E is the identity or the monary axis.
The combination of the identity element with all other elements must be commutative, i.e., (Fig. 4.5).

Fig. 4.5 The combination of reflection and mono-rotation causes reflection and the combination of monary rotation and reflection also causes reflection

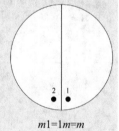

$m1=1m=m$

$$AE = EA = A$$

The combination of elements must be associative. It means that the following relationship must be fulfilled:

$$A(BC) = (AB)C$$

Where
A, B, and C are elements of the group.

Each of the elements of the group has the inverse element, so that the product of the element by its inverse is equal to the identity element.

$$AX = E$$

Where:
A is an element of the group,
X is its inverse element,
E is the identity.
Also, if X is the inverse of A, A must be the inverse of X:

$$AX = XA = E$$

4.3   Rules that Condition the Presence of Several Symmetry Elements in the Same Point Group

1. If there is an even-ordered axis of rotation and a plane of reflection perpendicular to it, there is a center of symmetry at its intersection.
2. If a series of symmetry planes are cut on a symmetry axis, there are as many planes as the order of the axis.
3. If a rotation axis of order n has binary axes perpendicular to it, there will be as many binary axes as the order of the axis.
4. If there is a binary axis perpendicular to a reversed axis of rotation, whose order n is even, there are n/2 planes intersecting the y-axis and n/2 binary axes perpendicular to it.

4.4   Crystalline System

Crystalline system is defined as the set of point groups compatible with Bravais lattices.

In the last century, point groups were grouped into classes that most authors call crystalline systems, although the terms singony and crystalline type have also been used.

Two or more point groups are said to belong to the same crystalline system if they support the same Bravais lattices. In this way, seven crystalline systems appear (Table 4.1).

Table 4.1 Crystalline systems and Bravais lattices

Crystalline systems	Bravais lattices
Triclinic	P
Monoclinic	P, A (B, C)
Orthorhombic	P, I, F, A (B, C)
Tetragonal	P, I
Hexagonal	P
Rhombohedral	P
Cubic	P, I, F

Table 4.2 Axial cross and axial cross angles of each crystalline system

Crystalline system	Axial cross	Axial cross angles
Triclinic	$a \neq b \neq c$	$\alpha \neq \beta \neq \gamma \neq 90°$
Monoclinic	$a \neq b \neq c$	$\alpha = \gamma = 90° \neq \beta$
Orthorhombic	$a \neq b \neq c$	$\alpha = \beta = \gamma = 90°$
Tetragonal	$a = b \neq c$	$\alpha = \beta = \gamma = 90°$
Hexagonal	$a = b \neq c$	$\alpha = \beta = 90° \; \gamma = 60°$ or $120°$
Rhombohedral	$a = b \neq c$	$\alpha = \beta = 90° \; \gamma = 60°$ or $120°$
Cubic	$a = b = c$	$\alpha = \beta = \gamma = 90°$

The axial cross in each crystalline system (Table 4.2) is chosen, taking into account the following considerations:

– The axial crossings of each system coincide with the seven primitive Bravais cells (P).
– They are constructed so that the coordinate axes coincide with the symmetry elements of the material in the crystalline state.
– When there is a rotation axis of higher order than binary, the direction of the *c-axis* is chosen according to the direction of that axis.
– The directions of the *a* and *b* axes are chosen according to the binary axes if they exist.

In the orthorhombic system, the directions of *a*, *b* and *c* are chosen according to the binary axes when they exist.
In the monoclinic system, the direction of *b* is chosen according to the single binary axis if it exists.
In the triclinic system, the choice of *a*, *b*, and *c* is made by selecting the three smallest and non-coplanar edges.

– The choice of these axes on the crystal would be made in the same way as for the choice of the lattice constants *a*, *b*, and *c*.
– It is customary to call *x*, *y*, and *z* the coordinates coinciding with the directions of the lattice constants *a*, *b*, and *c*.

4.5 Point Group Symbol

There are two types of symbols.

1. Schoenflies notation is the oldest. It consists of a capital letter, characteristic of the point group type. It can be accompanied by one or more subscripts: One of them is numerical. The other one is a small letter. When both exist, they are written in this order.
2. Hermann–Mauguin notation (or international notation). It consists of a sequence of numbers and the letter *m* (reflection plane). They correspond to the symbols that represent the different elements of symmetry. They can include the following: Slash. Denominator is the letter *m*. Numerator is a number that refers to the order of a rotation axis. Some symbols can be simplified if this does not lead to confusion with other symbols.

Box 4.2. Symmetry directions
To obtain the symbol of a point group, according to the international notation, the directions of symmetry of the plane lattices (Table 4.3) or three-dimensional lattices (Table 4.4) must be taken into account, depending on whether the point group is two-dimensional (Table 4.3) or three-dimensional (Table 4.4).

There are three directions of symmetry: Primary, secondary, and tertiary.

- In triclinic lattices, there is no direction of symmetry.
- In monoclinic lattices, there is a direction of symmetry.
- In rhombohedral lattices, there are two directions of symmetry, but it must be taken into account that either hexagonal or rhombohedral axes are chosen.
- In the orthorhombic, tetragonal, hexagonal, and cubic lattices, there are three directions.
- In the symbol for the point groups of the triclinic and monoclinic systems, it is only necessary to specify the existing symmetry element (1 or $\bar{1}$) in the triclinic and the existing symmetry element in the only symmetry direction of the monoclinic.

Table 4.3 Plane lattices symmetry directions

Lattice	Symmetry directions Position in Hermann–Mauguin notation		
Oblique	Rotation point in the plane	Secondary	Tertiary
Rectangular		[10]	[01]
Square		[10] [01]	[1$\bar{1}$] [11]
Hexagonal		[10] [01] [$\bar{1}$1]	[1$\bar{1}$] [1$\bar{2}$] [$\bar{2}$1]

Table 4.4 Three-dimensional lattices symmetry directions

Lattices	Symmetry directions		
	Primary	Secondary	Tertiary
Triclinic	Ninguna		
Monoclinic	[010] or [001] or [100]		
Orthorhombic	[100]	[010]	[001]
Tetragonal	[001]	[100] [010]	$[1\bar{1}0][110]$
Hexagonal	[001]	[100] [010] $[\bar{1}\bar{1}0]$	$[1\bar{1}0][120][\bar{2}\bar{1}0]$
Rhombohedral	[001]	$[100][001][\bar{1}\bar{1}0]$	
Rhombohedral	[111]	$[1\bar{1}0]$ $[01\bar{1}]$ $[\bar{1}01]$	
Cubic	[100] [010][001]	[111] $[1\bar{1}\bar{1}]$ $[\bar{1}1\bar{1}]$ $[\bar{1}\bar{1}1]$	$[1\bar{1}0]$ $[01\bar{1}]$ $[\bar{1}01]$ [110] [011][101]

- In the case of point groups of the rhombohedral system, the existing symmetry elements in the two symmetry directions must be specified.
- In the case of the point groups of the orthorhombic, tetragonal, hexagonal, and cubic systems, the symmetry elements existing in the three symmetry directions must be specified.
- If, in any of the symmetry directions, there is no symmetry element, nothing is written.

Note: When there is more than one direction of symmetry in the column of primary, secondary or tertiary directions for a given crystalline system, it means they are equivalent.

4.6 Symmetry Operations of Point Groups

The point group symmetry operations are as follows:

- Proper rotations: 1, 2, 3, 4, 6
- Improper rotations: $\bar{1}$, $\bar{2}$ (reflection), $\bar{3}$, $\bar{4}$, $\bar{6}$.

The total number of three-dimensional point groups is 32.

They are given various names, some derived from geometric shapes that possess the symmetry of the point group, while other names describe the characteristics of the group.

4.7 Point Groups and Crystal Classes

A crystal class is a set of crystals that possess the same point group. According to the International Tables of Crystallography[4] a "geometrical" crystal class classifies the symmetry groups of the external shape of macroscopic crystals.

Their names come from the geometric shapes that have the symmetry of the point group or the characteristics of the group.

Depending on the elements of symmetry, they are distinguished by class:

- Holohedry is the crystalline class that has the highest number of symmetry operations.
- Hemihedry is the class containing half the symmetry operations. In turn, it may be:

  Paramorphic: It is characterized by the preservation of the center of symmetry.
  Enantiomorphic: There are no planes of symmetry in it.
  Hemimorphic: The symmetry axes are polars.

- Tetartohedry is the crystalline class that has the fourth part of the symmetry operations.

Tables 4.5, 4.6, 4.7, 4.8, 4.9, 4.10 and 4.11 present, for each of the seven crystalline systems, the point groups with Hermann–Mauguin notation and Schoenflies notation in brackets, the respective crystalline classes named according to[5] symmetry operations and symmetry elements expressed by the formula described in Box 4.3.

Table 4.5 Point groups, crystalline classes, symmetry operations, and elements of the triclinic crystalline system

Point group	Crystalline class	Number of operations	Symmetry elements	Symmetry elements diagram
$\bar{1}$ (C_i)	Pinacoidal (holoedry)	2	C	
1 (C_1)	Pedial (hemiedry)	1	1	

---

[4] Hahn [4].

[5] Friedel [5].

Table 4.6 Point groups, crystalline classes, symmetry operations, and elements of the monoclinic crystalline system

Point group	Crystalline class	Number of operations	Symmetry elements	Symmetry elements diagram
$2/m$ (C_{2h})	Prismatic (holoedry)	4	$1E^2\ m\ C$	
2 (C_2)	Sphenoidal (enantiomorphic hemiedry)	2	$1E^2$	
m (C_s)	Domatic (hemimorphic hemiedry)	2	m	

Table 4.7 Point groups, crystalline classes, symmetry operations, and elements of the orthorhombic crystalline system

Point group	Crystalline class	Number of operations	Symmetry elements	Symmetry elements diagram
mmm (D_{2h})	Dipyramidal (holoedry)	8	$3E^2\ 3m\ C$	
222 (D_2)	Disphenoidal (enantiomorphic hemiedry)	4	$3E^2$	
$mm2$ (D_{2v})	Pyramidal (hemimorphic hemiedry)	4	$2m\ 1E^2$	

Table 4.8 Point groups, crystalline classes, symmetry operations, and elements of the tetragonal crystalline system

Point group	Crystalline class	Number of operations	Symmetry elements	Symmetry elements diagram
4/mmm (D_{4h})	Ditetragonal-dipyramidal (holoedry)	16	$1E^4$ $4E^2$ $5m$ C	
4mm (C_{4v})	Ditetragonal-pyramidal (enantiomorphic hemiedry)	8	$1E^4$ $4m$	
$\bar{4}$ 2m (D_{2d})	Scalenohedral (hemiedry with inversion)	8	$1E^4$ $2E^2$ $2m$ C	
422 (D_4)	Trapezohedral (enantiomorphic hemiedry)	8	$1E^4$ $4E^2$	
4/m (C_{4h})	Dipyramidal (paramorphic hemiedry)	8	$1E^4$ $1m$ C	
$\bar{4}$ (S_4)	Disphenoidal (tetartoedry with inversion)	4	$1E^4$ C	
4 (C_4)	Pyramidal (tetartoedry)	4	$1E^4$	

Table 4.9 Point groups, crystalline classes, symmetry operations, and elements of the rhombohedral crystalline system

Point group	Crystalline class	Number of operations	Symmetry elements	Symmetry elements diagram
$\bar{3}m$ (D_{3d})	Ditrigonal scalenohedral (holoedry)	12	$1E^3$ $3E^2$ $3m$ C	
$3m$ (C_{3v})	Ditrigonal pyramidal (hemimorphic hemiedry)	6	$1E^3$ $3m$	
32 (D_3)	Trapezohedral (enantiomorphic hemiedry)	6	$1E^3$ $3E^2$	
$\bar{3}(C_{3i})$	Rhombohedral (paramorphic hemiedry)	6	$1E^3$ C	
3 (C_3)	Pyramidal tetartohedral (tetartoedry)	3	$1E^3$	

Table 4.10 Point groups, crystalline classes, symmetry operations, and elements of the hexagonal crystalline system

Point group	Crystalline class	Number of operations	Symmetry elements	Symmetry elements diagram
6/mmm (D_{6h})	Dihexagonal dipyramidal (holoedry)	24	$1E^6$ $6E^2$ $7m$ C	
$\bar{6}$ 2m (D_{3h})	Ditrigonal dipyramidal (hemiedry with inversion)	12	$1E^3$ $3E^2$ $4m$ C	
6mm (C_{6v})	Dihexagonal pyramidal (hemimorphic hemiedry)	12	$1E^6$ $6m$	
622 (D_6)	Trapezohedral (enantiomorphic hemiedry)	12	$1E^6$ $6E^2$	
6/m (C_{6h})	Dipyramidal (paramorphic hemiedry)	12	$1E^6$ $1m$ C	
$\bar{6}$ (C_{3h})	Trigonal dipyramidal (tetartohedry with inversion)	6	$1E^3$ $1m$ C	
6 (C_6)	Pyramidal (tetartohedry)	6	$1E^6$	

Table 4.11 Point groups, crystalline classes, symmetry operations, and elements of the cubic crystalline system

Point group	Crystalline class	Number of operations	Symmetry elements	Symmetry elements diagram
$m\,\bar{3}m$ (O_h)	Hexakisoctahedral (holoedry)	48	$3E^4\ 4E^3\ 6E^2$ $9m$ C	
$\bar{4}\,3m$ (T_d)	Hexakistetrahedral (hemiedry with inversion)	24	$3E^4\ 4E^3\ 6m$	
432 (O)	Pentagon-icositetrahedral (enantiomorphic hemiedry)	24	$3E^4\ 4E^3\ 6E^2$	
$m\,\bar{3}$ (T_h)	Disdodecahedral (paramorphic hemiedry)	24	$3E^2\ 4E^3\ 3m,$ C	
23 (T)	Tetrahedral-Pentagondodecahedral (tetartohedry)	12	$3E^2\ 4E^3$	

Box 4.3. Formula for expressing the elements of symmetry of a given point group

The formula consists of a series of characters expressing the elements of symmetry. The symbols used are as follows:

- C indicates the center of symmetry
- E indicates the symmetry axis

The number of axes of symmetry of a given type is expressed by placing the number before the letter E.

The type of symmetry axis is expressed by placing the symbol of the corresponding axis, according to the Hermann–Mauguin notation, in the form of a superscript to the right of the letter E.

Example: The formula to indicate that there are 4 ternary axes of rotation inversion would be as follows: $4E^{\bar{3}}$.

- m indicates reflexion plane

Example: The formula to indicate that there are 3 binary and 4 ternary axes would be as follows: $3E^2$, $4E^3$.

Appendix I, Tables 1, 2, 3, 4, 5, 6 and 7, present point groups and crystalline classes, equivalent positions, and the stereographic projection of the symmetry elements and face poles of the general form corresponding to each crystalline system.

4.8 Two-Dimensional Point Groups and Point Groups of the Plane Lattices

There are 10 two-dimensional point groups and they are:
1, *m*, 2, *2mm*, 3, *3m*, 4, *4mm*, 6, *6mm*.
Examples can be seen in Box 4.4.

Box 4.4. Examples of two-dimensional point groups

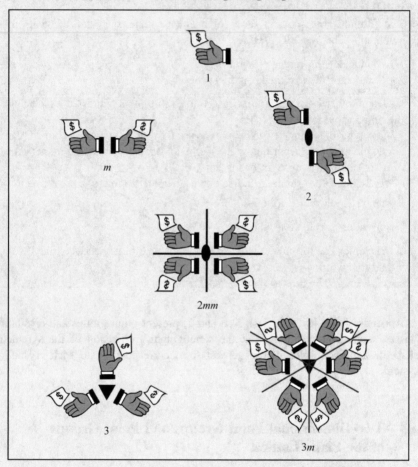

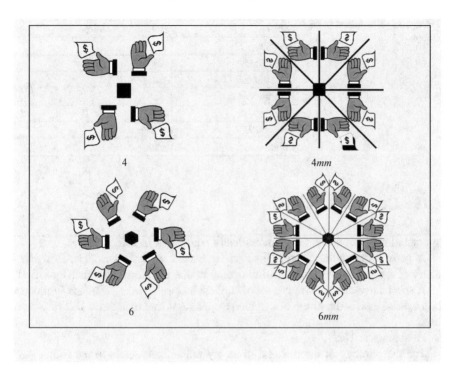

The following groups characterize plane lattices:

Point group	Plane lattice
2	Oblique
2*mm*	Rectangular
2*mm*	Orthorhombic
4*mm*	Square
6*mm*	Hexagonal

4.9   Three-Dimensional Point Groups and Three-Dimensional Lattices Point Groups

There are 32 three-dimensional point groups. Their symbols (Table 4.3 are obtained by taking into account the symmetry direction of the lattices), according to international notation. The point groups that characterize three-dimensional lattices are the following (Table 4.12).

Table 4.12 Holohedral point groups

Lattices	Holohedral point group
Triclinic	$\bar{1}$
Monoclinic	$2/m$
Orthorhombic	mmm
Rhombohedral	$\bar{3}m$
Hexagonal	$6/mmm$
Tetragonal	$4/mmm$
Cubic	$m\bar{3}m$

4.10 Crystalline Forms

A crystalline form is a set of symmetrically equivalent faces.

A general crystalline form is a set of equivalent general faces. The face symmetry of a general face is the identity operation that transforms this face into itself.

A special crystalline form is a set of equivalent special faces. The face symmetry of a special face is the group of symmetry operations that transforms this face onto itself.

Its symbol is $\{hkl\}$.

The morphology of the material in its crystalline state refers to the forms generated by natural processes.

The number, aspect, and distribution of the faces of a crystal is governed by the symmetry of the crystal.

Crystalline forms can be:

–   Open: crystalline forms do not limit a space (Fig. 4.6).

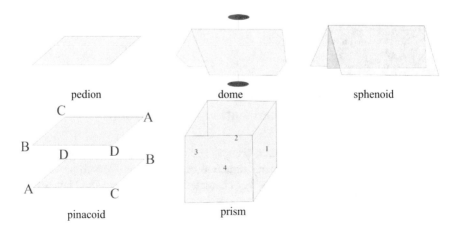

Fig. 4.6 Open crystalline forms

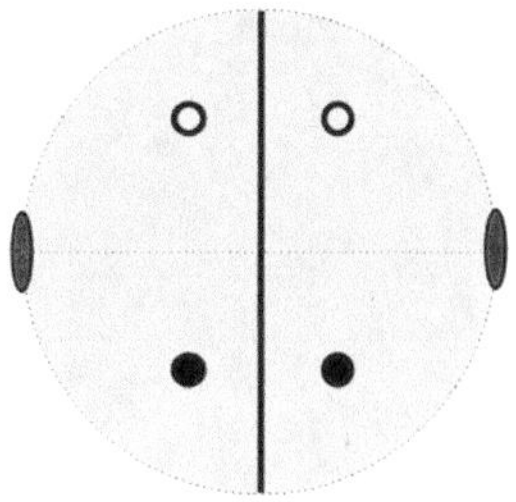

Fig. 4.7 Faces of a monoclinic crystal (2/*m*) in a general position, with multiplicity
4

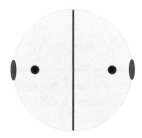

Fig. 4.8 Faces of a monoclinic crystal (2/*m*) in a particular   position,   with
multiplicity 2

– Closed: crystalline forms that limit a space.
– Simple: crystalline forms made up of a single form.
– Composite: crystalline forms made up of several forms.

Multiplicity is the number of faces generated by the symmetry elements.

A face is said to be in a *general position* when it is not placed on any symmetrical element (Fig. 4.7).

A face is said to be in a *special* or *particular position* when it is placed on a
symmetrical element (Fig. 4.8).

4.11   Zone and Zone Axis

A zone is defined as the set of crystalline planes with a common crystallographic
direction, called the zone axis.

A zone axis is the crystallographic direction common to a series of crystalline
planes.

Its symbol is [*uvw*] and it is obtained as explained in paragraph 13 of Theme 2.

Fig. 4.9 Bundle of normal to
faces

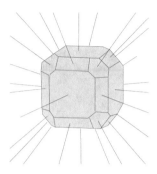

4.12   Bundle of Normals to Faces

The bundle of perpendiculars is the set of perpendiculars traced from the origin of
coordinates to different crystalline faces (Fig. 4.9). It is characterized by the fact
that it contains the angles between the faces.

4.13   Crystalographic Projections

4.13.1   *Spherical Projection*

Spherical projection is the three-dimensional projection of the perpendiculars to the
faces and the elements of symmetry in a sphere, called *sphere of poles*. The projection of a perpendicular to a face on the poles sphere is a point called pole.
Figure 4.10 shows the poles (dots) of the faces of the crystal inscribed on the poles
sphere.

In the poles sphere, the angles between the faces, the zones (the maximum
circles joining the poles of the faces. In Fig. 4.11, they have a common direction),
and the angles between the edges, are preserved.

– Spherical coordinates of a pole.

The spherical coordinates of a pole are the coordinates, φ and ρ, that determine the
position of a pole on the pole sphere. The angle φ is the distance between two
meridians, one of them is taken as the origin of coordinates and passes through the
pole N; the other goes through the N pole, the S pole, and the face pole. The angle
is equal to the arc between the point N and the pole P, drawn on the meridian
passing through N, P, and S (Fig. 4.12).

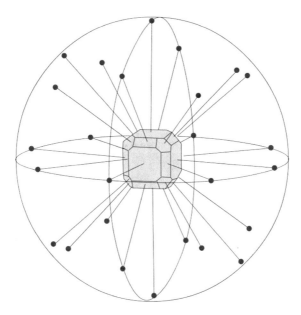

Fig. 4.10 Poles sphere

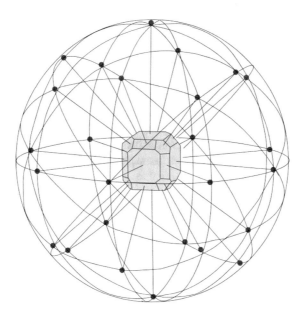

Fig. 4.11 Zones

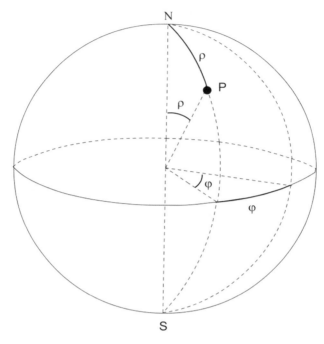

Fig. 4.12 Spherical coordinates of a pole, φ and ρ, that determine the position of a pole on the pole sphere

4.13.2 Stereographic Projection

The stereographic projection is a two-dimensional projection in which the elements of symmetry and the bundle of normal are projected onto the faces of a crystal.

The projection plane used is usually the equatorial plane. The point of view is the southern pole for the poles of the upper hemisphere of the polar sphere; and the northern pole for the poles of the lower hemisphere of the polar sphere. The angle φ is retained but not the ρ angle, whose value is (Fig. 4.13).

$$\rho_{\text{projection}} = \text{Rtg}(\rho/2) \tag{4.1}$$

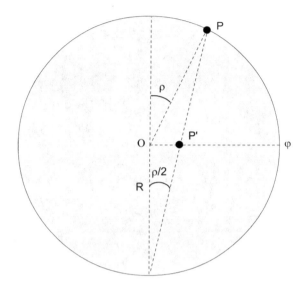

Fig. 4.13 Stereographic projection of a pole P. ρ in the projection is the distance OP and is equal to Rtg (ρ/2), where R is the radius of the sphere

Examples of stereographic projections of minerals can be seen in Appendix II.

Box 4.5. Stereographic projection of the crystallographic axes, zones and poles of the fundamental faces of the different crystalline systems

- Orthogonal systems: Cubic, tetragonal, orthorhombic (Fig. 4.14).
- Rhombohedral and hexagonal systems (Fig. 4.15).
- Monoclinic system (Fig. 4.16).
- Triclinic system (Fig. 4.17).

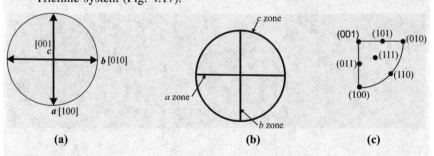

Fig. 4.14 Orthogonal systems: **a** Projection of crystallographic axes; **b** projection of zones of crystallographic axes; **c** projection of fundamental faces

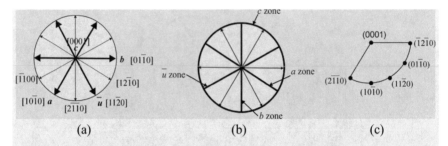

Fig. 4.15 Rhombohedral and hexagonal systems: **a** Projection of crystallographic axes; **b** projection of zones of crystallographic axes; **c** projection of fundamental faces

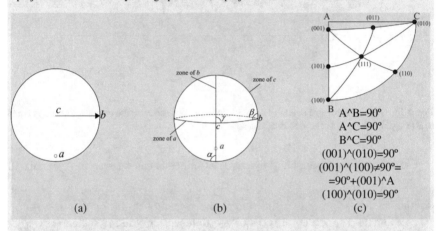

Fig. 4.16 Monoclinic system: **a** Projection of crystallographic axes; **b** projection of zones of crystallographic axes; **c** projection of fundamental faces

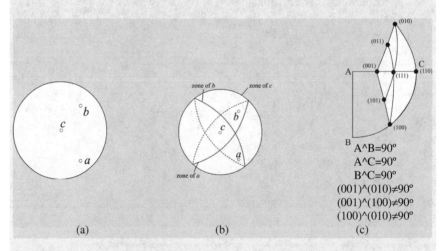

Fig. 4.17 Triclinic system: **a** Projection of crystallographic axes; **b** projection of zones of crystallographic axes; **c** projection of fundamental faces

4.14   Crystallographic Calculations

The stereographic projection is useful because it allows us to obtain the point group and the crystalline system from the representation of the poles of their faces.

For the representation of these poles, it is necessary to know the spherical coordinates obtained by measuring the angles with a *goniometer* (Box 4.6).

Box 4.6. Goniometer

It is a device to measure the interfacial angles* of the crystals. Two main types are used:

1. The contact goniometer, for large crystals. It consists of an angle conveyor, with an oscillating arm that is placed in contact with the crystalline faces. In general, the results are not very accurate.
2. The optical goniometer, suitable for small crystals with reflective and shiny faces.

   There are several versions of this type, depending on the ability of the crystal to reflect a beam of light directed at it from a collimator. The reflection is detected by an observing telescope. The crystal is rotated from one reflection position to the next, and the angle of rotation is measured. Optical goniometers are very useful, due to their high degree of precision and accuracy.

   Knowing these coordinates, parametric relationship (*a/b:b/b: c/b* or *a/b*:1:*c/b*) and angles of the axial cross (*α, β, γ*) can be obtained. To do this, it is necessary to know the faces (100), (010), (001), and (110) or (101) or (011), which are part of a spherical triangle, called the fundamental triangle.

   * Interfacial angle is the angle between the normals on both sides of a crystal

---

In an orthogonal crystal (cubic, tetragonal or orthorhombic), the parametric relationship is obtained by Eqs. 4.2 and the angles α, β and γ by expressions (4.3).

$$\frac{a}{b} = \frac{\sin(100) \wedge (110)}{\sin(110) \wedge (010)} \cdot \frac{\sin(010) \wedge (001)}{\sin(001) \wedge (100)}$$

$$\frac{c}{b} = \frac{\sin(001) \wedge (011)}{\sin(011) \wedge (010)} \cdot \frac{\sin(010) \wedge (100)}{\sin(100) \wedge (001)}$$

$$\frac{c}{a} = \frac{\sin(001) \wedge (101)}{\sin(101) \wedge (100)} \cdot \frac{\sin(100) \wedge (010)}{\sin(010) \wedge (001)}$$

(4.2)

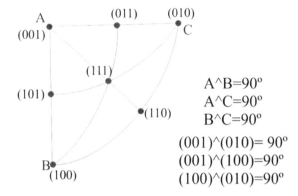

Fig. 4.18 Stereographic projection of the fundamental faces of an orthogonal crystal

$$\alpha = 180 - (001) \wedge (010)$$
$$\beta = 180 - (001) \wedge (100) \qquad\qquad (4.3)$$
$$\gamma = 180 - (100) \wedge (010)$$

It is necessary to know the angles among the fundamental crystalline faces (Fig. 4.18) and to use spherical trigonometry (Box 4.7) to obtain the exact values of the parametric ratio.

Box 4.7. Spherical trigonometry

Spherical trigonometry is of great importance for stereographic projection theory and geodesy. It is also the basis for astronomical calculations. For example, the solution of the so-called astronomical triangle is used to find the latitude and longitude of a point, the time of day, the position of a star, and other magnitudes.

We start with a radio unity sphere. If we cut this sphere with a plane that passes through the center of the sphere, we obtain what is called a *maximum circle*. If, on the other hand, the cutting plane does not pass through the center of the sphere, what we'll get is a *smaller circle*.

Let us now consider a sphere and a full circle. If we draw a line perpendicular to the plane that defines the maximum circle and that passes through the center of the sphere, what we obtain is two points on the sphere that are called *poles*. In addition, the maximum circle will divide the sphere into two hemispheres.

The *dihedral angle* is the angle between two maximum circles.

At this point, we can define a *spherical triangle* as a portion of a spherical surface limited by three maximum circles, with the condition that the measure of each of the arcs is less than 180°.

To solve a spherical triangle, it is enough to know at least three of the six data, sides and/or angles, of that triangle.

Relationships that satisfy the sides and angles of a triangle:

- One side of a spherical triangle is less than the sum of the other two sides and greater than its difference.
- The sum of the three sides of a spherical triangle is less than 360°.
- The sum of the three angles is greater than 180° and less than 540°.
- If a spherical triangle has two equal angles, the opposite sides are also equal to each other.
- If a spherical triangle has two unequal angles, the greater the angle, the greater the side.

After seeing these relationships, it is interesting to note that, for the resolution of spherical triangles, there are a number of formulas, such as Bessel formulas, cotangent formulas, and Borda formulas. In the case of a right-angled spherical triangle (one angle is 90°), or a straight triangle (one side is 90°), the resolution is simplified with the Neper pentagon rule.

Sines theorem: In a spherical triangle, the sides and their opposite angles fulfill the relations (first Bessel group).

$$(\mathrm{sen}a/\mathrm{sen}A) = (\mathrm{sen}b)/(\mathrm{sen}B) = (\mathrm{sen}c)/(\mathrm{sen}C)$$

Cosine theorem: In a spherical triangle, each side and its opposite angle satisfy the equality (second Bessel group).

$$\cos a = \cos b \cos c + \sin b \sin c \cos A$$

$$\cos b = \cos c \cos a + \sin c \sin a \cos B$$

$$\cos c = \cos a \cos b + \sin a \sin b \cos C$$

Miller indices of a single-sided pole, other than one of the fundamentals, can be calculated by one of the following methods:

- Wulff director cosine method:

$$(a/h) \cos \varphi = (b/k) \cos \chi = (c/l) \cos \omega \qquad (4.4)$$

The angles (φ, χ, ω) are those that form the normal to the face with the three coordinate axes (x, y, and z), respectively.

- Method of sines ratio of Miller consists of calculating the indices of the pole of an unknown face from the indices of the poles of the other three known faces with which it is in a zone.

Exercises

1. Assemble the paper solids shown in Table E1, and for each:

1. Observe the elements of symmetry.
2. Obtain the point group.
3. Orient the solid to perform a stereographic projection of the symmetry elements and the face poles.
4. Project stereographically the elements of symmetry and the poles of the faces (using the zone concept) on diagrams in Table E1.
5. Give notation to the face poles.
6. Indicate the crystalline forms present (using Appendix I).

Guidelines:

Observe the elements of symmetry.

1. Obtain the crystalline system from said elements, since the axis of symmetry of the highest order characterizes said system.

- Axis of order 1 characterizes the triclinic system.
- Axis of order 2 characterizes the monoclinic and orthorhombic systems. The difference between them is that in the monoclinic there is only one other element of symmetry at most, which is a plane perpendicular to the axis.
- The axis of order 3 characterizes the rhombohedral system.
- The axis of order 4 characterizes the tetragonal system.
- The axis of order 6 characterizes the hexagonal system.
- Axes of order 4, 3, and 2 characterize the cubic system, although some distinction must be made.
- When there are axes of order 4, 3, and 2, the only possible system is the cubic system.
- The axis of order 4 also characterizes the tetragonal system, but in this, there is only one and in the cubic system there are three.
- Axis of order 3 also characterizes the rhombohedral system, but there is only one and in the cubic system there are four.
2. Obtaining the specific group.

To obtain the specific group, the symmetry elements must be associated with symmetry directions, for which you must do the following:

Orient the solid, essential to project the elements of symmetry and poles of the faces. For this, it is necessary to know the axial cross and the symmetry directions of said system. As the crystalline system is known, the symmetry elements found must be related to the symmetry directions (that coincide with the directions of the crystallographic axes or intermediate directions) of said system. Remember that normally the axis of maximum symmetry coincides with the *c* crystallographic axis. The axes of symmetry coincide (are parallel)

Table E1 In the first column of the table are the crystalline solids of which the corresponding crystalline forms must be indicated (second column) and the stereographic projection of the symmetry elements and poles of the faces in the diagrams of the third and fourth columns, respectively, must be performed

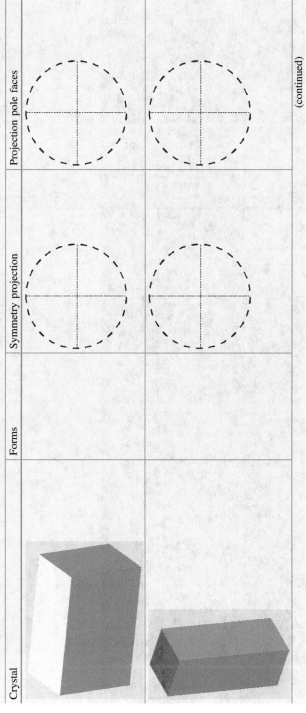

(continued)

Table E1 (continued)

Crystal	Forms	Symmetry projection	Projection pole faces
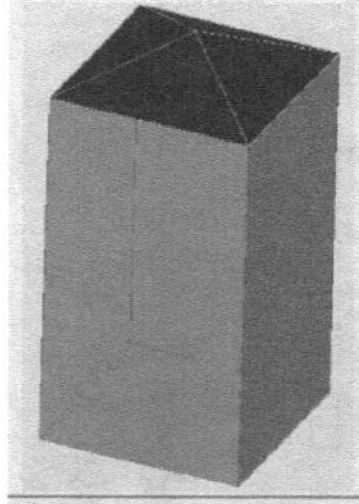		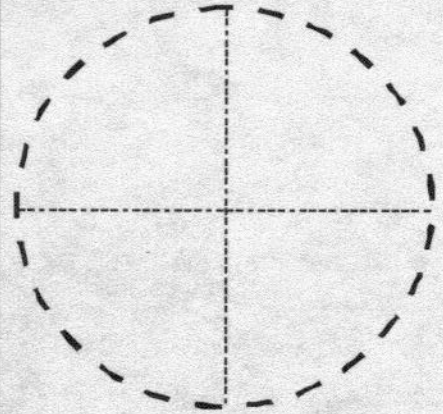	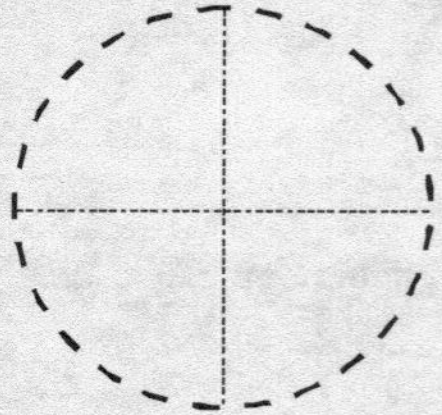
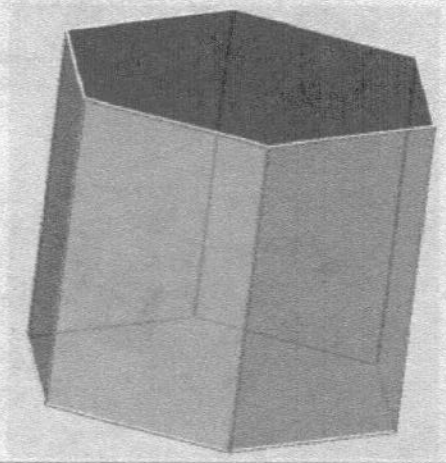		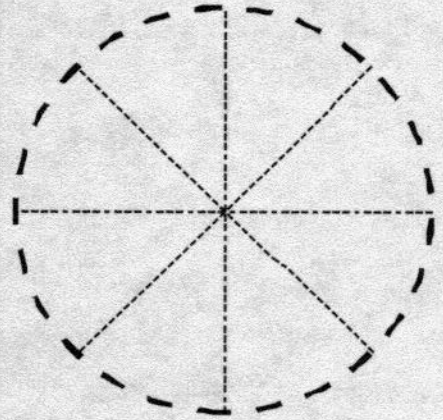	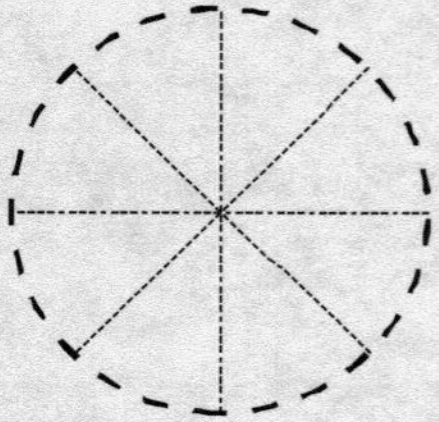

(continued)

Table E1 (continued)

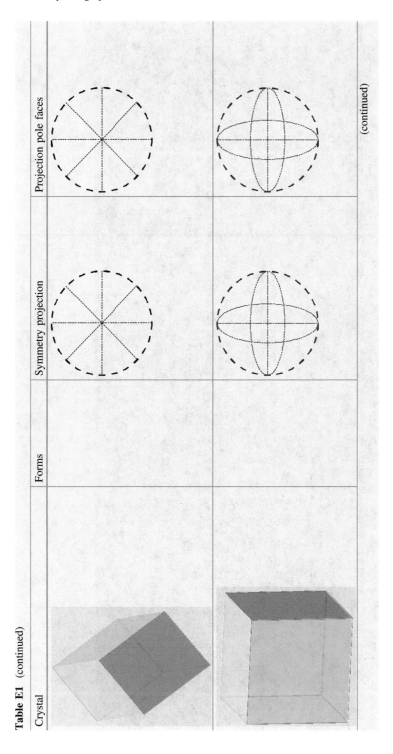

Crystal	Forms	Symmetry projection	Projection pole faces

(continued)

Table E1 (continued)

Crystal	Forms	Symmetry projection	Projection pole faces

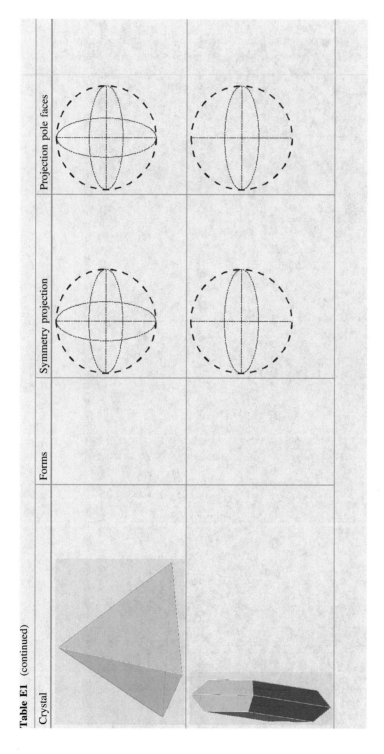

with the symmetry directions and the symmetry planes are perpendicular to the symmetry directions.

3. Stereographic projection of the elements of symmetry and poles of the faces.

After crystal orientation, the elements of symmetry and the poles of the faces are projected.

For this, it must be remembered how the crystallographic axes are projected for each crystalline system and the zones of these axes.

4. Assign generic Miller or Bravais Miller indices (hexagonal and rhombohedral systems) to the poles of the faces.

An easy way to do this is using the zone concept and remembering the following:

– For faces that are in the zone of one of the crystallographic axes, the corresponding Miller index is zero.
– Faces that do not fall in the zone of any of the crystallographic axes cut all three axes, and the Miller indices will take different values according to the distance at which they cut to those axes. The greater the distance, the smaller the Miller indices because they are the inverse of the Weiss coefficients.

The crystalline forms are obtained through the Miller indices since there is correspondence between them.

Paper-based crystalline models to be assembled from the first column of Table E1 are shown in Table E2.

Table E2 Paper-based crystalline models

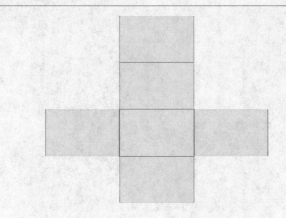

(continued)

Table E2 (continued)

Table E2 (continued)

Table E2 (continued)

Table E2 (continued)

Questions

1. The 4/*mmm* point group characterizes the crystalline system

   ○ a. tetragonal and consists of 3 quaternary axes
   ○ b. tetragonal and consists of a single quaternary axis
   ○ c. cubic and consists of 3 quaternary axes
   ○ d. cubic and consists of a single quaternary axis

2. The hemihedral crystalline class has

   ○ a. a fifth of the elements of symmetry in relation to the holohedral class
   ○ b. a quarter of the elements of symmetry in relation to the holohedral class
   ○ c. half of the elements of symmetry in relation to the holohedral class
   ○ d. one third part of the elements of symmetry in relation to the holohedral class

3. The point group 4*mm*

   ○ a. is a group that can belong to both cubic space lattices and flat squares
   ○ b. is a group that can belong to both tetragonal and rectangular flat space lattices
   ○ c. is a group that can belong to both tetragonal space lattices and flat squares
   ○ d. is a group that can belong to both cubic and flat rectangular space lattices

4. The axial cross with $a \neq b \neq c$ corresponds to

   ○ a. the monoclinic and orthorhombic crystal systems
   ○ b. the triclinic and monoclinic crystal systems
   ○ c. the triclinic and orthorhombic crystalline systems
   ○ d. the triclinic, monoclinic and orthorhombic crystal systems

5. The Holohedral Point Group of the Tetragonal System is 4/*m*2/*m*2/*m* and Simplified *mmm*.

   ○ True
   ○ False

6. A point group is defined as the set of symmetry operations existing in a crystal lattice. It is a mathematical group because A, B, C, E (identity element), and X (inverse element) are elements of the group it fulfills

   AB = E
   AE = E
   AE = EA = A
   A(BC) = (AB)C
   AX = E
   ○ True
   ○ False

7. The rhombohedral system is characterized as:

$a = b = c$ and $\alpha = \beta = 90°$ $\gamma = 60°$ ó $120°$

○ True
○ False

8. In Hermann Mauguin's notation (or international notation) the point group symbol consists of a succession of characters that correspond to the symbols that represent the different elements of symmetry. They can include the letter corresponding to the type of lattice; and break bar with the letter m as the denominator, and a number that refers to the order of an axis of rotation as the numerator. Some symbols can be simplified if it does not lead to confusion with other symbols

○ True
○ False

9. The monoclinic system has only one direction of symmetry.

○ True
○ False

10. In the holohedral point group of the hexagonal system, the number of axes of order 6 is 1, of order 2 is 6, has 6 planes of reflection, and has no center of symmetry

○ True
○ False

11. Match each point group with its corresponding crystalline system

$m\bar{3}m$	Orthorhombic system
422	Cubic system
222	Tetragonal system

12. Match each space group with its corresponding crystal lattice

$P2/d\,\bar{3}$	Rombohedral P
P312	Orthorhombic P
Imma	Monoclinic P

13. How many space groups are there?

Answer: [   ]

14. Write the maximum number (using numbers) of characters to be specified in the symbol of the spatial or point group, without considering the translations

Answer: [   ]

15. The number of symmetry elements in a spatial group is finite.

   ○ True
   ○ False

16. Match each element of symmetry (only that with maximum symmetry) with the type of crystal lattice (three-dimensional) that it characterizes.

Axis of order 4	Triclinic
Binary axis	Tetragonal
Center of symmetry	Monoclínic

17. Match each space group with its corresponding point group of which it is a subgroup

I2/b₂/c₂/a₂	4mm
Bb	m
P4mm	mmm

18. The spatial groups reflect the internal and external symmetry of the material in a crystalline state, and the point groups only reflect the internal symmetry.

   ○ True
   ○ False

19. If you think that there are axes of order 5 in space groups, answer with the term **yes**, and otherwise answer with **no**.

Answer: [   ]

20. If you consider that, in the definition of space groups, crystalline materials must be considered as discrete media, write **discrete**. If you think they must be defined as continuous media, write **continuous**.

Answer: [   ]

References

1. de Romé Delisle JBL (1783) Cristallographie ou Description des Formes Propres à tous le Corps du Regne Minéral 4 vols. Paris
2. Haüy RJ (1803) Traité Élémentaire de Physique I-II. 426 + 447 p., Paris
3. Haüy RJ (1784) Essai d'une Théorie sur la Structure des Crystaux. Gogué et Née de La Rochelle
4. Hahn T (2005) International tables for crystallography, vol A: Space group symmetry: space group symmetry (5th revised Edition. 2002. 2nd printing Edition). Springer
5. Friedel G (1926) Leçons de crisrallographie. Blanchard, Paris (Reprint 1964)

Chapter 5
Space Groups

Abstract The concept of spatial group is defined, differentiating it from point group, and its symbol is described. The concept of equivalent position is introduced, and the difference between general and particular equivalent positions is shown. How the spatial groups are represented in the International Crystallography Tables is described. The concepts of multiplicity, Wyckoff's symbol, position symmetry, and asymmetric unit are defined.

5.1 Space Groups Definitions

Spatial groups can be defined as groups of transformations of the homogeneous and discrete three-dimensional space itself.

The principle of homogeneity of a substance in its crystalline state, considered at a microscopic level, i.e., considering the atomicity of the crystalline substance, includes the principles of symmetry (the crystalline substance contains an infinite number of equal points per symmetry) and of discretion (not all points of a crystalline substance are identical).

These principles are realized simultaneously in the crystal lattice. The conditions of homogeneity and discretion determine that all the spatial groups are periodic groups and therefore crystallographic groups, with symmetry axes of orders 1, 2, 3, 4 and 6.

Space groups contain the translation group of the three-dimensional lattice as a subgroup and, therefore, space groups can also be defined as groups in which proper and improper rotations are accompanied by the translations.

© The Author(s), under exclusive license to Springer Nature Switzerland AG 2022
C. Marcos, *Crystallography*, Springer Textbooks in Earth Sciences,
Geography and Environment, https://doi.org/10.1007/978-3-030-96783-3_5

5.2   Space Group Symmetry Operations

The symmetry operations contained in the space groups are as follows:

- Proper rotations
- Improper rotations
- Translations
- Reflexions with translations
- Rotations with translations.

The number of elements of symmetry that exist in a space group is infinite because the translations repeat the elemental cell infinitely but, for the same reason, the space group is perfectly defined from this cell.

The total number of space groups is 230, which were obtained in 1890 almost simultaneously by Federov[1] and Schoenflies.[2]

5.3   Derivation of Space Groups

The methods for deriving the space groups can be geometric, arithmetic, combinatorial or group theory.

In any case, we must bear in mind the following:

1. A Bravais lattice is an arrangement of mathematical points that have position but not magnitude or shape. For each crystalline system, the possible space lattices have the symmetry of the holohedral point group.
2. Matter in the crystalline state is made up of ions, atoms or molecules, which are associated with each of the nodes of the Bravais lattice, i.e., the motif with point group symmetry that is repeated by the group of translations of the lattice must be that of the holohedral point group of the lattice or a subgroup thereof.

Figure 5.1 shows an example of spatial symmetry, corresponding to the *Imm2* space group (Fig. 5.1a). It is based on an interior-centered orthorhombic lattice I (Fig. 5.1b), whose point symmetry, *mmm,* is that of the holohedral point group of orthorhombic lattices (Fig. 5.1c). The repetition motif in the example is a star with symmetry 6 *mm*, which is a point group subgroup of the *mmm* (holohedral point group of the orthorhombic lattices). An example is the structure of the hemimorphite ($Zn_4Si_2O_7(OH)_2 \cdot H_2O$).

- **Types of space groups**

There are two types of space groups:—Symmorphic space groups and non-symmorphic space groups.

---

[1] Fedorov [1].

[2] Schoenflies [2].

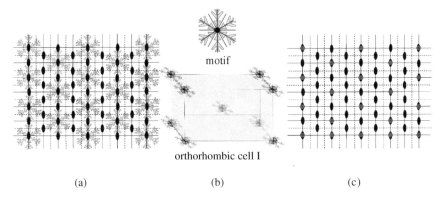

motif

orthorhombic cell I

(a)   (b)   (c)

Fig. 5.1 Spatial symmetry: **a** *Imm2* space group, **b** motif and cell, **c** *mmm* holohedric point group of the orthorhombic lattices

Table 5.1 Number of symmorphic groups by crystalline system

Crystalline system	Specific groups	Bravais lattices	Symmorphic groups
Cubic	5	3	15
Tetragonal	7	2	14 + 2
Rhombohedral	5	2	10 + 3
Hexagonal	7	1	7 + 1
Orthorhombic	3	4	12 + 1
Monoclinic	3	2	6
Triclinic	3	1	2

– *Symmorphic space groups*

Symmorphic space groups are space groups obtained in a simple way by combining each of the 32 point groups with each of the Bravais lattices compatible with them. In total, there are 66 plus 7 symmorphic space groups. Table 5.1 shows the number of symmorphic space groups corresponding to each crystalline system.

These groups appear because the geometrical relations between the symmetry elements of the group and the lattice vary, as a consequence of which Bravais lattices centered on the bases (A, B or C) have a special direction (Fig. 5.2).

The point symmetry of the group is not the holohedry and the lattice presents different orientations with respect to the symmetry elements of the group (Fig. 5.3).

In the three-dimensional space, the geometric idea of a symmetric group is equivalent to placing the symmetry elements of a given point group on the nodes of the Bravais lattice compatible with that symmetry (Fig. 5.4). Symmetry elements with associated translation are obtained.

Figure 5.5 show the motif (Fig. 5.5a) and how, by combining a one-dimensional lattice (Fig. 5.5b) with a point group *m* (Fig. 5.5c), the symmorphic space group

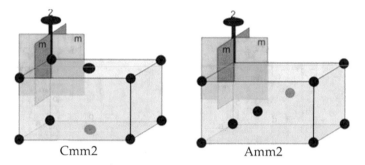

Fig. 5.2   Appearance of space groups due to the centering of lattices

Fig. 5.3   Appearance of space groups due to the different orientation of symmetry elements

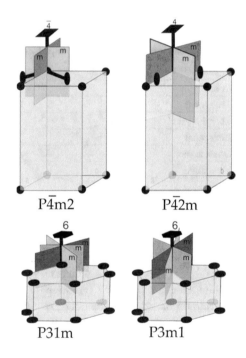

Fig. 5.4 Geometric idea of a symmorphic group

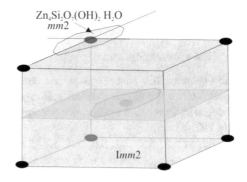

Fig. 5.5 Obtaining a symmorphic space group: motif (**a**), translation (**b**), point group m (**c**), translation b + reflexion m = symmorphic space group (**d**), glide line g (**e**)

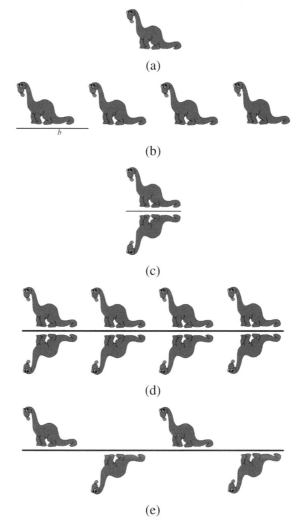

(Fig. 5.5d) is obtained, in which the glide line g appears (Fig. 5.5e), in addition to translation, identity and reflexion.

– Non-symmorphic space groups

Non-symmorphic space groups are those derived from the symmorphic space groups when a multiple translation is considered, since elements of symmetry with associated translation appear, not present in the symmorphic groups. In total there are 157.

Tables 5.2, 5.3, 5.4, 5.5, 5.6. 5.7, 5.8 present the symmorphic and non- symmorphic space groups for each crystalline system.

Table 5.2 Symmorphic and non-symmorphic space groups of triclinic crystalline system

Symmorphic groups	Non-symmorphic groups	
	Hemisymmorphic	Asymmorphic
1 $P1$		
2 $P\bar{1}$		

Table 5.3 Symmorphic and non-symmorphic space groups of monoclinic crystalline system

Symmorphic groups	Non-symmorphic groups	
	Hemisymmorphic	Asymmorphic
1st orientation: c = 2 and/or −2		
3 $P2$	7 Pc	4 $P2_1$
5 $C2$	9 Cc	11 $P2_1/m$
6 Pm	13 $P2/c$	14 $P2_1/c$
8 Cm	15 $C2/c$	
10 $P2/m$		
12 $C2/m$		

Symmorphic groups	Non-symmorphic groups	
	Hemisymmorphic	Asymmorphic
2nd orientation: b = 2 and/or −2		
3 $P2$	7 Pb	4 $P2_1$
5 $B2$	9 Bb	11 $P2_1/m$
6 Pm	13 $P2/b$	14 $P2_1/b$
8 Bm	15 $B2/b$	
10 $P2/m$		
12 $B2/m$		

Table 5.4 Symmorphic and non-symmorphic space groups of orthorhombic crystalline system

Symmorphic groups	Non-symmorphic groups	
	Hemisymmorphic	Asymmorphic
16 *P222*	27 *Pcc2*	17 *P222$_1$*
21 *C222*	28 *Pma2*	18 *P2$_1$2$_1$2*
22 *F222*	30 *Pnc2*	19 *P2$_1$2$_1$2$_1$*
23 *I222*	32 *Pba2*	20 *C222$_1$*
25 *Pmm2*	34 *Pnn2*	24 *I2$_1$2$_1$2$_1$*
35 *Cmm2*	37 *Ccc2*	26 *Pmc2$_1$*
38 *Amm2*	39 *Abm2*	29 *Pca2$_1$*
42 *Fmm2*	40 *Ama2*	31 *Pmn2$_1$*
44 *Imm2*	41 *Aba2*	33 *Pna2$_1$*
47 *Pmmm*	43 *Fdd2*	36 *Cmc2$_1$*
65 *Cmmm*	45 *Iba2*	51 *Pmma*
69 *Fmmm*	46 *Ima2*	52 *Pnna*
71 *Imm*	48 *Pnnn*	53 *Pmna*
	49 *Pccm*	54 *Pcca*
	50 *Pban*	55 *Pbam*
	66 *Cccm*	56 *Pccn*
	67 *Cmma*	57 *Pbcm*
	68 *Ccca*	58 *Pnnm*
	70 *Fddd*	59 *Pmmn*
	72 *Ibam*	60 *Pbcn*
		61 *Pbca*
		62 *Pnma*
		63 *Cmcm*
		64 *Cmca*
		73 *Ibca*
		74 *Imma*

Table 5.5 Symmorphic and non-symmorphic space groups of rhombohedral crystalline system

Symmorphic groups	Non-symmorphic groups	
	Hemisymmorphic	Asymmorphic
143 *P3*	158 *P3c*1	144 *P3$_1$*
146 *R3*	159 *P31c*	145 *P3$_2$*
147 *P$\bar{3}$*	161 *R3c*	151 *P3$_1$12*
148 *R$\bar{3}$*	163 *P$\bar{3}$1c*	152 *P3$_1$2$_1$*
149 *P312*	165 *P$\bar{3}$c1*	153 *P3$_2$12*
150 *P321*	167 *R$\bar{3}$c*	154 *P3$_2$21*
155 *R32*		
156 *P3m1*		
157 *P31m*		
160 *R3m*		
162 *P$\bar{3}$1m*		
164 *P$\bar{3}$m1*		
166 *R$\bar{3}$m*		

Table 5.6 Symmorphic and non-symmorphic space groups of hexagonal crystalline system

Symmorphic groups	Non-symmorphic groups	
	Hemisymmorphic	Asymmorphic
168 *P6*	184 *P6cc*	169 $P6_1$
174 *P$\bar{6}$*	188 *P$\bar{6}$c2*	170 $P6_5$
175 *P6/m*	190 *P$\bar{6}$2c*	171 $P6_2$
177 *P622*	192 P6/*mcc*	172 $P6_4$
183 *P6mm*		173 $P6_3$
187 *P$\bar{6}$m2*		176 $P6_3/m$
189 *P$\bar{6}$2m*		178 $P6_122$
191 *P6/mmm*		179 $P6_522$
		180 $P6_222$
		181 $P6_422$
		182 $P6_322$
		185 $P6_3cm$
		186 $P6_3mc$
		193 $P6_3/mcm$
		194 $P6_3/mmc$

Table 5.7 Symmorphic and non-symmorphic space groups of tetragonal crystalline system

Symmorphic groups	Non-symmorphic groups	
	Hemisymmorphic	Asymmorphic
75 *P4*	85 *P4/n*	76 $P4_1$
79 *I4*	100 *P4bm*	77 $P4_2$
81 *P$\bar{4}$*	103 *P4cc*	78 $P4_3$
82 *I$\bar{4}$*	104 *P4nc*	80 $I4_1$
83 *P4/m*	108 *I4cm*	84 $P4_2/m$
87 *I4/m*	112 *P$\bar{4}$ 2c*	85 *P4/n*
89 P*422*	116 *P$\bar{4}$ c2*	86 $P4_2/n$
97 *I422*	117 *P$\bar{4}$ b2*	88 $I4_1/a$
99 *P4mm*	118 *P$\bar{4}$ n2*	90 $P42_12$
107 *I4mm*	120 *I$\bar{4}$ c2*	91 $P4_122$
111 *P$\bar{4}$2m*	124 *P4/mcc*	92 $P4_12_12$
115 *P$\bar{4}$m2*	125 *P4/nbm*	93 $P4_222$
119 *I$\bar{4}$m2*	126 *P4/nnc*	94 $P4_22_12$
121 *I$\bar{4}$2m*	140 *I4/mcm*	95 $P4_322$
123 *P4/mmm*		96 $P4_32_12$
139 *I4/mmm*		98 $I4_122$
		101 $P4_2cm$
		102 $P4_2nm$
		105 $P4_2mc$
		106 $P4_2bc$
		109 $I4_1md$
		110 $I4_1cd$
		113 *P$\bar{4}$ 2_1m*
		114 *P$\bar{4}$ 2_1c*
		122 *I$\bar{4}$ 2d*
		127 P4/*mbm*
		129 *P4/nmm*
		130 *P4/ncc*

Table 5.7 (continued)

Symmorphic groups	Non-symmorphic groups	
	Hemisymmorphic	Asymmorphic
		131 $P4_2/mmc$
		132 $P4_2/mcm$
		133 $P4_2/nbc$
		134 $P4_2/nnm$
		135 $P4_2/mbc$
		136 $P4_2/mnm$
		137 $P4_2/nmc$
		138 $P4_2/ncm$
		141 $I4_1/amd$
		142 $I4_1/acd$
		128 $P4/mnc$

Table 5.8 Symmorphic and non-symmorphic space groups of cubic crystalline system

Symmorphic group	Non-symmorphic groups	
	Hemisymmorphic	Asymmorphic
195 $P23$	201 $Pn\bar{3}$	198 $P2_13$
196 $F23$	203 $Fd\bar{3}$	199 $I2_13$
197 $I23$	218 $P\bar{4}3n$	205 $Pa\bar{3}$
200 $Pm\bar{3}$	219 $F\bar{4}3c$	206 $Ia\bar{3}$
202 $Fm\bar{3}$	222 $Pn\bar{3}n$	208 $P4_232$
204 $Im\bar{3}$	226 $Fm\bar{3}c$	209 $F432$
207 $P432$		210 $F4_132$
209 $F432$		212 $P4_332$
211 $I432$		213 $P4_132$
215 $P\bar{4}3m$		214 $I4_132$
216 $F\bar{4}3m$		220 $I\bar{4}3d$
217 $I\bar{4}3m$		223 $Pm\bar{3}n$
221 $Pm\bar{3}m$		224 $Pn\bar{3}m$
225 $Fm\bar{3}m$		225 $Fm\bar{3}m$
229 $Im\bar{3}m$		227 $Fd\bar{3}m$
		228 $Fd\bar{3}c$
		230 $Ia\bar{3}d$

5.4   Space Group Symbol

There are two types of symbols.

1. Schoenflies notation is the oldest of all and consists of a capital letter, characteristic of the point group type. It can be accompanied by one or more subscripts, one numerical and one small letter. When both exist, they are written in this order.

2. Hermann Mauguin notation (or international notation) consists of a capital letter that indicates the type of Bravais lattice, and a set of characters, after the capital letter, indicating elements of symmetry referring to the symmetry directions of the lattice (Box 2 Chap. 4). They can include a slash and a denominator, the letter m. The numerator is a number that refers to the order of a rotation axis. Some symbols can be simplified if this does not lead to confusion with other symbols.

Table 5.9 shows the 230 space groups. The number corresponds to the order number in the International Tables for Crystallography.[3]

Table 5.9 Space groups

No	Symbol	No	Symbol	No	Symbol	No	Symbol	No	Symbol
1	$P1$	2	$P\bar{1}$	3	$P2$	4	$P2_1$	5	$C2$
6	Pm	7	Pc	8	Cm	9	Cc	10	$P2/m$
11	$P2_1/m$	12	$C2/m$	13	$P2/c$	14	$P2_1/c$	15	$C2/c$
16	$P222$	17	$P222_1$	18	$P2_12_12$	19	$P2_12_12_1$	20	$C222_1$
21	$C222$	22	$F222$	23	$I222$	24	$I2_12_12_1$	25	$Pmm2$
26	$Pmc2_1$	27	$Pcc2$	28	$Pma2$	29	$Pca2_1$	30	$Pnc2$
31	$Pmn2_1$	32	$Pba2$	33	$Pna2_1$	34	$Pnn2$	35	$Cmm2$
36	$Cmc2_1$	37	$Ccc2$	38	$Amm2$	39	$Abm2$	40	$Ama2$
41	$Aba2$	42	$Fmm2$	43	$Fdd2$	44	$Imm2$	45	$Iba2$
46	$Ima2$	47	$Pmmm$	48	$Pnnn$	49	$Pccm$	50	$Pban$
51	$Pmma$	52	$Pnna$	53	$Pmna$	54	$Pcca$	55	$Pbam$
56	$Pccn$	57	$Pbcm$	58	$Pnnm$	59	$Pmmn$	60	$Pbcn$
61	$Pbca$	62	$Pnma$	63	$Cmcm$	64	$Cmca$	65	$Cmmm$
66	$Cccm$	67	$Cmma$	68	$Ccca$	69	$Fmmm$	70	$Fddd$
71	$Immm$	72	$Ibam$	73	$Ibca$	74	$Imma$	75	$P4$
76	$P4_1$	77	$P42$	78	$P43$	79	$I4$	80	$I41$
81	$P\bar{4}$	82	$I\bar{4}$	83	$P4/m$	84	$P4_2/m$	85	$P4/n$
86	$P42/n$	87	$I4/m$	88	$I41/a$	89	$P422$	90	$P42_12$
91	$P4_122$	92	$P4_12_12$	93	$P4_222$	94	$P4_22_12$	95	$P4_322$
96	$P4_32_12$	97	$I422$	98	$I4_122$	99	$P4mm$	100	$P4bm$
101	$P4_2cm$	102	$P4_2nm$	103	$P4cc$	104	$P4nc$	105	$P4_2mc$
106	$P4_2bc$	107	$I4mm$	108	$I4cm$	109	$I41md$	110	$I41cd$
111	$P\bar{4}2m$	112	$P\bar{4}2c$	113	$P\bar{4}2_1m$	114	$P\bar{4}2_1c$	115	$P\bar{4}m2$
116	$P\bar{4}c2$	117	$P\bar{4}b2$	118	$P\bar{4}n2$	119	$I\bar{4}m2$	120	$I\bar{4}c2$
121	$I\bar{4}2m$	122	$I\bar{4}2d$	123	$P4/mmm$	124	$P4/mcc$	125	$P4/nbm$

(continued)

---

[3] Hahn [3].

Table 5.9 (continued)

No	Symbol	No	Symbol	No	Symbol	No	Symbol	No	Symbol
126	$P4/nnc$	127	$P4/mbm$	128	$P4/mnc$	129	$P4/nmm$	130	$P4/ncc$
131	$P4_2/mmc$	132	$P4_2/mcm$	133	$P4_2/nbc$	134	$P4_2/nnm$	135	$P4_2/mbc$
136	$P4_2/mnm$	137	$P4_2/nmc$	138	$P4_2/ncm$	139	$I4/mmm$	140	$I4/mcm$
141	$I4_1/amd$	142	$I4_1/acd$	143	$P3$	144	$P3_1$	145	$P3_2$
146	$R3$	147	$P\bar{3}$	148	$R\bar{3}$	149	$P312$	150	$P321$
151	$P3_112$	152	$P3_121$	153	$P3_212$	154	$P3_221$	155	$R32$
156	$P3m1$	157	$P31m$	158	$P3c1$	159	$P31c$	160	$R3m$
161	$R3c$	162	$P\bar{3}1m$	163	$P\bar{3}1c$	164	$P\bar{3}m1$	165	$P\bar{3}c1$
166	$R\bar{3}m$	167	$R\bar{3}c$	168	$P6$	169	$P6_1$	170	$P6_5$
171	$P6_2$	172	$P6_4$	173	$P6_3$	174	$P\bar{6}$	175	$P6/m$
176	$P6_3/m$	177	$P622$	178	$P6_122$	179	$P6_522$	180	$P6_222$
181	$P6_422$	182	$P6_322$	183	$P6mm$	184	$P6cc$	185	$P6_3cm$
186	$P6_3mc$	187	$P\bar{6}m2$	188	$P\bar{6}c2$	189	$P\bar{6}2m$	190	$P\bar{6}2c$
191	$P6/mmm$	192	$P6/mcc$	193	$P6_3/mcm$	194	$P6_3/mmc$	195	$P23$
196	$F23$	197	$I23$	198	$P213$	199	$I2_13$	200	$Pm\bar{3}$
201	$Pn\bar{3}$	202	$Fm\bar{3}$	203	$Fd\bar{3}$	204	$Im\bar{3}$	205	$Pa\bar{3}$
206	$Ia\bar{3}$	207	$P432$	208	$P4_232$	209	$F432$	210	$F4_132$
211	$I432$	212	$P4_332$	213	$P4_132$	214	$I4_132$	215	$P\bar{4}3m$
216	$F\bar{4}3m$	217	$I\bar{4}3m$	218	$P\bar{4}3n$	219	$F\bar{4}3c$	220	$I\bar{4}3d$
221	$Pm\bar{3}m$	222	$Pn\bar{3}n$	223	$Pm\bar{3}n$	224	$Pn\bar{3}m$	225	$Fm\bar{3}m$
226	$Fm\bar{3}c$	227	$Fd\bar{3}m$	228	$Fd\bar{3}c$	229	$Im\bar{3}m$	230	$Ia\bar{3}d$

In Tables 5.10, 5.11, 5.12, 5.13, 5.14, 5.15 and 5.16, the space groups ordered by crystalline systems and the Laue group symmetry are presented. The Laue group describes the symmetry of the diffraction pattern. The Laue symmetry can be lower than the metric symmetry of the unit cell but never higher.

Table 5.10 Space groups of the triclinic system

Point groups	1	$\bar{1}$
Space groups	1 $P1$	2 $P\bar{1}$
Laue group symmetry	$\bar{1}$	

Table 5.11 Space groups of the monoclinic system

Point groups	1^a orientation: $c = 2$ and/or -2		
	2	m	2/m
Space groups	3 P2	6 Pm	10 P2/m
	4 P2₁	7 Pb	11 P2₁/m
	5 B2	8 Bm	12 B2/m
		9 Bb	13 P2/b
			14 P2₁/b
			15 B2/b
Laue group symmetry	2/m		
Point groups	2^a orientation: $b = 2$ and/or -2		
	2	m	2/m
Space groups	3 P2	6 Pm	10 P2/m
	4 P2₁	7 Pc	11 P2₁/m
	5 C2	8 Cm	12 C2/m
		9 Cc	13 P2/c
			14 P2₁/c
			15 C2
Laue group symmetry	2/m		

Table 5.12 Space groups of the orthorhombic system

Point groups	222	mm2	mmm
Space groups	16 P222	25 Pmm2	47 Pmmm
	17 P222₁	26 Pmc2₁	48 Pnnn
	18 P2₁2₁2	27 Pcc2	49 Pccm
	19 P2₁2₁2₁	28 Pma2	50 Pban
	20 C222₁	29 Pca2₁	51 Pmma
	21 C222	30 Pnc2	52 Pnna
	22 F222	31 Pmn2₁	53 Pmna
	23 I222	32 Pba2	54 Pcca
	24 I2₁2₁2₁	33 Pna2₁	55 Pbam
		34 Pnn2	56 Pccn
		35 Cmm2	57 Pbcm
		36 Cmc2₁	58 Pnnm
		37 Ccc2	59 Pmmn
		38 Amm2	60 Pbcn
		39 Abm2	61 Pbca
		40 Ama2	62 Pnma
		41 Aba2	63 Cmcm
		42 Fmm2	64 Cmca
		43 Fdd2	65 Cmmm
		44 Imm2	66 Cccm
		45 Iba2	67 Cmma
		46 Ima2	68 Ccca
			69 Fmmm
			70 Fddd
			71 Immm
			72 Ibam
			73 Ibca
			74 Imma
Laue group symmetry	Mmm		

Table 5.13 Space groups of the rhombohedral system

Point groups	3	$\bar{3}$	32	3m	$\bar{3}m$
Space groups	143 $P3$ 144 $P3_1$ 145 $P3_2$ 146 $R3$	147 $P\,\bar{3}$ 148 $R\,\bar{3}$	149 $P312$ 150 $P321$ 151 $P3_112$ 152 $P3_121$ 153 $P3_212$ 154 $P3_221$ 155 $R32$	156 $P3m1$ 157 $P31m$ 158 $P3c1$ 159 $P31c$ 160 $R3m$ 161 $R3c$	162 $P\,\bar{3}1m$ 163 $P\,\bar{3}1c$ 164 $P\,\bar{3}m1$ 165 $P\,\bar{3}c1$ 166 $R\,\bar{3}m$ 167 $R\,\bar{3}c$
Laue group symmetry	$\bar{3}\,m$				

Table 5.14 Space groups of the hexagonal system

Point groups	6	$\bar{6}$	6/m	622	6mm	$\bar{6}m2$	6/mmm
Space groups	168 $P6$ 169 $P6_1$ 170 $P6_5$ 171 $P6_2$	174 $P\,\bar{6}$	175 $P6/m$ 176 $P6_3/m$	177 $P622$ 178 $P6_122$ 179 $P6_522$ 180 $P6_222$ 181 $P6_422$ 182 $P6_322$	183 $P6mm$ 184 $P6cc$ 185 $P6_3cm$ 186 $P6_3mc$	187 $P\,\bar{6}m2$ 188 $P\,\bar{6}c2$ 189 $P\,\bar{6}2m$ 190 $P\,\bar{6}2c$	191 $P6/mmm$ 192 $P6/mcc$ 193 $P6_3/mcm$ 194 $P6_3/mmc$
Laue group symmetry	6/mmm						

Table 5.15 Space groups of the tetragonal system

Point groups	4	$\bar{4}$	4/m	422	4mm	$\bar{4}2m$	4/mmm
Space groups	75 $P4$ 76 $P4_1$ 77 $P4_2$ 78 $P4_3$ 79 $I4$ 80 $I4_1$	81 $P\,\bar{4}$ 82 $I\,\bar{4}$	83 $P4/m$ 84 $P4_2/m$ 85 $P4/n$ 86 $P4_2/n$ 87 $I4/m$ 88 $I4_1/a$	89 $P422$ 90 $P42_12$ 91 $P4_122$ 92 $P4_12_12$ 93 $P4_222$ 94 $P4_22_12$ 95 $P4_322$ 96 $P4_32_12$ 97 $I422$ 98 $I4_122$	99 $P4mm$ 100 $P4bm$ 101 $P4_2cm$ 102 $P4_2nm$ 103 $P4cc$ 104 $P4nc$ 105 $P4_2mc$ 106 $P4_2bc$ 107 $I4mm$ 108 $I4cm$ 109 $I4_1md$ 110 $I4_1cd$	111 $P\,\bar{4}2m$ 112 $P\,\bar{4}2c$ 113 $P\,\bar{4}2_1m$ 114 $P\,\bar{4}2_1c$ 115 $P\,\bar{4}m2$ 116 $P\,\bar{4}c2$ 117 $P\,\bar{4}b2$ 118 $P\,\bar{4}n2$ 119 $I\,\bar{4}m2$ 120 $I\,\bar{4}c2$ 121 $I\,\bar{4}2m$	123 $P4/mmm$ 124 $P4/mcc$ 125 $P4/nbm$ 126 $P4/nnc$ 127 $P4/mbm$ 128 $P4/mnc$ 129 $P4/nmm$ 130 $P4/ncc$ 131 $P4_2/mmc$ 132 $P4_2/mcm$ 133 $P4_2/nbc$ 134 $P4_2/nnm$ 135 $P4_2/mbc$

(continued)

Table 5.15 (continued)

Point groups	4	$\overline{4}$	4/m	422	4mm	$\overline{4}2m$	4/mmm
						122 $I\,\overline{4}2d$	136 $P4_2/mnm$
							137 $P4_2/nmc$
							138 $P4_2/ncm$
							139 $I4/mmm$
							140 $I4/mcm$
							141 $I4_1/amd$
							142 $I4_1/acd$
Laue group symmetry	4/mmm						

Table 5.16 Space groups of the cubic system

Point groups	23	$m\,\overline{3}$	432	$\overline{4}3m$	$m\,\overline{3}m$
Space groups	195 $P23$	200 $Pm\,\overline{3}$	207 $P432$	215 $P\,\overline{4}3m$	221 $Pm\,\overline{3}m$
	196 $F23$	201 $Pn\,\overline{3}$	208	216 $F\,\overline{4}3m$	222 $Pn\,\overline{3}n$
	197 $I23$	202 $Fm\,\overline{3}$	$P4_232$	217 $I\,\overline{4}3m$	223 $Pm\,\overline{3}n$
	198	203 $Fd\,\overline{3}$	209 $F432$	218 $P\,\overline{4}3n$	224 $Pn\,\overline{3}m$
	$P2_13$	204 $Im\,\overline{3}$	210	219 $F\,\overline{4}3c$	225 $Fm\,\overline{3}m$
	199 $I2_13$	205 $Pa\,\overline{3}$	$F4_132$	$220I\,\overline{4}3d$	226 $Fm\,\overline{3}c$
		206 $Ia\,\overline{3}$	211 $I432$		227 $Fd\,\overline{3}m$
			212		228 $Fd\,\overline{3}c$
			$P4_332$		229 $Im\,\overline{3}m$
			213		230 $Ia\,\overline{3}d$
			$P4_132$		
			214 $I4_132$		
Laue group symmetry	$m\,\overline{3}m$				

5.5 Plane Space Groups and Symbol

There are 17 plane space groups, and they correspond to the spatial symmetry of plane cells.

The symbol consists of a lowercase letter that indicates the type of plane cell (p = primitive and c = centered), followed by a series of characters that consist of the symbol of the symmetry elements associated with the symmetry directions of the plane cells (Table 4.3).

Representation of the plane space groups are presented in Table 5.17.

Table 5.17 Plane space groups

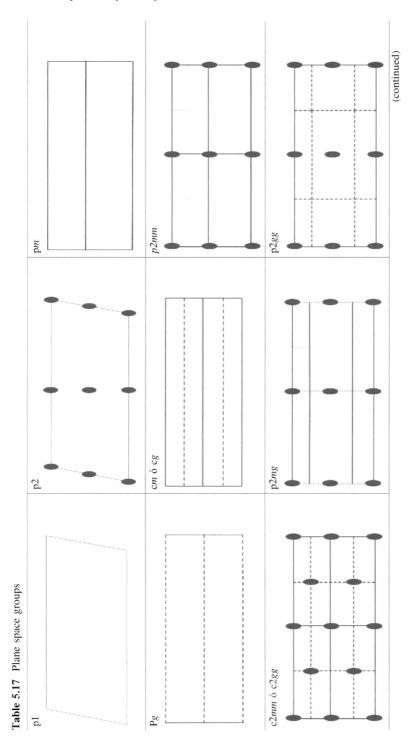

(continued)

Table 5.17 (continued)

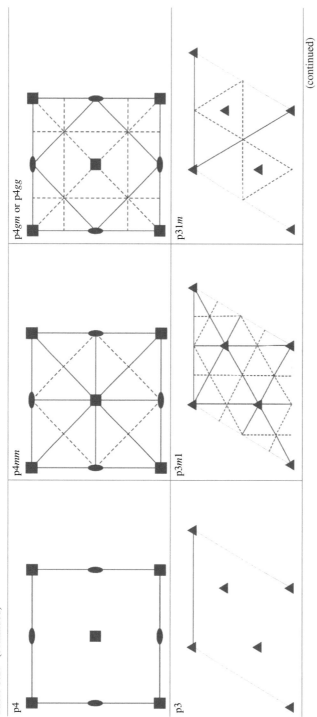

(continued)

Table 5.17 (continued)

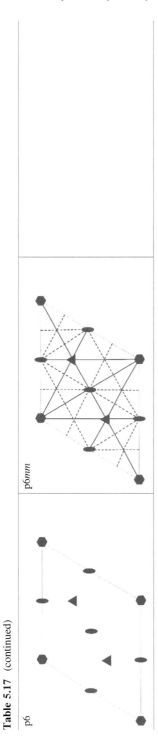

p6

p6mm

Examples

p1

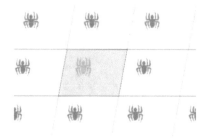

p2

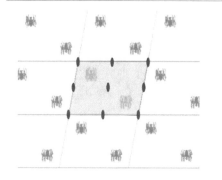

Egyptian motif

P*m*

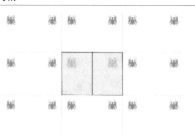

Egyptian motif

(continued)

(continued)

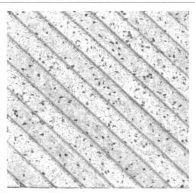

Pavement tile

Pg

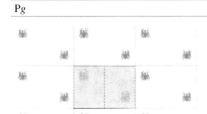

Kent Damask (England)

cm or *cg*

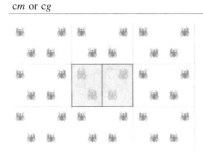

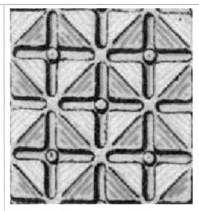

St. Denis (France, century XII)

(continued)

(continued)

---

p2mm

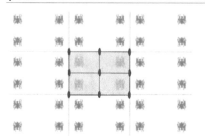

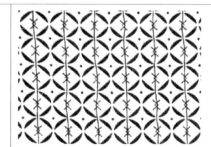

Egyptian motif

Pavement tile

---

c2*mm* or c2*gg*

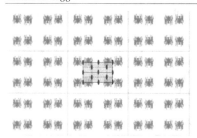

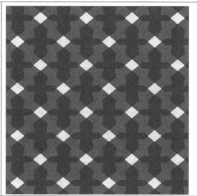

(continued)

(continued)

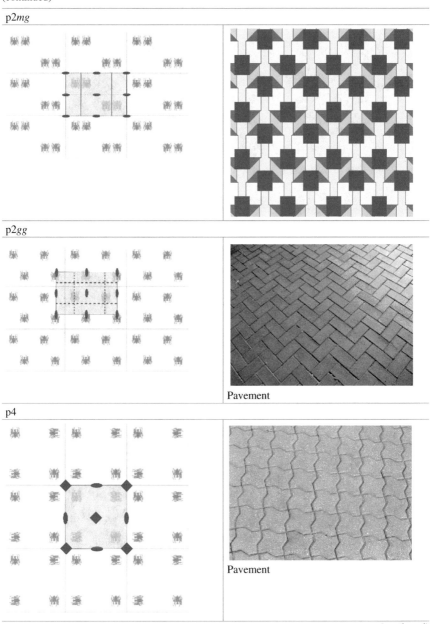

p2*mg*

p2*gg*

Pavement

p4

Pavement

(continued)

(continued)

p4*mm*

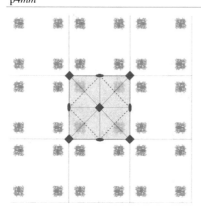

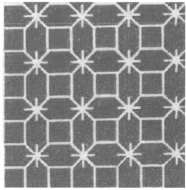

Persian motif

Pavement

p4*gm* or p4*gg*

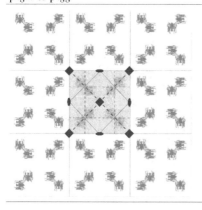

Fence

(continued)

(continued)

Pavement tile

p3

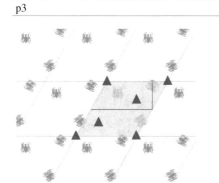

Arabian motif

p3*m*1

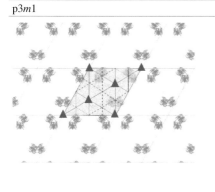

Persian motif

(continued)

p31*m*	
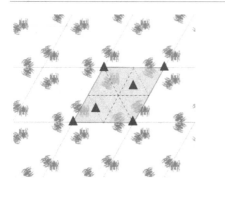	 Chinese motif
p6	
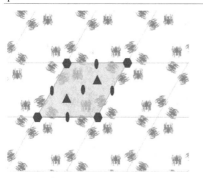	 Persian motif
p6*mm*	
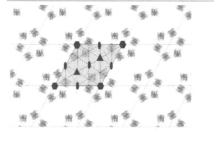	 Chinese motif

5.6 Equivalent Positions

The existence of symmetry operations in a space group produces a series of equivalent points per symmetry.

The number of points equivalent per symmetry is called *multiplicity*.

Knowing all the operations of a given space group, it is possible to obtain, from any point, all the points symmetrically equal to it.

The set of points thus generated is called the *regular point system of the group*. The regular point system is described by the set of coordinates of each of the starting points and the coordinates of the derived points expressed in terms of the coordinates of the starting point.

– *General position*

Point is not located on any symmetrical element.

The set of points derived from it is called a *regular point system of general position*.

The points are asymmetrical and are assigned the symbol of identity—1—because they are equivalent to themselves along the monary axis.

– *Special (or particular) position*

Point is located on some element of symmetry.

The number of points generated by it is called a *special position regular point system*.

The multiplicity is less than in the case of the general position.

The points in special position are assigned the symmetry of the element on which they are located, called position symmetry.

– Asymmetric unit

The asymmetric unit, also called the *fundamental region*, is a part of space that does not contain any element of symmetry and, by application of the elements of the space group, the elementary cell is obtained.

It is useful for interpreting and describing the crystalline structure.

5.7 Graphic Description of Space Groups

The graphic representation of the spatial groups is generally made by two projections on the plane (*001*):

– Elements of symmetry

For representation of the symmetry elements (Fig. 5.6), the following convention is adopted:

The orientation of the *b*-axis in the projection plane is taken from left to right, while that of the *a*-axis goes from top to bottom in the projection plane.

Fig. 5.6 Representation of symmetry elements

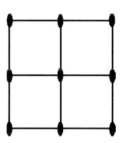

The angle between the **a** and **b** axes is 90° in the orthogonal systems, and 120° in the hexagonal and rhombohedral systems and any value in the triclinic system. In the monoclinic system, there are two alternatives for orientation of the axes, depending on whether b or c is the direction of symmetry of the network. Thus, in the first case, the projected cell axes will be a and c, and the angle between them (b) will be different from 90°, while in the second case, the projected axes will be a and b, and the angle between them (γ) will be 90°.

The symbols of the symmetry elements can be seen in Tables 5.18, 5.19, 5.20, 5.21 (adapted from the International Crystallography Tables).

Table 5.18 Planes of symmetry perpendicular to the plane of projection (3 dimensions) and lines of symmetry in the plane of the figure (2 dimensions)

Plane or line of symmetry	Graphic symbol	Glide vector in units of lattice translation vectors parallel and perpendicular to the projection plane axis	Printed symbol
Reflexion plane or reflexion line	———	None	m
Axial glide plane or glide line	– – – –	1/2 of the lattice vector along the line in the projection plane 1/2 of the lattice vector along the line	a, b or c g
Axial glide plane	··········	1/2 perpendicular to the projection plane	a, b or c
"*Double*" glide plane (only in centered cells)	– ·· – ·· – ·	2 glide vectors: 1/2 along the line parallel to the projection plane 1/2 perpendicular to the projection plane	e
Diagonal glide plane	– · – · – ··	1 glide vector with two components: 1/2 along the line parallel to the projection plane 1/2 perpendicular to the projection plane	n
Diamond glide plane (pairs of planes; in centered cells only)	· – ⟶ – · · ⟵ · –	1/4 along the line parallel to the projection plane, combined with 1/4 perpendicular to the projection plane (the arrows indicate the direction parallel to the projection plane for which the perpendicular component is positive)	d

Table 5.19 Planes of symmetry parallel to the projection plane

Plane symmetry	Graphic symbol	Glide vector in lattice translation vector units parallel to the projection plane	Printed symbol
Reflexion plane or reflexion line		None	m
Axial glide plane		2 glide vectors: 1/2 lattice vector in the direction of the arrow	a, b or c
"*Double*" glide plane (only in centered cells)		2 glide vectors: 1/2 in either direction of the two arrows	e
Diagonal glide plane		1/2 in the direction of the arrow	n
Diamond glide plane (pairs of planes; in centered cells only)	3/8 1/8	1/2 in the direction of the arrow; the glide vector is 1/4 of one of the diagonals of the cell centered on the faces	d

Table 5.20 Symmetry axes parallel to the projection plane

Axis of symmetry	Graphic symbol	Glide vector in grid vector units parallel to the axis	Printed symbol
Binary axis		None	2
Binary screw axis "2 sub 1"		½	2_1

2. General equivalent positions

Beside the diagram representing the symmetry elements of a given spatial group, another diagram appears representing the general equivalent positions on the projection plane (*001*). In it, + and − refer to the heights along the crystallographic axis perpendicular to the plane of projection. The symbols 1/2 + or 1/2−, refer to the height along the axis being 1/2 more than the position indicated only with + or with −. The enantiomorphic positions are symbolized by a circle with a comma in the center of it. When two enantiomorphic positions are related by a plane of symmetry parallel to the plane of projection, it is symbolized by a circle split in half and with a comma in one of the halves (Fig. 5.7).

Other data that are shown in the International Tables for Crystallography beside these diagrams are the following:

Table 5.21 Symmetry axes perpendicular to the projection plane

Axis of symmetry or point of symmetry	Graphic symbol	Glide vector parallel to the helical axis	Printed symbol (partial elements in parentheses)
Identity		None	1
Binary axis Binary rotation point		None	2
Ternary axis Ternary rotation point		None	3
Quaternary axis Quaternary rotation point		None	4 (2)
Senary axis Senary rotation point		None	6 (3, 2)
Center of symmetry or inversion: 1 bar		None	$\bar{1}$
Ternary rotation-inversion axis: 3 bar		None	$\underline{3} \equiv 3 + \underline{1}$ (3,$\underline{1}$)
Quaternary axis of rotation-inversion: 4 bar		None	$\underline{2}(2)$
Senary axis of rotation-inversion: 6 bar		None	$\underline{3} \equiv 3/m$
Ternary screw axis: "2 sub 1"		1/2	2_1
Ternary screw axis: "3 sub 1"		1/3	3_1
Ternary screw axis: "3 sub 2"		2/3	3_2
Quaternary screw axis: "4 sub 1"		1/4	4_1 (2_1)
Quaternary screw axis: "4 sub 2"		1/2	4_2 (2)
Quaternary screw axis: "4 sub 3"		3/4	4_3 (2_1)
Senary screw axis: "6 sub 1"		1/6	6_1 $(3_1, 2_1)$
Senary screw axis: "6 sub 2"		1/3	6_2 $(3_2, 2)$
Senary screw axis: "6 sub 3"		1/2	6_3 $(3, 2_1)$
Senary screw axis: "6 sub 4"		2/3	6_4 $(3_1, 2)$

(continued)

Table 5.21 (continued)

Axis of symmetry or point of symmetry	Graphic symbol	Glide vector parallel to the helical axis	Printed symbol (partial elements in parentheses)
Senary screw axis: "6 sub 5"		5/6	6_5 (3_2, 2_1)
Binary axis with center of symmetry		None	$2/m$ ($\underline{1}$)
Binary screw axis with center of symmetry		1/2	$2_1/m$ ($\underline{1}$)
Quaternary axis with center of symmetry		None	$4/m$ ($\underline{4}$, 2,$\underline{1}$)
Quaternary screw axis: "4 sub 2" with center of symmetry		1/2	$4_2/m$ ($\underline{4}$, 2,$\underline{1}$)
Senary axis with center of symmetry		None	$6/m$ ($\underline{6}$, $\underline{3}$, 3, 2, $\underline{1}$)
Senary screw axis: "6 sub 3"¾ with center of symmetry		1/2	$6_3/m$ ($\underline{6}$, $\underline{3}$, 3, 2_1,$\underline{1}$)

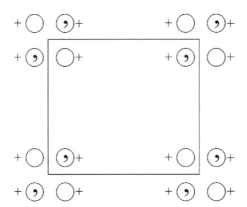

Fig. 5.7 Diagram of the general equivalent positions

- Symbol of the corresponding space group with the international notation and the number it occupies within the Tables.
- Symbol of the plane space group with the Schoenflies notation and below with the international notation.
- Symbol of the corresponding point group, according to the international notation.

- Crystalline system.
- The point symmetry of the origin, with a symbol similar to that used for point groups.
- The multiplicity of equivalent positions, expressed by a number.
- The Wyckoff symbol, which is expressed by a letter of the alphabet indicating the position symmetry of the corresponding equivalent position. The first letters of the alphabet are associated with higher position symmetry, while lower position symmetry is assigned more advanced letters of the alphabet.
- Position symmetry, which expresses point symmetry and is symbolized in a similar way to point groups.
- Coordinates of equivalent positions, both general and special or particular.

- Origin of the coordinate system

The origin of the coordinate system in the centroid groups is located in the center of symmetry.

In the case of non-centrosymmetric groups, there is no special rule for the group. The position of the origin is located on a point with the highest position symmetry.

The determination and description of crystalline structures is facilitated by the choice of a suitable origin and their own identification.

There are several ways to determine the location and position symmetry of the origin. One is to inspect it directly on the diagrams of the space groups, in the International Tables for Crystallography. Another is to look for a special equivalent position with 0, 0, 0 coordinates.

Example

Table 5.22 Data of space group P*mm*2

P*mm*2	C12v		*mm*2	Orthorhombic
No. 25	p*mm*2			
Origin in *mm*2				
Multiplicity	Wyckoff symbol	Position symmetry	Coordinates			
4	i	.	x, y, z	−x, −y, z	x, −y, z	−x, y, z
2	h	*m*. .	1/2, y, z	1/2, −y, z		
2	g	*m*. .	0, y, z	0, −y, z		
2	f	.*m*.	x, 1/2, z	−x, 1/2, z		
2	g	.*m*.	x, 0, z	−x, 0, z		
1	d	*mm*2	1/2, 1/2, z			
1	c	*mm*2	1/2, 0, z			
1	b	*mm*2	0, 1/2, z			
1	a	*mm*2	0, 0, nz			

Exercises

1. In the periodic models of Fig. E1:

   a. Draw the elements of symmetry (proper, improper and with associated translation).
   b. Select the elementary cell based on the elements of symmetry.
   c. Symbolize the symmetry of each model using the plane space group, taking into account section (b) and the symmetry directions of the lattice.

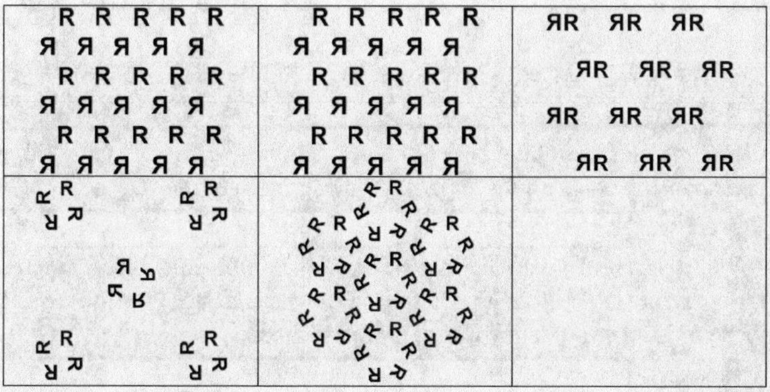

R R R R R Я Я Я Я Я R R R R R Я Я Я Я Я R R R R R Я Я Я Я Я	R R R R R Я Я Я Я Я R R R R R Я Я Я Я Я R R R R R Я Я Я Я Я	ЯR ЯR ЯR ЯR ЯR ЯR ЯR ЯR ЯR ЯR ЯR ЯR

Fig. E1

2. Obtain the equivalent positions to that indicated in Fig. E2 by the symmetry elements of the space group $Pcm2_1$.

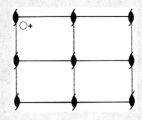

Fig. E2

3. In Fig. E3:

   (a) Draw the symmetry elements, based on the indicated equivalent positions.
   (b) Write the space group symbol.

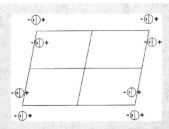

Fig. E3

Questions

1. Write the maximum number of characters to be specified in the symbol of the space or point group, without considering the translations.

   Response: [   ]

2. Write **discrete** if you believe crystalline materials must be considered as discrete media. Write in the definition of space groups **continuous** if you believe it must be defined as continuous media.

   Response: [   ]

3. Spatial groups reflect the internal and external symmetry of a material in a crystalline state, and the point groups only reflect internal symmetry.
   - ○ True
   - ○ False

4. General position is the position located on the identity.
   - ○ True
   - ○ False

5. Match each element of symmetry (only with maximum symmetry) with the type of crystal lattice (three-dimensional) that it characterizes.

Axis of order 4	Triclinic
Binary axis	Orthorhombic
Symmetry center	Tetragonal

6. Match each space group with its corresponding crystal lattice.

Imma	Cubic
P2/d $\underline{3}$	Orthorhombic
P312	Rhombohedral

7. Do you believe there are axes of order 5 in space groups? (Yes or No).

   Response: [   ]

8. How many space groups are there?

    Response: [   ]

9. The number of symmetry elements in a space group is finite.
    ○ True
    ○ False

10. What is the symmetry element (s) after the lattice type in a plane space group?
    ○ a. The least symmetrical rotation point
    ○ b. The axis of rotation of maximum symmetry
    ○ c. The point of rotation of maximum symmetry
    ○ d. The axis of rotation with the least symmetry

11. Write the symmetry elements that you see in Fig. E1 corresponding to a plane cell.

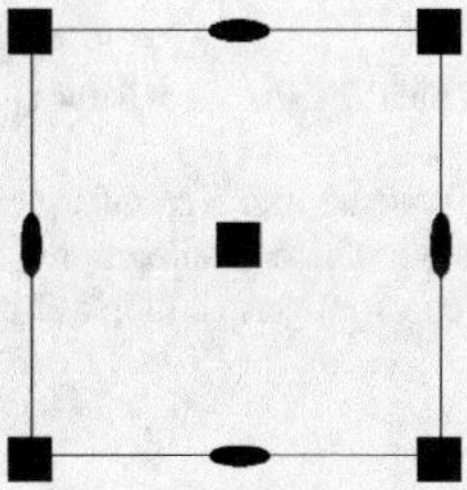

Fig. E1

    Response: [   ]

12. Write the directions of symmetry associated with the second line of symmetry *m* of the plane space group p2*mm* to which the Fig. E2 corresponds.

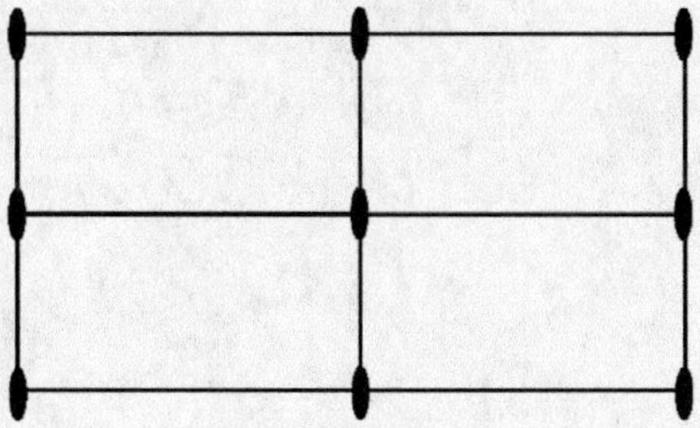

Fig. E2

    Response: [   ]

13. Points in special position are assigned symmetry of the
    ○ a. translation group
    ○ b. element on which they are, called position symmetry
    ○ c. space group
    ○ d. point group

14. An enantiomorphic equivalent position is related by:
   ○ a. any symmetry element associated with rotation
   ○ b. any symmetry element associated with the inversion
   ○ c. any element of symmetry associated with identity
   ○ d. any symmetry element with associated translation
15. General and special equivalent positions have different multiplicity and the same symmetry position
   ○ True
   ○ False

References

1. Fedorov ES (1890) Protokol'naia Zapis. Zap Min Obshch 26:454. (Official report Trans Mineral Soc)
2. Schoenflies A (1891) Krystallsysteme und Krystallstructur
3. Hahn T (2005) International tables for crystallography, vol. A: space group symmetry: space group symmetry, 5th revised ed. Springer. (2002, 2nd printing edition)

Part II
Crystallochemistry

In this part, Chaps. 6 to 9, we study the arrangement of atoms in the crystalline matter; that is, their structure. Up to this point, the student has learned to see the crystal as a periodic medium in which order prevails at all times, as a consequence of the reticular, structural, and energetic postulates. The concept of real crystal is then introduced, and its imperfections must be considered, contrary to what was considered in geometric crystallography.

Chapter 6
Crystal Structures, Compact Packing, Coordination

Abstract Underlying any structural criteria based on chemical bonding, there is one fact: the main condition for a crystalline structure to be s is that its free energy is minimal; therefore, the structures most likely to be energetically stable are those whose atoms or molecules are arranged in space in the most compact way possible. Despite the chemical diversity observed in the large number of resolved structures, all these structures can refer to a rather reduced number of fixed structural types which, in turn, are characterized by the fact that their atoms or molecules are arranged in space in the most compact way possible. The structure depends on the type of bond, the number and diversity of atoms and the repetition associated with the network. However, although the types of bonds are included in the program, they are not explained, since the student has knowledge of them through chemistry studied in pre-university courses and even in this first course of Geology. It is simply to remind them of this dependence on the structure. Covalent, metallic, and ionic crystals will be briefly explained. Depending on the different nature of the chemical bonding forces, crystalline structures can be grouped into two large categories: structures in whose space molecules are energetically individualized, and structures that can be considered constituted by a single molecule extending to the entire crystalline space. The mentioned crystals belong to this category. Therefore, the types of packing will be explained based on the hypothesis that atoms can be represented by hard and impenetrable spheres. In this explanation, it is necessary to introduce the concepts of coordination polyhedron and coordination number and, since this depends on the relative sizes of the coordinated ions, we will talk about the ratio of radii. Finally, the Pauling rules are presented, which fundamentally consider the conditions of maximum compactness and markedly ionic character of the bond presented by many crystalline structures.

© The Author(s), under exclusive license to Springer Nature Switzerland AG 2022   149
C. Marcos, *Crystallography*, Springer Textbooks in Earth Sciences,
Geography and Environment, https://doi.org/10.1007/978-3-030-96783-3_6

6.1   Introduction

All matter consists of ions, atoms, or molecules. The atom and its interaction forces determine the chemical and physical properties of matter. This depends on chemical composition, geometric arrangement of the constituent atoms or ions, and the nature of the electrical forces that bind them.

Crystallochemistry studies the relationships between the chemical composition of crystalline materials and their structures, as well as their effects on physical properties. It includes the study of chemical bonds, morphology, and the formation of crystalline structures, according to the characteristics of the atoms, ions or molecules, as well as their type of bond. This discipline is the link between crystallography, solid-state chemistry, and condensed matter physics.

Crystallochemistry was developed from mineralogy and, later, from crystallography.

The bases of crystallochemistry were established by Goldschmidt and Laves (1920–1930)[1] by means of the following postulates:

- *Principle of compact packing*: Atoms in a crystalline structure tend to be disposed of in a way that fills the space in the most efficient manner.
- *Principle of symmetry*: Crystalline structure atoms tend to achieve an environment with the highest possible symmetry.
- *Principle of interaction*: Atoms in a crystalline structure tend to surround themselves with as many neighboring atoms as possible, with which they can interact, i.e., they tend to achieve the highest coordination.

6.2   Crystalline Structures

A crystalline structure is the periodic and orderly arrangement in a three-dimensional space of matter constituents (ions, atoms, molecules, or sets therein). It provides information about the location of all atoms, binding positions and types, symmetry and chemical content of the elementary cell. Its study began with experiments in 1910 by Laue, Friedrich, and Knipping,[2] looking at the diffraction of X-rays by crystals. In 1913, Bragg[3] determined the first crystalline structure. Its knowledge is very important for faithfully, interpreting data on the chemical composition and physicochemical properties of minerals, and analyzing the conditions of formation and transformation of minerals in different environments. Then, hundreds of structures were solved and the general rules governing crystalline structures were formulated. The crystallochemistry of minerals was

---

[1] McSween et al. [1].

[2] Eckert [2].

[3] Bragg [3].

created, as part of a more general crystallochemistry, revealing even finer details of the mineral structure. The ionic model was developed. An ionic radius system and the additivity rule of such a radius was proposed. The rules governing ionic crystals were formulated. The concepts of solid solution and polymorphism were clarified. The concept of compact packing was developed and structural notions, stoichiometry, coordination number, and coordination polyhedrons were introduced; methods for the polyhedral representation of structures were also developed. Geometric analysis of the structure of minerals was completed in about 1960.

Stability of Crystalline Structures

For a crystalline structure to be stable, its free energy must be minimal.

This energy corresponds to Gibbs free energy, G, given by the expression (6.1):

$$G - U + PV - TS \tag{6.1}$$

where

G   is the free energy
U   is the internal energy
P   is the pressure
V   is the volume
T   is the temperature
S   is the entropy

A minimum energy corresponds to a minimum volume, or the structures most likely to stabilize energetically are those whose constituents are ordered in space in the most compact way possible.

6.3   Bonds in Crystalline Structures

Chemical bond is defined as the forces of attraction between atoms. It is determined by the outermost electrons or valence electrons. These electrons interact in ways that determine the symmetry of the atomic structure. There are three extreme types of chemical bond:

– *Ionic bond*

This bond is predominant in inorganic compounds. In this bond, the outermost layer is completed by transferring electrons from one atom to another (Fig. 6.1).

– *Covalent bond*

In the covalent bond, the outermost electrons are shared by predominant neighboring atoms (Fig. 6.2).

Fig. 6.1 Ionic bond

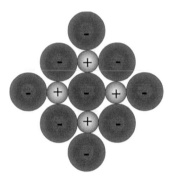

Fig. 6.2 C atom with four equivalent sp^3 orbitals, directed to the vertices of a tetrahedron

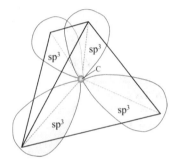

– *Metallic bond*

This bond is presented only by the elements (gold, silver, etc.). It is non-directional. Valence electrons can move freely (delocalized) (Fig. 6.3).

These bonds are not found pure in any crystal.

Electronegativity is a concept that expresses the relative measure of the force of attraction for electrons.

The ionic character of a bond is given by the difference between the electronegativities of the atoms that form it.

Fig. 6.3 Metallic bond

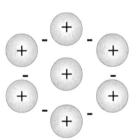

Table 6.1 Electronegativity values of some elements assigned by Pauling (1960)

F	O	N	Cl	C	S	B	Ca	Na
4	3.5	3	3	2.5	2.5	2	1	0.9

Pauling (1960) assigned a numerical value of electronegativity to each element. An example follows in descending order of the same (Table 6.1).

6.4   Ionic Crystals

The characteristics of ionic crystals are in general terms: To possess moderate hardness and specific weight, high melting and boiling points, very low electrical and thermal conductivity, high symmetry. To form ions in dissolution.

6.5   Covalent Crystals

The characteristics of covalent crystals are in general terms: insolubility, high stability, they do not form ions in dissolution, very high melting and boiling points, symmetry is lower than that of ion crystals.

6.6   Metal Crystals

The metal crystals are characterized by their great plasticity, tenacity, ductility, conductivity; and low hardness, melting and boiling points.

These crystals present high symmetry and very compact structures.

6.7   Compact Packing

Compact packing is defined as being available to atoms in space occupying the lowest volume and considering atoms to be hard and rigid spheres.

In the two-dimensional space, this is obtained when each atom surrounds itself with another six. The cell is hexagonal, and the lattice parameter is 2r (r is the radius of the sphere). Two types of voids can be distinguished: 1 and 2 (Fig. 6.4).

In the three-dimensional space, packing is achieved by stacking flat layers like the one described in the previous section. The second layer can be positioned so that the spheres rest over the 1 void or over the 2 voids. Both provisions are related by a 180° rotation, so they do not differ. If option B is chosen, this second layer is named B, and the layer sequence is AB.

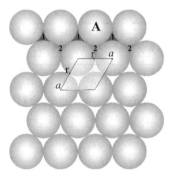

Fig. 6.4 Void types 1 and 2 in compact packing

When stacking the third layer, there are again two possibilities. When the spheres rest on the spheres of layer A, it is called layer A. The sequence is ABABAB...., and this type of packing is called *compact hexagonal packing* (hcp or hc). The occupied space is 74%. The atoms in layer B do not have the same environment as the atoms in layer A, as the orientation of the bonds is different (Fig. 6.5a). The atoms in layer A occupy the nodes of the hexagonal cell, with coordinates 0,0,0 and the atoms of layer B are located at 1/3,2/3,1/2 (Fig. 6.5b).

When the spheres are placed on void C, the layer is called layer C. The sequence is ABC ABC ABC...., and this type of packing is called *compact cubic packing* (ccp or cc) (Fig. 6.6a). The occupied space is 74%. The atoms in layers A, B, and C have the same environment (Fig. 6.6b). Layer A atoms occupy the nodes of the cubic cell, with coordinates 0,0,0; the atoms in B layer with 1/3,2/3,1/3; and in C layer the coordinates are 2/3,1/3,2/3. The elemental cell can be described as a cubic cell (Fig. 6.6c).

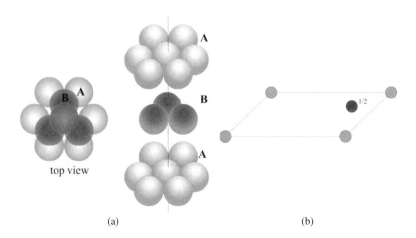

(a)   (b)

Fig. 6.5 Compact hexagonal packing

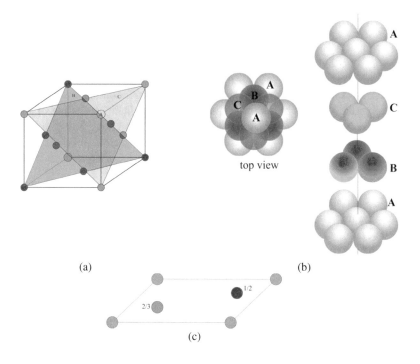

(a)   (b)

(c)

Fig. 6.6 Compact cubic packing

In the *interior-centered cubic packing* (bcc or bc), atoms are located at the vertices of an interior-centered cubic cell. The central atom has the same environment as the vertices. The volume occupied is 68%. Atoms have coordinates 0,0,0 and 1/2,1/2,1/2. The elemental cell can be described as F cubic cell (Fig. 6.7).

Fig. 6.7 Interior-centered cubic packing

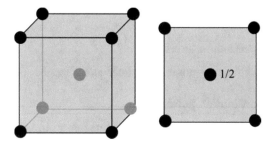

6.8   Size of Atoms

– Ionic radius

In the case of crystalline structures in which the ion bond predominates, the size of an ion is expressed in terms of its ion radius, which is defined as Radius of the sphere occupied by an ion in a particular structural environment (refers to the coordination of the ion).

Allows determination of the ratio of radii and is given by the expression (6.2):

$$RA : RX \tag{6.2}$$

RA being the radius of cation and RX being the radius of the anion, useful for systematic derivation of crystalline structures; understanding the replacement of one ion with another in the same kind of structure, and structural determination.

The interatomic distance is the sum of the radii of atoms.

– Covalent radius
Covalent radius is defined as the arithmetic mean of the interatomic distances of the crystals of elementary substances. It is used in covalent crystalline structures.

> **Example**
>
> – the C–C distance is 1.54,
> – the distance Si–Si is 2.34,
> – the distance C-Si, resulting from the union of C and Si to form the CSi, is 1.94; this value closely matches that obtained from X-ray diffraction equal to 1.930.

6.9   Coordination, Pauling Rules

– Coordination number

The coordination number is defined as the number of ions surrounding a given ion.

– Coordination polyhedron

The coordination polyhedron is defined as an imaginary polyhedron that arises by joining the lines of the neighboring ions or atoms closest to the central ion or atom (Table 6.2).

Table 6.2 Coordination polyhedra, radii ratio, and examples

n°	Configuration	Polyhedron	RA/RX	Example
12		cube-octahedron	1	Cu
8		cube	0.73	Ca^{2+} in fluorite

(continued)

Table 6.2 (continued)

n°	Configuration	Polyhedron	RA/RX	Example
6		trigonal prism	0.53	
6		octahedron	0.41	Cl⁻ in halite

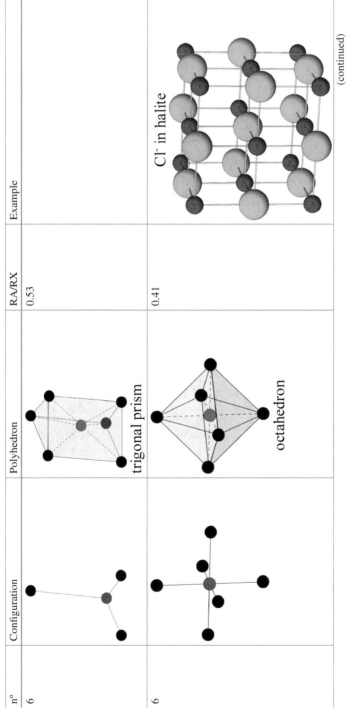

(continued)

Table 6.2 (continued)

n°	Configuration	Polyhedron	RA/RX	Example
4		tetrahedron	0.23	Si^{4+} tetrahedra in quartz
3		triangle	0.15	CO$_3^{2-}$ (C in figure) in calcite

– **Structural unit**

In a crystalline structure, the bonding characteristics allow differentiation of certain structural units; that is, certain atoms or groups of atoms that are ordered in the most compact way possible. They may consist of the following:

– An atom
– Finite groupings of atoms
– Chains of atoms
– Layers of atoms
– Three-dimensional framework of atoms

Each of these units can be packaged in many ways, resulting in a large number of structural types formed by the same structural unit. However, it is noted from the comparison of some structures that some of them may be derived from others. The latter are known as basic structures and the former as derived structures.

– **Pauling rules**

These rules basically consider the ionicity of the bond and the maximum compactness of the structure. These are as follows:

1. The maximum number of R radius ions that can coordinate with another with radius r < R is given by the r/R ratio (see Table 6.2).
2. An ion structure will be more stable with greater neutralization of the charges of anions and cations.
3. The stability of a crystal structure will be less stable with a greater number of edges and faces that are shared in the coordination polyhedron.
4. In a crystal containing different cations, those of higher valence and smaller coordination number tend not to share any element of the coordination polyhedron. When they share edges, they contract and the cations tend to move from the center of their coordination polyhedron.
5. Principle of parsimony: The number of different classes of constituents in a crystal tends to be small.

6.10 Positions in Compact Packaged Structures

Compact packing is of interest because many mineral structures can be described in terms of compact packing of anions, with cations occupying the spaces between them.

In compact packing, two types of spaces are distinguished between two layers (Fig. 6.8).

Fig. 6.8 Compact packing
holes: Tetrahedral holes, 1,
and octahedral holes, 2

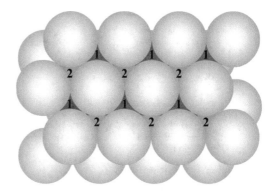

1. Tetrahedral holes: holes between three ions of one layer and one ion of another.
2. Octahedral holes: holes between three ions of one layer and three ions of
   another.

– **Tetrahedral positions**

Tetrahedral positions are positions occupied by cations surrounded by four anions
(Fig. 6.9). Its coordination number is 4. Its coordination polyhedron is a
tetrahedron.

In compact cubic compact packing, the maximum number of tetrahedral positions is eight, and in the compact hexagonal packing there are four (Fig. 6.10).

– **Octahedral positions**

Octahedral positions are positions occupied by cations surrounded by six anions
(Fig. 6.11). Its coordination number is 6. Its coordination polyhedron is an
octahedron.

In compact cubic compact packing, the maximum number of octahedral positions is four and in compact hexagonal compact packing there are two (Fig. 6.12).

Coordination will be stable, depending on radii ratio $R_A:R_X$.

Fig. 6.9 Tetrahedral position

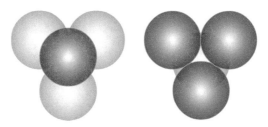

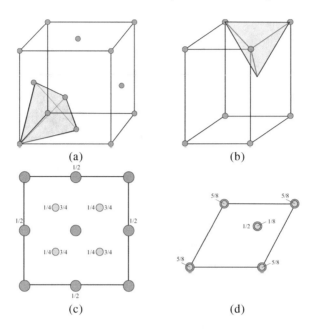

(a)   (b)

(c)   (d)

Fig. 6.10 Tetrahedral positions of the cubic (**a**) and hexagonal (**b**) compact packing and their respective projections (**c**) and (**d**)

Fig. 6.11 Octahedral position

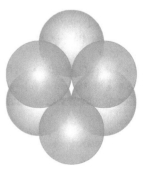

Fig. 6.12 Octahedral positions of the cubic (**a**) and hexagonal (**b**) compact packing and their respective projections into the basal plane (**c**) and (**d**)

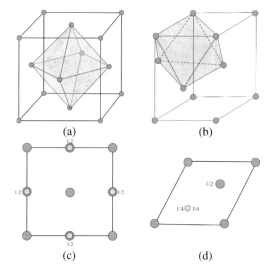

(a)   (b)

(c)   (d)

Exercises

1. (1°) Place a layer of ping pong balls (grey) on a flat surface adherent, layer A, as shown in Fig. E1. Count the number of balls surrounding a given. Look at the two types of voids, the so-called 1 point upwards and the so-called 2 point down. Do you think you could get another arrangement of the balls so that the voids were smaller? Draw the elemental cell on Fig. E1 and provide the values of a, b, and γ.

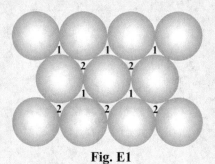

Fig. E1

(2°) Place another layer of ping pong balls (orange) over the voids 1 of the balls in layer A in Fig. E1, layer B, as shown in Fig. E2. Is the number of balls surrounding a given the same as in the previous case?

(3°) Place another layer of ping pong balls (grey) over voids 2 of the balls in layer B (Fig. E2), layer A, as shown in Fig. E3. Notice that the balls in this layer overlap those of the first layer A. This is the *compact hexagonal packing* (hcp or hc) AB AB AB.

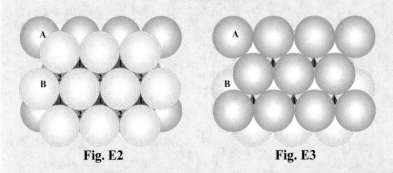

Fig. E2   Fig. E3

(4°) Place another layer of ping pong balls (red) over voids 1 of the balls in layer B (Fig. E3), layer C, as shown in Fig. E4. Notice that the balls in this layer C do not overlap those of the first layer A.

(5°) Place another layer of ping pong balls (grey) over voids 2 of the balls in layer B (Fig. E4), layer A, as shown in Fig. E5. Notice that the balls in this layer overlap those of the first layer A. This is the *compact cubic packing* (ccp or cc) ABC ABC ABC.

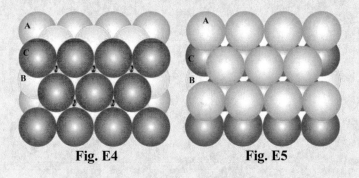

Fig. E4   Fig. E5

Questions

1. Relate the number of faces to the coordination polyhedron

8	Tetrahedron
6	Cube
4	Octahedron

2. Relate the value of the cation–anion radius ratio to coordination

1 and greater than 1	Tetrahedron
0.15–0.22	Cube
0.41–0.53	Trigonal prism

3. Interatomic distance is defined for ionic crystals as
   - a. the semi-halflife of the radii of the atoms considered
   - b. the sum of the radii of the atoms considered
   - c. the arithmetic mean of the interatomic distances of atoms when considered separately forming crystals
   - d. the arithmetic mean of the interatomic distances of the atoms when each individually considered forms crystals with other atoms

4. What is the name of the compact packing in which the maximum number of tetrahedral positions is 4?
   Response:

5. How many octahedral positions at most does compact cubic packing support?
   Response:

6. Metallic crystals are characterized by their low hardness and high conductivity and plasticity
   - True
   - False

7. In compact packing, the maximum occupancy of space is 66%
   - True
   - False

8. In octahedral coordination, each ion is surrounded by six nearest neighbors
   ○ True
   ○ False

9. In compact hexagonal packing, there are a maximum of eight tetrahedral positions and four octahedral positions
   ○ True
   ○ False

10. In compact cubic packing, there are at most
    ○ a. four octahedral and two tetrahedral positions
    ○ b. eight tetrahedral and four octahedral positions
    ○ c. eight octahedral and four tetrahedral positions
    ○ d. four tetrahedral and two octahedral positions.

References

1. McSween HY, Richardson SM, Uhle ME (2003) Geochemistry pathways and processes, 2nd edn. Columbia University, New York
2. Eckert M (2012) Max von Laue and the discovery of X-ray diffraction in 1912. Ann Phys (Berlin) 524(5):A83–A85
3. Bragg WH (1913) X-rays and crystals. Nature 90:572

Chapter 7
Crystal Structures

Abstract The relative pressure and temperature stability of the minerals and, therefore, their value in reconstructing geological history depends very much on the crystal structure. In the case of polymorphism, the same chemical constituents can be arranged in two or more structures, depending on the formation conditions. Based on the spatial principle and considering also that there are practically ionic crystals and practically covalent crystals, the concepts of structural unit and derived structure will be explained, which have helped to rationalize the structural characteristics of crystals. Cubic and hexagonal compact structures will be explained, and then those derived from these packages. Other structures (CsCl, spinel, calcite, etc.) will also be explained. The structures of silicates will also be shown.

7.1 Introduction

The number of structures in minerals is so great that it is necessary to use classifications for each class of minerals in order to manipulate them. The classification must be chosen according to a specific purpose. There are, therefore, several classifications.

Due to the close interdependence between chemical composition, atomic structure, thermodynamic stability, and chemical reactivity, the most appropriate classification is typically the classification based on crystallochemical principles. According to these principles, the minerals are grouped into the following classes, according to Strunz (1941)[1]: Elements (e.g., gold); sulfides (e.g., pyrite); halides (e.g., fluorite); oxides and hydroxides (e.g., cassiterite); nitrates, carbonates and borates (e.g., calcite); sulfates (e.g., gypsum); phosphates (e.g., apatite); silicates (e.g., quartz); organic substances (e.g., amber).

---

[1] Strunz [1].

© The Author(s), under exclusive license to Springer Nature Switzerland AG 2022
C. Marcos, *Crystallography*, Springer Textbooks in Earth Sciences,
Geography and Environment, https://doi.org/10.1007/978-3-030-96783-3_7

The structure figures of this chapter were obtained with the software BS,[2] considering the model of hard and rigid spheres for atoms, from the American Mineralogist Crystal Structure Database,[3] and Atoms[4] software for the silicate structures. In all figures, the parallel orthogonal projection (based on the axonometric parallel projection by Haldinger, 1826[5]) has been used, taking as a basis the considerations of Naumann[6], used in most textbooks.

7.2 Compact Cubic Structures

They are typical structures of metals such as gold (Au), silver (Ag), copper (Cu), or aluminum (Al).

7.2.1 Gold Structure

- Space group $P2_1/c$.
- Gold atoms are arranged at the vertices and centers of the faces of a cubic cell (F).
- The total number of gold atoms is four.
- The coordinates are: 0,0,0; 1/2,1/2.0; 1/2,0,1/2; 0,1/2,1/2 (Fig. 7.1).

7.3 Compact Hexagonal Structures

They are typical structures of many metals, including magnesium (Mg), zinc (Zn), and titanium (Ti).

7.3.1 Magnesium Structure

- Space group $P6_3/mmc$.
- Magnesium atoms are arranged at the vertices and centers of the faces of a cubic cell (F).
- The total number of magnesium atoms is two.
- The coordinates are 2/3,1/3,3/4; 1/3, 2/3,1/4 (Fig. 7.2).

---

[2] Ozawa [2].

[3] Downs [3].

[4] Atoms V6.1.2 © 2004 by Shape Software.

[5] Haldinger [4].

[6] Naumann [5].

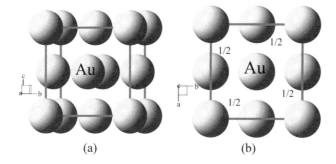

Fig. 7.1 Gold structure: **a** three-dimensional view, **b** projection on the plane (001)

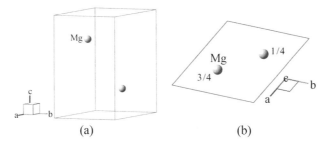

Fig. 7.2 Magnesium structure: **a** three-dimensional view, **b** projection on the plane (001)

7.4 Body-Centered Cubic Structures

Body-centered cubic, bcc, is another type of structure adopted by many metals such as iron (Fe), chromium (Cr), or sodium (Na).

7.4.1 Iron Structure

– Space group $Im\,\bar{3}m$.
– Iron atoms are arranged at the vertices and center of a cubic cell (I).
– The total number of iron atoms is two.
– The coordinates are 0,0,0; (1/2,1/2,1/2) (Fig. 7.3).

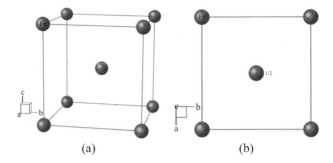

Fig. 7.3 Iron structure: **a** three-dimensional view, **b** projection on the plane (001)

7.5   Structures Derived from Compact Cubic Packing

7.5.1   Halita (NaCl)

– It is an AX structure.
– Space group $Fm\bar{3}m$.
– X anions form the compact packing.
– Cations A occupy all octahedral positions.
– Anions and cations have octahedral coordination.

   This structure can be described as consisting of two interpenetrated lattices, one of anions and one of cations and displaced from each other 1/2,1/2,1/2.

   The coordinates of the ion that forms the compact packing are 0,0,0; 1/2,1/2,0; 1/2,0,1/2; 0,1/2,1/2.

   The coordinates of the octahedral positions are 0,0,1/2; 0,1/2,0; 1/2,0,0; 1/2,1/2,0 (Fig. 7.4).

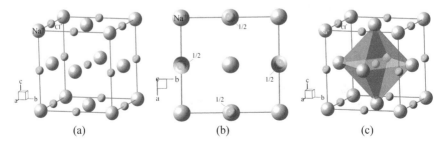

Fig. 7.4 Halite structure: **a** three-dimensional view, **b** projection on the plane (001), **c** structure with chlorine octahedra

7.5.2 Fluorite-Type Structures (CaF₂)

- These structures are AX_2 structures.
- Fluorite space group $Fm\bar{3}m$.
- Cations form compact cubic packing.
- Ca^{2+} in the fluorite structure has hexahedral coordination and occupies the vertices and face centers of the cubic cell.
- There are four Ca^{2+} per cell.
- The coordinates are 0,0,0; 1/2,1/2,0; 1/2,0,1/2; 0,1/2,1/2.
- Anions (F^-) occupy all tetrahedral positions.
- There are eight F^- per cell.
- The coordinates are 1/4,1/4,1/4; 1/4,1/4,3/4; 3/4,1/4,1/4; 3/4,1/4,3/4; 1/4,3/4,1/4; 1/4,3/4,3/4; 3/4,3/4.1/4; 3/4,3/4,3/4 (Fig. 7.5).

Antifluorite-type structures

They are structures A_2X. Example:

Na₂O structure

- Space group $Fm\bar{3}m$.
- Na atoms are coordinated to four oxygen atoms, while oxygen atoms are coordinated to eight sodium atoms.
- Anions (O^{2-}) occupy the positions of compact cubic packing.
- They are coordinated to eight Na^+.
- Their coordinates are 0,0,0; 1/2,1/2,0; 1/2,0,1/2; 0,1/2,1/2.
- Cations Na^+ are coordinated to 4 O^{2-}.
- Their coordinates are 1/4,1/4,1/4; 1/4,1/4,3/4; 3/4,1/4,1/4; 3/4,1/4,3/4; 1/4,3/4,1/4; 1/4,3/4,3/4; 3/4,3/4,1/4; 3/4,3/4,3/4 (Fig. 7.6).

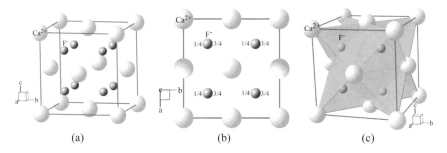

(a)   (b)   (c)

Fig. 7.5 Fluorite structure: **a** three-dimensional view, **b** projection on the plane (001), **c** structure with fluorine tetrahedra

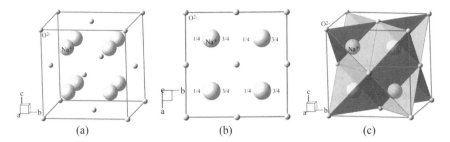

Fig. 7.6 Sodium oxide structure: **a** three-dimensional view, **b** projection on the plane (001), **c** structure with sodium tetrahedra

7.5.3   Sphalerite—Type Structures (ZnS)

- These structures are AX-type structures.
- Sphalerite space group F $\bar{4}3m$.
- Cations (Zn^{2+}) occupy the vertices and centers of the faces of the cubic cell centered on the faces (F).
- The number of Zn^{2+} in the cell is four.
- The coordinates are 0,0,0; 1/2,1/2,0; 1/2,0,1/2; 0,1/2,1/2.
- Anions (S^{2-}) occupy half of the tetrahedral positions.
- The number of S^{2-} in the cell is four.
- The coordinates are 1/4,1/4,1/4; 3/4,1/4,3/4; 1/4,3/4,3/4; 3/4,3/4,1/4 (Fig. 7.7).

7.5.4   Diamond

- Space group F$d\,\bar{3}m$.
- In diamond, the C^{4+} occupy two types of positions.
- Some C^{4+} occupy the vertices and the centers of faces of the cubic cell centered on the faces.
- The coordinates are 0,0,0; 1/2,1/2,0; 1/2,0,1/2; 0,1/2,1/2.
- Other C^{4+} occupy half of the tetrahedral positions.

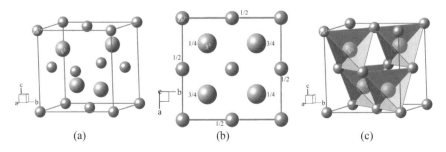

Fig. 7.7 Structure of the sphalerite: **a** three-dimensional view, **b** projection on the plane (001), **c** structure with sulfur tetrahedra

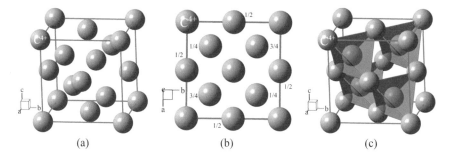

Fig. 7.8 Diamond structure: **a** three-dimensional view, **b** projection on the plane (001), **c** structure with carbon tetrahedra

– The coordinates are 1/4,1/4,1/4; 3/4,1/4,3/4; 1/4,3/4,3/4; 3/4,3/4,1/4.
– The content of C^{4+} per cell is eight (Fig. 7.8).

7.6 Structures Derived from Compact Hexagonal Packing

7.6.1 Nickelite Structure (NiAs)

– This structure is an AX-type structure.
– Space group $P6_3/mmc$.
– Cations (As^{2-}) occupy the positions of the compact hexagonal packing.
– There are two As^{2-} per cell.
– The coordinates are 0,0,0; 0,0,1/2.
– Anions (Ni^{2+}) occupy all octahedral positions.
– There are two Ni^{2+} per cell.
– The coordinates are 2/3,1/3,1/4; 2/3,1/3,3/4 (Fig. 7.9).

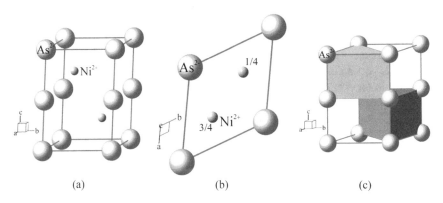

Fig. 7.9 Nickelite structure: **a** three-dimensional view, **b** projection on the plane (001), **c** structure with arsenic octahedra

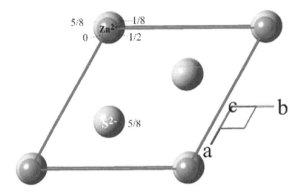

Fig. 7.10 Structure of the wurtzite

7.6.2   Wurtzite Structure (ZnS)

– It is an AX-type structure.
– Space group P6$_3$mc.
– The anions (Zn^{2+}) occupy the positions of the compact hexagonal packing.
– There are two Zn^{2+} per cell.
– The coordinates are 0,0,0; 1/3,2/3,1/2.
– Cations (S^{2-}) occupy half of the tetrahedral positions.
– There are two S^{2-} per cell.
– The coordinates are 0,0,5/8; 1/3,2/3,1/8 (Fig. 7.10).

7.7   Other Structural Types

7.7.1   CsCl Type Structure

– The CsCl structure is an AX-type structure.
– Space group Pm$\overline{3}$m .
– Cations A occupy the positions of a primitive cubic cell (P).
– X anions are located in the center of that cell.
– The Cs$^+$ occupies the vertices of a single cubic cell (sc).
– There is one per cell.
– The coordinates are 0,0,0.
– The Cl$^+$ occupy the center of the cubic cell.
– There is one per cell.
– The coordinates are 1/2,1/2,1/2.
– The coordination of Cs$^+$ and Cl$^-$ is hexahedral or cubic (the coordination number is 8) (Fig. 7.11).

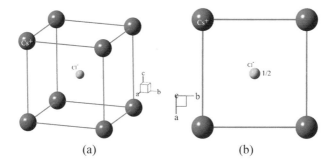

Fig. 7.11 Cesium chloride structure: **a** three-dimensional view, **b** projection on the plane (001)

7.7.2   Calcite Structure (CaCO₃)

– Space group $R\bar{3}c$.
– Calcite structure can be described as an NaCl structure distorted, compressed along the ternary axis of inversion rotation, until the 90° angle of the cube passes at an angle of 101°55′, typical of a rhombohedron.
– The Na^+ have been replaced by Ca^{2+}, which occupy the vertices and face centers of that rhombohedron.
– The Cl^- has been replaced by flat triangular groups of the carbonate ion CO_3^{2-}, perpendicular to the ternary axis of rotation-inversion.
– Most carbonates belong to this structural type (Fig. 7.12).

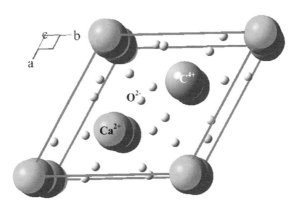

Fig. 7.12 Calcite structure

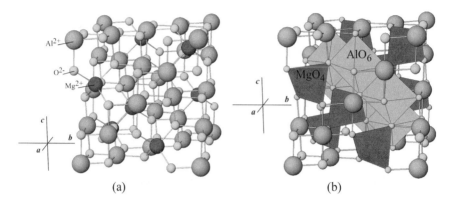

Fig. 7.13 Spinel structure: **a** three-dimensional view, **b** structure with magnesium tetrahedra and aluminum octahedral

7.7.3   Spinel Structure (AB₂O₄)

The spinel group is a group of oxide minerals that crystallizes in cubic systems, with a general composition XY_2O_4, where X can be Mg^{2+}, Fe^{2+}, Mn^{2+} and Zn^{2+}; and Y, Al^{3+}, Fe^{3+}, Cr^{3+}. The most common spinel minerals are magnetite (Fe_3O_4) and spinel *s.s.* ($MgAl_2O_4$).

– Space group Fd $\bar{3}m$.
– The spinel structure consists of a compact cubic package of oxygen when arranged in layers parallel to {111}, in which one-eighth of the tetrahedral positions and half of the octahedral positions are occupied by cations.
– Tetrahedra and octahedra appear alternating.
– The tetrahedral position is smaller and is occupied by Mg^{2+}, Fe^{2+}, Mn^{2+}, Zn^{2+}.
– The octahedral position is larger and is occupied by Al^{3+}, Cr^{3+} or Fe^{3+}.
– They are classified into *normal* and *inverse spinels*.

Normal spinels are those in which octahedral positions are occupied. Example: Spinel ($MgAl_2O_4$) (Fig. 7.13).
Inverse spinels are those in which the tetrahedral positions are occupied. Example: Magnetite (Fe_3O_4).

7.8   Silicate Structures

Silicates are minerals in which silicon (Si) and oxygen (O) are basic and fundamental elements in their composition and crystalline structure.

Silicate minerals are very abundant and make up most rocks. This is because silicon is the second most abundant element in the earth's crust and mantle, after

Fig. 7.14 Silicon tetrahedron

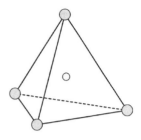

Table 7.1 Classification of silicates

	Number of oxygen atoms shared by the Si	Repetition unity	Ratio Si:O	Examples
Nesosilicate	0	$(SiO_4)^{4-}$	1:4	Olivine
Sorosilicates	1	$(Si_2O_7)^{6-}$	1:3.5	Epidote
Ciclosilicates	2	$(SiO_3)_8^{2-}$	1:3	Tourmaline
Single Chain Inosilicates	2	$(SiO_3)^{2-}$	1:3	Pyroxene
Double Chain Inosilicates	2.5	$(Si_4O_{11})^{3-}$	1:2.75	Amphybole
Phyllosilicates	3	$(Si_2O5_7)^{2-}$	1:2.5	Mica
Tectosilicates	4	$(SiO_2)^0$	1:2	Quartz

oxygen. The Si–O bond is considerably stronger than that between any other element and oxygen. The fundamental unit of the crystalline structure of silicates is formed by a regular tetrahedron in whose vertices the oxygen are located and in the center of the silicon tetrahedron (Fig. 7.14). Tetrahedral coordination is the most stable, because the radius ratio is 0.278.

The wide variety of structures existing in silicates is due to the polymerization capacity. Polymerization is the bond of silicon tetrahedrons that share apical oxygen atoms.

Based on the different tetrahedron bonds, silicates are classified by the number of oxygen atoms shared by the Si^{4+} (Table 7.1).

Examples of silicate structures

- **Nesosilicates**
- **Forsterite**

Forsterite (Mg_2SiO_4) is the extreme magnesium member of the olivine series that crystallizes in the orthorhombic system. Space group P*bnm*. The structure can be described as follows:

The silicon tetrahedra are arranged parallel to the c axis, forming rows, some pointing upwards and others pointing downwards, alternately. Some tetrahedra are at one level ($a = 0$) and others at another level ($a = 1/2$).

Within each level, the tetrahedra are joined by the octahedra, containing the M cations and also forming rows along the c axis.

There are two kinds of octahedra, the M1 and the M2.

– M1 octahedra form chains along c. These octahedra are attached to another chain above (along axis a) by means of M2 octahedra.
– M2 octahedra are slightly larger and slightly more distorted than the M1.

These octahedra in the actual structure are somewhat distorted. In general, in any structure, octahedra and tetrahedra are somewhat distorted. This distortion is given by a parameter called mean quadratic elongation.[7] (Fig. 7.15)

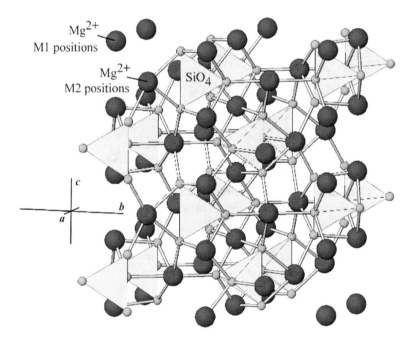

Fig. 7.15 Olivine structure

---

[7] Quadratic elongation and the variance of bond angles are linearly correlated for distorted octahedral and tetrahedral coordination complexes, both of which show variations in bond length and bond angle. The quadratic elongation is dimensionless, giving a quantitative measure of polyhedral distortion which is independent of the effective size of the polyhedron Robinson [6].

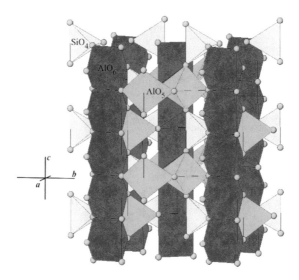

Fig. 7.16 Andalusite structure

– **Andalusite**

Andalusite (Al_2SiO_5) is a nesosilicate that crystallizes in the orthorhombic system. Space group P*nnm*. Its crystalline structure consists of aluminum octahedra (AlO_6) chains parallel to *c*, with shared edges and contains half of the aluminum with coordination number 6, and the other half with coordination number 5 (Fig. 7.16).

The structure of kyanite, a polymorph of andalusite, consists of aluminum octahedra (AlO_6) chains parallel to *c*, with shared edges and contains half of aluminum with coordination number 6, and the other half with coordination number 6.

The structure of sillimanite, the other polymorph of andalusite, consists of aluminum octahedra (AlO_6) chains parallel to *c* axis, with shared edges and contains half of aluminum with coordination number 6, and the other half with coordination number 4.

– **Garnets**

Garnets are nesosilicates that crystallize in the cubic system, space group I*a* $\overline{3}$ *d*. The general formula is

$A_3B_2(SiO_4)_3$,

where

A is Ca^{2+}, Mg^{2+}, Fe^{2+} or Mn^{2+}

B is At^{3+}, Fe^{3+} or Cr^{3+}

In the structure of the garnet, silicon tetrahedra alternate with BO_6 octahedra with those that share vertices. Cations A are large, with coordination 8, and the coordination polyhedra they form are distorted cubes.

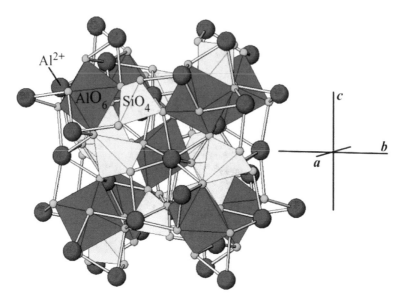

Fig. 7.17 Garnet structure

Silicon tetrahedra are distorted. The distortion depends on the size of the cation in the distorted AO_8 cubes with which silicon tetrahedra share two opposite vertices (Fig. 7.17).

– **Sorosilicates**
– **Epidote**

The epidote is a sorosilicate mineral of calcium, aluminum and iron, its formula is expressed by $Ca_2(Al,Fe)_3(SiO_4)_3(OH)$ and crystallizes in the monoclinic system, space group $P2_1/m$. Iron and aluminum are in octahedral coordination (Fig. 7.18).

– **Cyclosilicates**
– **Beryl**

The structure of the beryl, space group $P6/mcc$, is formed by rings of six silicon tetrahedrons. These rings are perpendicular to the crystallographic c axis and, in the different layers along it, they are rotated, one with respect to the other. The holes of these rings along the c axis are large and can accommodate a variety of ions and molecules. Among the rings and perpendicular to their planes, there are rows of octahedral positions of aluminum and rows of tetrahedral positions of beryllium (Fig. 7.19).

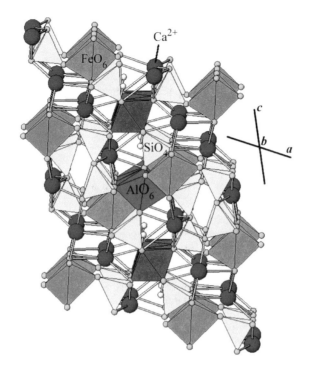

Fig. 7.18 Epidote structure

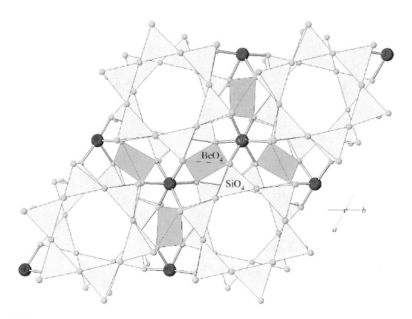

Fig. 7.19 Beryl structure

- **Inosilicates**
- **Diopside**

Diopside belongs to the pyroxenes group and crystallizes in the monoclinic system, space group C2/c. The chemical composition is represented by the general chemical formula:

$$XYZ_2OR_6$$

where

X is Na^+, Ca^{2+}, Mn^{2+}, Fe^{2+}, Mg^{2+} and Li^+, in M2 positions.

Y is Fe^{2+}, Mn^{2+}, Mg^{2+}, Fe^{3+}, Al^{3+}, Cr^{3+}, Ti^{4+}, in M1 positions.

Z is Si^{4+}, Al^{3+}, in tetrahedral positions.

The structure of the diopside can be described as follows: Silicon is tetrahedrically coordinated to oxygen. Tetrahedra are joined by forming chains parallel to the crystallographic axis c. Some tetrahedra have their vertices facing each other and others have their basis facing each other. The M cationic positions form chains that share edges and are parallel to silica tetrahedra chains. They are of two types:

- M1 positions lie among the opposing tetrahedra apexes. These positions are smaller and are almost regular octahedra.
- M2 positions lie between the bases of these tetrahedra. These positions are larger and more distorted (Fig. 7.20).

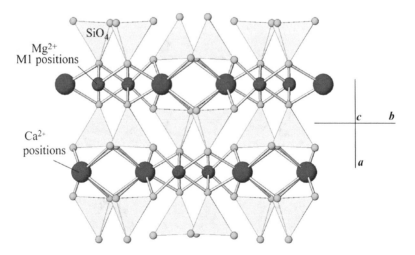

Fig. 7.20 Diopside structure

- **Phyllosilicates**
- **Muscovite**

Muscovite ($KAl_2(AlSi_3O_{10})(OH)_2$) is a phyllosilicate that crystallizes in the monoclinic system, space group $C2/c$. Its structure (Fig. 7.21) consists of stacks of the two basic layers, trioctahedral and dioctahedral.

The repetition structural unit includes one octahedral layer between two tetrahedral layers. The three layers are joined by monovalent alkaline ions (Na^+, K^+) by weak bonds. The repeating distance of the layers is 10 Å.

- **Tectosilicates**
- **Quartz**

In the structure of quartz silicon, tetrahedra share their oxygen vertices with other neighbors, resulting in a structure with strong bonds. The composition is SiO_2, and there is charge neutrality.

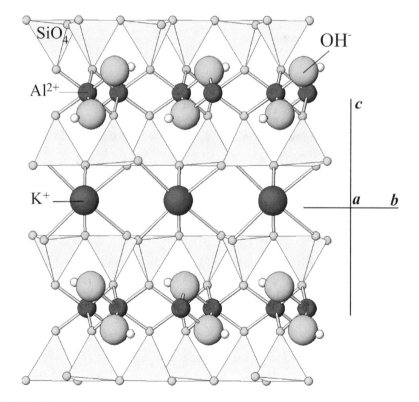

Fig. 7.21 Muscovite structure

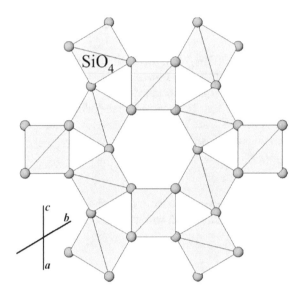

Fig. 7.22 High quartz structure

– Quartz β or high T

Quartz high has a structure (Fig. 7.22) based on spiral-shaped tetrahedra chains around a screw ternary axis, parallel to the crystallographic c axis. Its space group is $P6_222$. The positions of the tetrahedra of these spirals are located at 0, 1/3, and 2/3 of the crystallographic c axis.

In the unit cell, there are two spirals, each around a screw axis, and both are joined by vertices of the tetrahedra. These spirals can be right-handed or left-handed because the ternary screw axis is enantiomorph, so there are enantiomorph quartz crystals (Fig. 7.22).

– Quartz α or low T

The structure (Fig. 7.23) and properties of low quartz are similar to high quartz. Its space group is $P3_221$. The silicon tetrahedra in the structure are rotated: the senary axis of the high T quartz (hexagonal) is reduced to a ternary axis in the low T (rhombohedral).

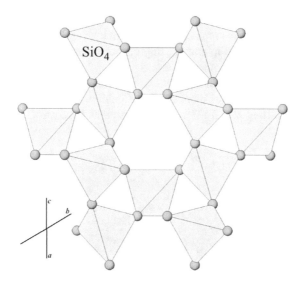

Fig. 7.23 Low quartz structure

– **Feldspars**
– **Sanidine**

Sanidine $((\underline{K},\underline{Na})(Si,Al)_4O_8)$ crystallizes in the monoclinic system, space group $C2/m$. Its structure (Fig. 7.24) can be described as follows: The frame of sanidine, as in the rest of feldspars, consists of rings of four tetrahedra, two T1 and two T2, alternating and occupied by Si^{4+} and Al^{3+}, with a disordered distribution, but the average occupancy of each tetrahedron is 50% silicon and 25% aluminum. In this

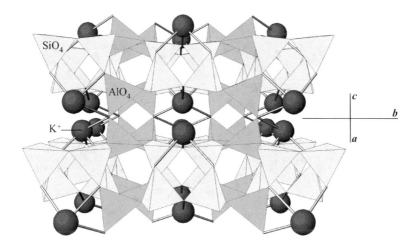

Fig. 7.24 Sanidine structure

structure, two tetrahedra have vertices pointing upwards and another two have them pointing down. These rings come together to form a layer. In this layer, the rings of four tetrahedrons are joined to another ring of four other tetrahedrons, so that both rings are related by a plane of symmetry.

The tetrahedra are joined along axis a, resulting in a crankshaft-shaped chain. Among the tetrahedron rings, there are large voids occupied by K^+ ions, on the symmetry planes. They also form a chain. These ions have a coordination number of 10.

Exercises

(1) Calculate the atomic radius of the following:

   (a) A metal whose structure consists of a cubic package centered in the interior with cell parameter $a = 3.294$ Å
   (b) The sodium that, in the structure of the halite, has a cell parameter of $a = 4.0862$ Å, and forms the cubic packing centered in the faces.

   For interior-centered cubic structures, the atomic radius is given by the expression:

   $$r = \frac{(\sqrt{3})a}{4}$$

   For faces-centered cubic structures, the atomic radius is given by the expression:

   $$r = \frac{(\sqrt{2})a}{4}$$

(2) Rutile structure (TiO_2)

   (a) Observe carefully the graphic representation of the symmetry elements of the spatial group $P4_2/mnm$ (Fig. E1) to which the rutile belongs.

   Symbols:

   – Projection on the x–y plane of a general position $\bigcirc$
   – Projection on the x–y plane of two positions with the same x and y coordinates but different z coordinates $\bigcirc\!\!|$
   – Projection of an enantiomorphic position (derived from another by some element of symmetry that implies the inversion) $\bigodot$
   – Projection on the x–y plane of two enantiomorphic positions with the same x and y coordinates but different z coordinates $\bigodot\!\!|$

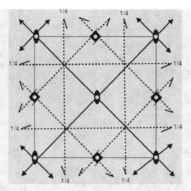

Fig. E1 Graphic representation of the symmetry elements of the spatial group P4$_2$/*mnm*

(b) Observe the rutile structure (Fig. E2). The titanium cations have octahedral coordination.
(c) Project in Scheme 1 of Table 1, the atoms Ti and O, whose coordinates are provided in Table 1.

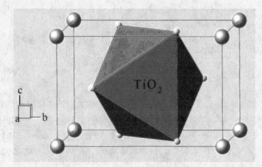

Fig. E2 Rutile structure. The cell parameters are a = 4.594 Å and c = 2.959 Å

In the same table, the position symmetry and Wyckoff letter of the different positions of the Ti and O atoms are provided. Both the position symmetry and the Wyckoff letter represent the symmetry elements on which the atoms are located. The symbol used to specify this symmetry is similar to that of the point groups, expressed by the symbols of the elements of symmetry associated with the directions of symmetry (primary, secondary, tertiary), depending on the crystalline system. The position symmetry group is isomorphic to a subgroup of the point group to which the space group under consideration belongs. Example: The symmetry group *m.2m* is iso-morphic to the point group *mm2*, subgroup of the point group 4/*mmm* to which the rutile space group P4$_2$/*mnm* belongs.

Observe that the titanium atoms occupy positions whose Wyckoff letter is a, position symmetry *m.mm*, while the oxygen atoms occupy positions whose Wyckoff letter is f and position symmetry is *m.2m*. In both cases, a point appears in the second position of the symbol, and it refers to the second direction of symmetry. So, for Wyckoff letter to position symmetry m.mm, it indicates that the atom's posi-tion is on three planes of symmetry, the first perpendicular to the *c* axis (i.e., plane parallel to the paper), the other two perpendicular to the tertiary directions [110] (i.e., to the intermediate directions between the *a* and *b* axes) and [1 $\bar{1}$ 0]. For Wyckoff letter f, the first *m* of the position symmetry *m.2m* is perpendicular to *c*, the binary axis parallel to the *b* axis and the other plane is perpendicular to the tertiary direction [110] and equivalent [1 $\bar{1}$ 0].

(d) Relate the position of the atoms projected in the scheme, both tita-nium and oxygen, with the position of the elements of symmetry, and observe that all the atoms are on some element of symmetry (special or particular positions).

(e) Observe in the projection of the equivalent positions of Fig. E1, one of the aspects that your projection could show if there were general positions (both titanium and oxygen); that is, those that are not located on any element of symmetry.

(f) Look at one atom and count the number of them generated by the symmetry elements of the space group to obtain the multiplicity. As an example, look at the general position labeled 1 and observe that the number of positions obtained by applying the symmetry elements of the spatial group is eight.

(g) Count the number of titanium and/or oxygen atoms that are in a special position to obtain the multiplicity of both types. To do this, look at one atom and count the number of them generated by the symmetry elements of the space group (Table E1).

Table E1 Table with the atomic coordinates of Ti and O (first and second columns) that compose rutile to be represented in the third column diagram 1; the fourth column is reserved to indicate the multiplicity of each atom and the fifth and sixth columns show the Wyckoff letter and position symmetry of the mentioned atoms

Atom	Coordinates	Multiplicity	Wyckoff letter	Site symmetry
Ti	0,0,0 1/2,1/2,1/2	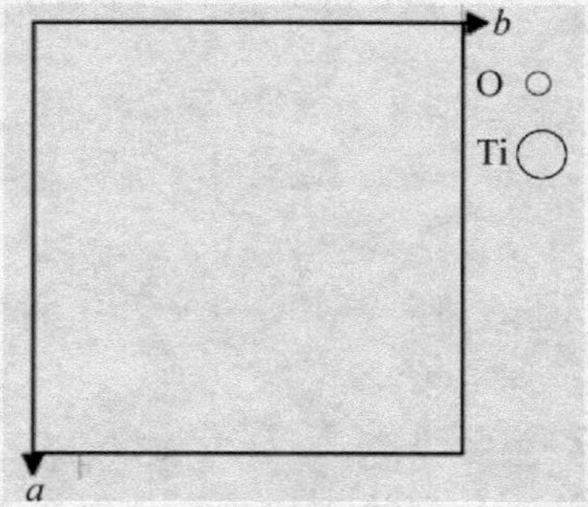	a	*m.mm*
O	x,x,0 1/2 + x,1/2−x,1/2 1/2−x, 1/2+x,1/2 −x,−x,0 x = 0.3		f	*m.2m*

Diagram 1

Questions

1. Relate each compound to the type of structure that corresponds to it.

Nickelite	Structure derived from compact cubic packing
Fluorite	Structure derived from compact hexagonal packing
Sodium	Interior-centered cubic

2. Write the name of the structure of Fig. Q1.

Fig. Q1 Type of structure to be named

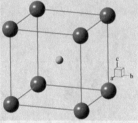

Response: [   ]

3. Gold has which structure?
   - ○ a. derived from the compact cubic
   - ○ b. compact cubic
   - ○ c. derived from the compact hexagonal
   - ○ d. compact hexagonal
4. Magnesium structure is
   - ○ a. cubic compact
   - ○ b. compact hexagonal
   - ○ c. derived from the compact cubic
   - ○ d. derived from the compact hexagonal
5. What type of structure corresponds to that of Fig. Q2? (projection on the plane (001) of the elemental cell)

Fig. Q2 Type of structure to be named

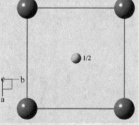

Response: [   ]

6. Write the name of the wurtzite ion to which the coordinates are (0,0,5/8), (1/3,2/3,1/8)

   Response: [   ]

7. Write the coordination of calcium in the calcite

   Response: [   ]

8. Carbon in calcite has which coordination?
   - ○ a. triangular
   - ○ b. tetrahedral
   - ○ c. octahedral
   - ○ d. hexahedral

9. The polyhedron coordinating chlorine ions in the halite is a
   - ○ a. cube, just like sodium
   - ○ b. cube-octahedron, just like sodium
   - ○ c. tetrahedron, just like sodium
   - ○ d. octahedron, just like sodium

10. In the structure of the sphalerite
   - ○ a. half of the tetrahedral positions are occupied
   - ○ b. all tetrahedral positions are occupied
   - ○ c. all octahedral positions are occupied
   - ○ d. half of the octahedral positions are occupied

References

1. Strunz KH (1941) Mineralogische Tabellen
2. Ozawa TC, Kang SJ (2004) Balls & sticks: easy-to-use structure visualization and animation creating program. J Appl Cryst 37:679
3. Downs RT, Hall-Wallace M (2003) The American mineralogist crystal structure database. Am Miner 88:247–250
4. Haldinger (1826) 1822/25, Poggendorffs Annalen 5:507
5. Naumann CF (1826) Grundriß der Kristallographie
6. Robinson KF, Gibbs GV, Ribbe PH (1971) Quadratic elongation: a quantitative measure of distortion in coordination Polyhedra. Science 172(3983):567–70

Chapter 8
Defects

Abstract It becomes clear that real crystal is full of defects that affect different conditions of the glass. First and foremost, the different types of imperfections will be discussed, and some information about the concepts of order and disorder will be given. Later, the crystalline defects will be considered, depending on whether they affect the crystal at an atomic level, atomic row, atomic plane, or at a three-dimensional level. The importance of the punctual defects in the diffusion in solid state, color of the minerals, and chemical composition will be explained. The importance of helical defects in the growth of crystals, the clusters, and the inclusions will also be discussed.

8.1 Introduction

Crystalline theory defines crystal as a perfect entity, according to the following postulates [1]:

- *Reticular*: The crystal is an infinite periodic medium, defined by a lattice corresponding to one of 14 types of Bravais.
- *Structural*: The crystal has an atomic structure whose symmetry corresponds to one of 230 spatial groups.
- *Energy*: Atoms in the crystalline structure occupy equilibrium positions for which energy is minimal.

From the moment a crystal or mineral is formed, it is subject to changes in its physical and chemical environment.

The response to such changes is the adaptation of its structure and composition to the new environment. Such changes may be

- Minor changes in bond lengths and angles or major structural transformations.
- Atomic-scale chemical changes or reactions that cause new mineral species.

© The Author(s), under exclusive license to Springer Nature Switzerland AG 2022   193
C. Marcos, *Crystallography*, Springer Textbooks in Earth Sciences,
Geography and Environment, https://doi.org/10.1007/978-3-030-96783-3_8

– Many minerals have formed at relatively high temperatures or pressures.
– The high-temperature state is characterized by chemical variability and a more generalized structure.
– The high-pressure state is characterized by higher density.

8.2   Order and Disorder

The order in a crystalline material is understood to be the regular and geometric distribution of the atoms that form it.

The degree of disorder is defined by a statistical factor called entropy. The entropy of a system, in this case a particular distribution of atoms and their vibrational energy levels, is defined from the statistical point of view as the probability of the existence of that state. This entropy is called configurational entropy. Mathematically this entropy S is expressed as

$$S = k \ln \omega \tag{8.1}$$

where

k is Boltzmann constant ($1.3 \cdot 10^{-23}$ JK^{-1}).
ω is the probability that a certain state will exist.

A distinction can be made between short-range order and long-range order:

– Short-range order refers to the order over distances comparable to interatomic distances. The short-range order coefficient is the ratio of atoms correctly positioned around a central atom.
– Long-range order refers to the order over distances infinitely great. Long-range order coefficient is the difference between the proportion of properly placed atoms and those placed incorrectly, relative to the perfectly ordered structure.

The concept of order can be seen in Fig. 8.1. A crystal with A and B atoms will have a completely ordered structure if all the 1 and 2 positions are appropriately occupied by A and B atoms, respectively.

Figure 8.2 shows that there are domains (zones) in which the order indicated in Fig. 8.1 exists.

The types of disorder are positional disorder, distortional disorder, substitutional disorder.

– *Positional disorder*

All atoms in a structure suffer thermal vibrations that can be described as position disorder over a base time. An example of positional disorder is that of potassium in feldspar (Fig. 8.3). At each moment, it can occupy a different position (1, 2, 3, 4, 5, 6).

Fig. 8.1 Schematic
illustration of long-range
order

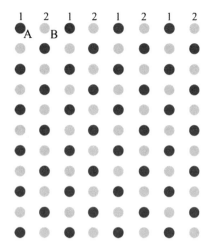

Fig. 8.2 Structure scheme
with ordered domains

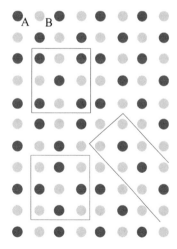

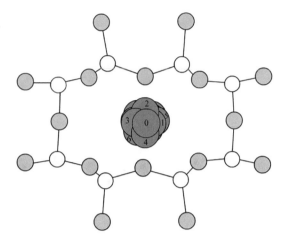

Fig. 8.3 Position disorder scheme

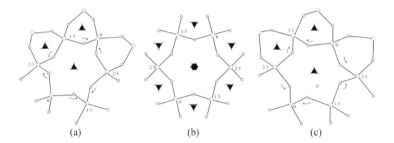

(a)   (b)   (c)

Fig. 8.4 Distortional disorder. Structure representations of low (or α) quartz, with distortion of the bonds left (**a**) and right (**c**), and high (or β) quartz (**c**). Yellow circles represent Si atoms; tetrahedrals surrounded by small turquoise blue circles represent O atoms

– *Distortional disorder*

Distortional disorder occurs as a result of the distortion of the bonds. An example is the distortion of ±16.37°, with symmetry loss, presented by low quartz, left quartz, or right quartz after transformation of high quartz (Fig. 8.4b). The probability that the distortion of the bonds is in the direction of Fig. 8.4a is the same as that of Fig. 8.4c.

Note

Quartz can be low (β) or high (α), depending on whether it is formed at low or high temperature. Sometimes in literature its name is contradictory. Both low and high quartz can be right or left. Right quartz, from a structural point of view, is left from the morphological and optical point of view. Conversely, left quartz, from a structural point of view, is right from the morphological and optical point of view. The symmetry presented by these types of quartz is summarized in Table 8.1, according to the terminology used in the International Tables of Crystallography [2].

Table 8.1 Symmetry summary of high and low quartz

Type		Space group
High quartz (α)	Right	$P6_222$
	Left	$P6_422$
Low quartz (β)	Right	$P3_221$
	Left	$P3_121$

– *Substitutional disorder*

Substitutional disorder involves an exchange of atoms between two or more positions that become indistinguishable by increasing the temperature, resulting in a chemical disorder where the average chemical content of each position is the same (Fig. 8.5).

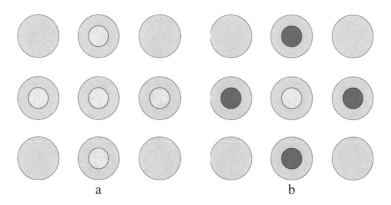

a   b

Fig. 8.5 Scheme showing substitutional disorder. In **a** some atoms, colored in light gray, are substituted in **b** by others colored in dark gray

8.3 Crystalline Defects

A crystalline material usually has imperfections that affect the orderly and geometric distribution of atomic constituents. One consequence of this disorder is the existence of defects.

A defect is the rupture of structural continuity in a crystalline material. Its importance lies in the effect this has on the properties of minerals.

Types of defects include point (zero-dimensional), linear (one-dimensional), two-dimensional, and three-dimensional.

8.4 Point Defects

Point defects are the result of an error in the occupation of an atomic position in the crystalline structure.

Point defects are important in processes such as solid-state diffusion, electrical conductivity, density, solid solutions, and mineral color.

Types of isolated point defects include vacancies, impurities, and interstitials (Fig. 8.6).

– *Vacancy*

Vacancy is an empty atomic position of the structure that should be occupied (Fig. 8.6a).

The number of vacancies increases with temperature and is given by the expression:

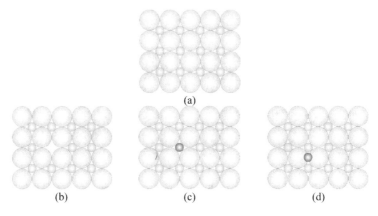

Fig. 8.6 Structure schemes ordered (**a**) and with specific defects: vacancy (**b**), impurity (**c**), interstitial atom (**d**)

$$N_V = N exp\left(-\frac{Q_V}{kT}\right)$$ (8.2)

where

N_V is the number of vacancies,
N   is the number of atomic positions,
Q_V is the energy to form a vacancy,
k is the Boltzman constant ($1.34 \cdot 10^{-23}$ J/K),
T is the temperature.

The activation energy can be obtained experimentally, knowing the concentration of vacancies N_V/N and its variation with temperature. The dependence of N_V/N versus T is exponential (Fig. 8.7).

The representation of $\ln N_V$ versus the inverse of T (Fig. 8.8) provides a line whose slope gives Q_V/K.

Fig. 8.7 Representation of N_V/N versus T

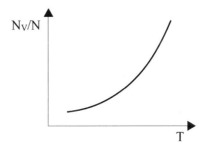

Fig. 8.8 Representation of $\ln N_V$ versus the inverse of T

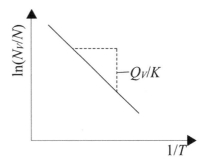

Fig. 8.9 Scheme showing the
Schottky defect

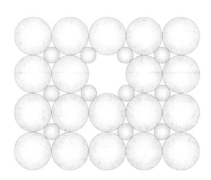

– *Impurity*

Impurity is an atom occupying the position of another characteristic of the crystal
structure.

– *Interstitial atom*

An interstitial atom occupies a space of the structure that does not correspond to it.
Point defects often appear not isolated but coupled because of charge balancing.

– *Schottky defect*

Schottky defect is the association of two vacancies of different signs (cationic
vacancy and anionic vacancy) (Fig. 8.9).

Example

1. Given a periclase crystal (MgO), with an NaCl-like structure (Fig. E1),
   with a cell parameter a = 3.96 Å, M_{Mg} = 24.31 g/mol, M_O = 16 g/mol,
   $R_{Mg}{}^{2+}$ = 0.072 nm, $R_O{}^{2+}$ = 0.140 nm, ρ_{exp} = 3.2 g/cm^3 and 1 Schottky
   defect for every 10 elementary cells, calculate the following: (a) the
   number of anion vacancies per cm^3; (b) the density of the MgO crystal;
   (c) the number of Schottky defects that will be in one elementary cell.

Fig. E1 Periclase structure

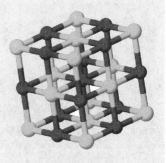

(a) In a unit cell of MgO, there are 4 Mg^{2+} and 4 O^{2-}; therefore, in 10 unit
cells there will be 40 Mg^{2+} and 40 O^{2-}. Due to the existence of a
Schottky defect, the number of ions is reduced to 39 Mg^{2+} and 39 O^{2-}.

(b) The volume of a periclase cell is a $= (3.96 \cdot 10^{-8}$ cm$)^3 = 62.1 \cdot 10^{-24}$
cm^3.
If, in 10 cells $(10 \cdot 62.1 \cdot 10^{-24}$ cm$^3)$, there is one anionic vacancy, the
number of vacancies by cubic centimeter is $1.61 \cdot 10^{21}$.

(c) If the crystal has defects, its experimental density, ρ_{exp}, will be lower
than the theoretical density, ρ_{theo}, so there will be fewer magnesium
and oxygen atoms in the cell than there should be.
Magnesium and oxygen atoms in the cell.

The number of Mg atoms in the cell is the same as the number of O atoms,
but their position is different.

Mg atoms: There are 12 in the center of the edges, 1 inside. Total: 12: 4
(each edge is shared by 3 other adjacent cells) + 1 = 4 Mg atoms/cell.

O atoms: There are 6 in the center of faces, 8 in vertices. Total: 6: 2 (each
face is shared by another adjacent cell) + 8: 8 (each vertex is shared by 7 other
adjacent cells) = 4 O atoms/cell.

$a = 2\,R_{Mg}{}^{2+} + R_O{}^{2+} = 0.424$ nm $= 4.24 \cdot 10^{-8}$ cm

$$\rho_{exp} = \frac{n\frac{at\,Mg}{cell} \times 24.31\frac{gMg}{mol} + n\frac{at\,O}{cell} \times 16\frac{gO}{mol}}{7.62 \times 10^{-23} cm^3 \times 6.023 \times 10^{23}\frac{at}{mol}}$$

$$3.2\frac{g}{cm^3} = n\frac{40.31}{45.89} \times \frac{g}{mol}$$

$$N = 3.64 \Rightarrow 0.36 \; vacancies/cell$$

$$\% \, vacancies = \frac{0.36}{4} \times 100 = 8.9.$$

2. Calculate the number of vacancies per cubic meter in iron at 850 °C. The
energy for vacancy formation is 1.08 eV/atom. The atomic weight of iron
is 55.85 g/mol and the density of iron is 7.65 g/cm^3.
The expressions to determine the number of vacancies (8.2) and the
density of a crystal (E.1) will be used.

$$\rho = \frac{Z \cdot M}{V \cdot A} \qquad\qquad\qquad (E.1)$$

where

M   is the molecular weight
Z   is the number of molecules in the unit cell. Z is normally referred to as
     the number of molecules in the unit cell; it is more strictly the number of
     formula units in the unit cell
V   is the volume
A   is the Avogadro number = $6.022 \cdot 10^{23}$ mol^{-1}

$$\frac{Z}{V} = \frac{\rho \cdot A}{M}$$

$$\frac{Z}{V} = N = \frac{\rho \cdot A}{M}.$$

Substituting the value of N and the known data in expression (8.2), the number of vacancies is obtained

$$N_V = \frac{\left(\frac{7.65g}{cm^3}\right)\left(\frac{6.023 \cdot 10^{23}}{mol}\right)}{\left(\frac{55.85g}{mol}\right)} exp \frac{\left(-1.08 \frac{eV}{atom}\right)}{\left(8.617 \cdot 10^{-5} \frac{eV}{K}\right)(850 + 273)K}$$

$$= \frac{1.17 \cdot 10^{18}}{cm^3}\left(\frac{100cm}{m^3}\right)^3 = 1.17 \cdot 10^{24} m^3.$$

– *Frenkel defect*

Frenkel defect is the association of a vacancy and an interstitial atom (Fig. 8.10).

Point defects play a very important role in processes such as solid-state diffusion, electrical conductivity, density, solid solutions, color, etc. These defects affect, perhaps, one cell in 10,000.

Fig. 8.10 Scheme showing the Frenkel defect

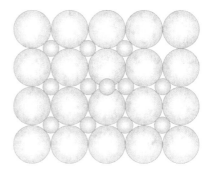

8.4.1   Point Defects and Solid-State Diffusion

Point defects allow diffusion in solid state, i.e., the transport of atoms in a crystalline material, as a result of their mobility.

When an atom moves to another position, it is because the potential energy is less.

Movement of Atoms
The movement of atoms can be by

1. Vacancies (Fig. 8.11)

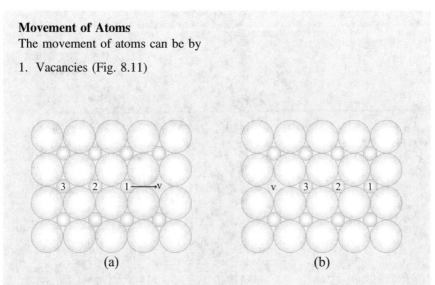

(a)   (b)

Fig. 8.11 Diffusion through a vacancy. The movement of atom 1 to the v vacancy (**a**) causes the vacancy to be subsequently occupied by atom 2 and so on (**b**)

2. Impurities (Fig. 8.12)

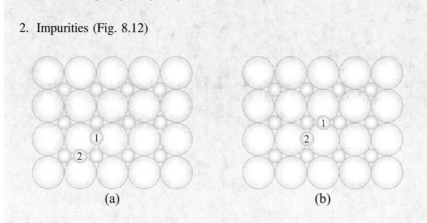

(a)   (b)

Fig. 8.12 Diffusion through impurities: As impurity 1 moves (**a**), its position would be occupied by impurity 2 and so on (**b**)

3. Exchange between atoms (Fig. 8.13)

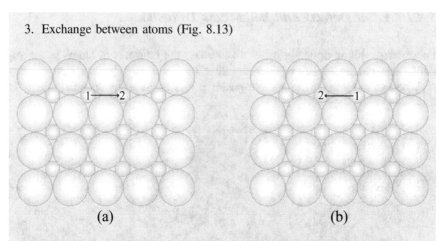

Fig. 8.13 Scheme showing diffusion by means of exchange between pairs of atoms. Atom 1 moves to position of atom 2 (**a**), then its position would be occupied by atom 2 (**b**)

4. Circular exchange between atoms (Fig. 8.14).

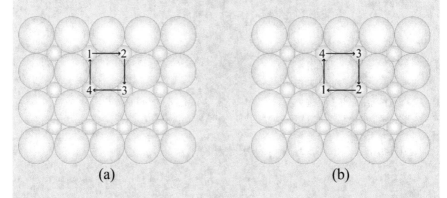

Fig. 8.14 Scheme showing diffusion through circular exchange between atoms. In **a** the positions of atoms 1–4 and the movement scheme are shown, and in **b** the position of the atoms after the circular exchange is shown

8.4.2 Point Defects and Color in Crystals and Minerals

All the sensations of light intensity and color reaching the human eye represent only a small part of the electromagnetic spectrum, ranging from wavelengths of 390 nm (violet) to 770 nm (red). Each wavelength is associated with a characteristic frequency (v), such that

$$hv = c \tag{8.3}$$

Being

h the Planck's constant,
v the frequency, and
c the rate of light propagation in a vacuum.

In turn, the wavelength carries associated energy, determined by

$$E = hv = h(c/\lambda) \tag{8.4}$$

where

λ is the wavelength

There are three main types of color causes in minerals:

1. Selective absorption of certain components of the visible spectrum and transmission of the rest.
2. Physical—optical effects related to scattering, refraction, and reflection of light.
3. Band theory, which explains the color cause, e.g., in the diamond.

1. *Selective absorption*
Considering selective absorption, when a crystal lets through all wavelengths of white light, it is colorless, as is the case in pure varieties of quartz or calcite. If all wavelengths are absorbed, the mineral will appear black.

The mineral will only be colored if the energy of any of the visible wavelengths matches the energy needed to raise an electron from a fundamental state to an excited state; that is, to produce an electronic transition.

The electronic transitions that most condition color are those that affect the elements of the first transition series: Ti, V, Cr, Fe, Co, Ni, and Cu; and also some rare earths—neodymium and praseodymium in lanthanides, and uranium in actinides. For these elements, the following electronic transitions are observed: (a) Crystal field transitions between the two groups of 3d orbitals in transition metals, (b) charge transfer transitions between neighboring ions, and (c) color centers.

– Crystalline field transitions
Crystal field theory is a chemical bond theory, very appropriate for phonic-type structures such as silicates and oxides. It considers ions as point loads which, being

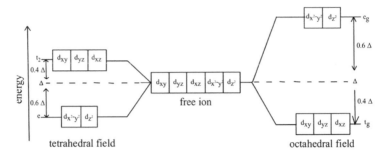

Fig. 8.15 Splitting of the d-orbitals of the Cr^{3+} ion with tetrahedral and octahedral coordination

part of a short-range structure (which means coordination polyhedra), create a field that interacts with the electronic levels of a transitional metal ion.

Transition metals have $3d$ orbitals which, in some cases, are incomplete and in the absence of electric and magnetic fields all have the same energy but correspond to different spatial orientations, the d_z^2 and d_x^2 and d_z^2 orbitals are projected along the cartesian axes, while the d_{xy} and d_{yz} orbitals are directed between the axes. The effect of the crystalline field created by the cation ligands in it is to produce splitting of the energies of $3d$ orbitals in two groups, with an energy difference called *splitting energy* (see the splitting of the Cr^{3+} ion with tetrahedral and octahedral coordination in Fig. 8.15).

In this situation, the distribution of electrons in the orbitals d is controlled by two factors:

1. Electrostatic and exchange interactions between electrons causes electrons to be distributed in as many orbitals with parallel spin as possible.
2. The splitting effect causes electrons to be placed in lower energy orbitals.
   This causes two possible configurations on the metal:

   1. *High spin*, corresponding to a weak crystalline field.
   2. *Low spin*, corresponding to strong crystalline field.

Based on these factors, when white light strikes a mineral with *chromophore elements* (elements causing color), if the energy of any of the visible wavelengths coincides with the splitting energy of the transitional metal cation, an electronic transition occurs, with absorption of that particular color. These chromophore elements can enter the chemical formula of the mineral, and these minerals are called *idiochromatic* (Table 8.2). If they enter as impurities, the minerals are called *allochromatic* (Table 8.3).

Table 8.2 Examples of idiochromatic minerals

Element	Example of color and mineral idiochromatic
Chromium	Green: Uvarovite
Manganese	Red: Rodocrosite; Pink: Rhodonite Orange: Spessartite
Iron	Green: Melanita; Red to Green: Hematites Yellow: Goethite
Copper	Green: Malaquite Blue: Azurite, Turquoise

Table 8.3 Examples of allochromatic minerals

Element	Example of color and mineral allochromatic
Chromium	Green: Emerald, Grosularia Red: Ruby
Manganese	Green and Red: Alexandria Red: Red Beryl; Pink: Morganite
Iron	Green to Blue: Aquamarine Yellow: Chrysoberyl, Citrine Quartz
Cobalt	Yellow + Green: Andalusite Blue: Synthetic spinel

The specific values of the crystalline field splitting energies, which determine the wavelengths to be absorbed, depend on the following:

– The transition metal ion and its oxidation state.
– Coordination of the anion and the type of anion–cation bond.
– Distortions of the ideal coordination polyhedron.

Transitions of this type can only occur in ions with half full $3d$ orbitals. For example, the Zn^{3+} and Cu^{2+} have complete $3d$ orbitals, so they do not have vacant orbitals to accommodate excited electrons. Similarly, the Ti^{4+} and Sc^{3+} ions do not have $3d$ electrons, so they cannot absorb it by this mechanism.

Only the transitions that keep the number of electrons unpaired should, theoretically, be observed.

Transitions that maintain the number of unpaired electrons are those that should, theoretically, be observed, because those that do not maintain it (called forbidden spin transitions) change the total spin of the cation, which is, theoretically, not allowed by magnetic properties.

However, these transitions are observed due to coupling effects between orbitals. Although they are generally weaker, and at higher energies than allowed spin transitions, they may be the main cause of color in some minerals.

An example is the typical red garnet color, which is not due to the main transition of allowed spin of ferrous iron, since it falls out of the visible (7640 cm^{-1}, about 1400 nm), but to weaker forbidden spin transitions in the shorter regions of the visible spectrum, which allow the passage of red.

Examples of the effect of cation coordination are garnet and olivine. In garnet, Fe^{2+} cations are in distorted cubic coordination, while in olivine they are in octahedral coordination, with anion–cation distance of 0.212 nm, relatively short compared to the distance of 0.222 nm in garnet. Consequently, electrons around Fe^{2+} in garnet experience a weaker crystalline field, with minimal splitting. This causes the absorption peak in olivine to be at a higher energy than garnet.

A clear example of the influence of the bond type on color is the color of ruby, emerald and alexandrite, which involve the same transitional metal, the Cr^{3+}, in the same octahedral position but have very different colors.

Example
Color in the ruby

Ruby is corundum, Al_2O_3, colorless when pure; however, with a small amount of Cr^{3+} replacing the Al^{3+} in octahedral positions (approximately 1% of Cr_2O_3) ruby acquires its characteristic red color.

The Cr^{3+} ion has three electrons in the 3d orbitals; the crystalline field induced in the ion by the effect of the six oxygen ligands causes the five orbitals to split, becoming different energies. As a result, two main absorptions can occur, one in which 2.2 eV are absorbed, and another in which 3.0 eV are absorbed, as seen in Fig. E1—Transitions (a) and ruby spectrum (b).

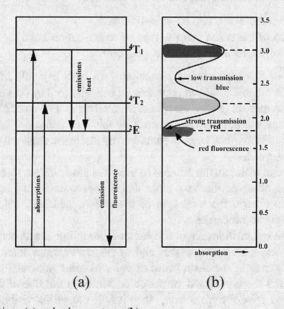

Fig. E1 Transitions (**a**) and ruby spectrum (**b**)

The first corresponds to the absorption of the green-yellow component of the spectrum.

The second corresponds to violet.

These absorptions allow a very important transmission of the red component, hence its color.

Ruby has another color phenomenon, called *dichroism* (see Figure):

If ruby is observed in polarized light, turning the glass results in a change in color.

In sections perpendicular to the optical axis, the color is purple red.

In sections parallel to the optical axis, that is, in the direction of the extraordinary ray, the color changes to an orange red. This is because one of the groups of unfolded levels ($^4T_1$ and $^4T_2$, according to group theory) is, in turn, unfolded in two others, because the octahedral coordination polyhedron in which chromium is placed is slightly distorted.

– Transitions by charge transfer

These transitions involve the transfer of valence electrons between neighboring cations in the structure of a mineral.

This phenomenon cannot be explained by crystalline field theory.

It is the cause of the blue color of sapphire and the dark colors of many transitional metal oxides, such as magnetite.

Example
Blue Sapphire color

Blue sapphire is corundum in which small amounts of Fe and Ti enter, replacing trivalent aluminum (see Fig. 8.16); iron can appear as Fe^{2+} or Fe^{3+}, and titanium is usually in its tetravalent state.

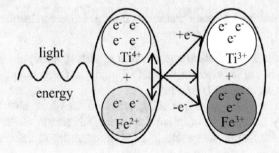

Fig. 8.16 Charge transfer between Fe^{2+} and Ti^{4+}

In order for there to be an interaction between the two, there must be a situation in which they are related, as occurs with octahedral sharing faces along the c axis. Corundum with only iron have a pale yellow color while, if only titanium appears, they are transparent. This indicates that the Fe^{2+}-Ti^{4+} interaction is responsible for producing the blue color.

In the arrangement above, the distance between the two cations is 2.65 Å, which produces sufficient overlap between the orbitals d for electrons to be transferred from one to another. In this case, the transfer absorbs 2.11 eV, which corresponds to a λ of 588 nm. There are also absorptions at both ends of the visible spectrum, so all colors except blue and blue-violet are absorbed. These conditions correspond to the ordinary ray.

There is another arrangement in which iron and titanium are adjacent in the corundum structure, in a direction perpendicular to the c axis, with octahedra sharing edges. In this situation the Fe-Ti distance is 2.79 Å, with less overlap of orbitals. The difference in energy is small, but the intensities difference is very high, which makes the absorptions different. These are the conditions that affect the extraordinary ray.

In general, charge transfer can be of two types:

(a) Heteronuclear between two different transition metal ions, for example, Fe^{2+-} + Ti^{4+}, as in sapphire. Similar transitions also occur in blue kyanite, benitoite, tanzanite, and brown andalusite.
(b) Homonuclear colors are derived from transfers that affect ions of the same transition metal, in different valence states, for example, Fe^{2+}-Fe^{3+} occupying different positions in the structure. They can be:

   – Idiochromatic—magnetite and vivianite
   – Allochromatic—green and blue tourmaline, brown and black micas, amphiboles and cordierite.

There is a third type of color cause that does not affect transition elements.

– Color centers

Color centers also involve electronic transitions, although they do not affect transition metals, but are an electron trapped in a gap in the structure or an anionic group in which an electron is missing.

These color centers appear naturally, although they can also be caused by irradiation or by the addition of impurities during the growth of the crystal.

Some are stable, while others are lost during exposure to light, or even in the dark.

There are two types of color centers: Electron and hole.

– Electron color centers: F-Centers
They occur when an electron displaced from an ion, usually after an energy input, reaches a relatively stable situation in a vacancy it finds when returning to its original ion or to a similar one.

They are common in halides and are usually related to excess metal or a halide deficiency in the crystal.

They originate from a halogen vacancy, and instead, an electron is trapped. They can be explained by Frenkel and Schottky defects. These defects, thermodynamically stable, at approximately 0.01%, can be caused by the addition of halide. The presence of divalent cation impurities produces vacancies that can be transformed into an F-color center.

A vacancy by itself produces no color. Light, as it passes through the crystal, transitions to a higher energy level of the electrons located in the vacancies, with absorption of a relatively wide range of the spectrum.

Depending on the conditions and even caused by exposure to light itself, other types of electron color centers may occur (Fig. 8.17). An F-center can pass to F', in which two electrons are trapped. An M-center, also called F2, consists of two interacting F-centers, absorbing into the infrared region. Finally, the R-color centers consist of three adjacent F-centers, arranged on the plane (111).

Crystals with electron color centers do not usually show pleochroism, because either they have cubic symmetry or are arranged according to all the equivalent directions of the crystal, thus maintaining isotropy.

– Hole color centers
In general, quartz grows with a small amount of Al^{3+}, which will replace the Si^{4+} but no color change will appear. However, when radiating the crystal with γ-rays or with X-rays it acquires the smoky color, which disappears if it is heated to 400 °C, recovering the color when it radiates again.

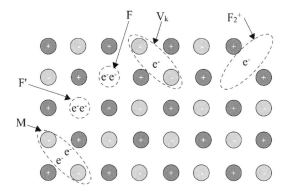

Fig. 8.17 Color centers types

The mechanism of formation of smoked quartz (Fig. 8.18) consists of some electrons escaping from O^{2-} when radiating the quartz, but they will quickly return to their original position or to a similar one.

If there is Al^{3+} replacing Si^{4+}, there will be a free proton H^+ as an interstitial atom for each substitution, to maintain the neutrality of the crystal. These protons can be combined with the released electrons to give a interstitial hydrogen. The resulting $[AlO_4]^{4-}$ group absorbs light, as it constitutes a hole (V_k) color center (see Fig. 8.17). A subsequent heating of the crystal releases the electron from hydrogen, returning to a nominal position, fading the crystal.

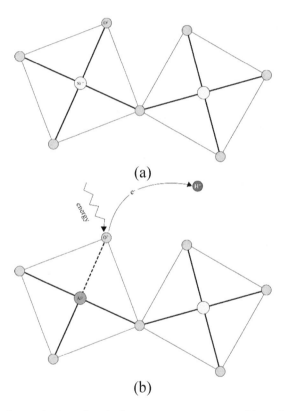

Fig. 8.18 Formation mechanism of smoked quartz: **a** quartz structure without aluminum **b** quartz structure with Al^{3+} replacing Si^{4+} and releasing an electron with H^+ formation, after irradiation

If, instead of Al^{3+}, $Fe^{3+}$is introduced, a light crystalline field on the cation colors the quartz a pale yellow color, obtaining citrine quartz. If citrine is irradiated, the center $[FeO_4]^{4-}$ is formed in a similar way to smoked quartz, quartz of purple color. Thus the amethyst is obtained (Fig. 8.19), which losses coloration with heat.

In general, any type of color center can be produced by irradiating the mineral.

2. Physico-optical effects

In this section, physico-optical effects are considered a series of phenomena that cause color. They are not related to selective absorption of visible light but to phenomena of dispersion and reflection, and interference with light affecting minerals with certain characteristics.

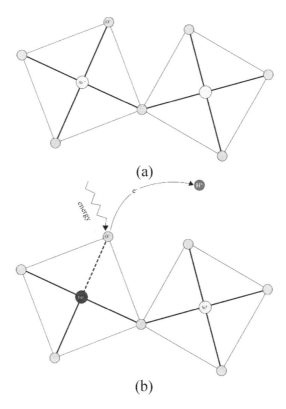

Fig. 8.19 Formation mechanism of amethyst quartz: **a** quartz structure without iron **b** quartz structure with Fe^{3+} replacing Si^{4+} and releasing an electron with H^+ formation, after irradiation

– Dispersion and reflexion

Dispersion is the separation of light at its different wavelengths. The variation of this property with the variation of the wavelength is what can be important when considering the color of the minerals. One of the most important effects is opalescence, which is the dispersion of light by a cloud of small particles placed in its path, similar to dust particles in a room's atmosphere. This effect is also called the *Tyndall effect*, giving rise to the milky appearance of some opals or contributing to the optical effect of moonstone.

The cat's eye effect consists of internal reflections originated by cavities or fibrous and fine inclusions, oriented regularly according to the symmetry of the crystal. Examples include quartz, sapphire, and diopside (Fig. 8.20).

If the inclusions are parallel to more than one crystalline face, the star effect can be obtained if the stone is carved in cabochon. This usually occurs in sapphires (star sapphire) (Fig. 8.21) and rubies.

Fig. 8.20 1–7 Quartz (Sri Lanka). 8–9 Smoky quartz (Brazil). 10: Black sapphire, Bang Kha Cha, Chanthaburi, Thailand. 11: Golden black sapphire, Bang Kha Cha, Thailand. 12: Golden black sapphire with 12 rays, Bang Kha Cha, Thailand. 13: Rubi, Burma. 14: Synthetic Ruby, Russia, 15: Synthetic ruby with diffuse star, Thailand. 16: Corundum with diffuse star (titanium). 17 y 18: Natural sapphire. 19: Synthetic sapphire, Russia. 20: Sunstone (orthoclase feldspar), Tanzania. 21: Green diopside, Burma. 22: Almandine garnet with multi-stars. 23: Black diopside. 24: Cat's eye sillimanite. Burma, 25: Cat's eye moonstone (orthoclase feldspar). 26: Pink scapolite cat's eye, Burma. 27: Cat's eye zircon, Sri Lanka. 28: Cat's eye sillimanite, Burma. 29: Tiger's eye quartz (taken from https://kaiajoyasuruguay.blogspot.com/2012/07/ciencia-el-asterismo-o-efecto-estrella. html, with permission)

Fig. 8.21 Asterism in sapphire (taken from https://kaiajoyasuruguay.blogspot.com/2012/07/ciencia-el-asterismo-o-efecto-estrella.html, with permission)

– Adularescence

Adularescence is an interesting effect affecting the adularia, or moonstone.

Adularia is potassium feldspar with intergrowths of albite in its interior, so that alternate layers of albite and orthoclase are arranged which, when traversed by light, produce a whitish or bluish glow by internal reflections on surfaces of different refractive index.

– Interferences

If the light waves from a single source are divided into two rays, and they combine again after traveling slightly different distances, they may be out of phase, and they produce interference (Fig. 8.22).

This phenomenon occurs, for example, in thin films of transparent substances, producing iridescent colors. The difference in trajectory will increase with the thickness of the film and the obliquity of the beam. If white light is used, different colors will be strengthened when the film thickness is varied.

This phenomenon has been observed with electron microscopy in the structure of the opal, in which by diffraction in the rows of spheres of its structure, colors by interference originate.

3. *Band theory*

Band theory is based on the fact that electrons, depending on their energy state, can occupy certain energy bands. This theory classifies solids into three groups: Insulators, semiconductors, and metals.

Fig. 8.22 Wave interference

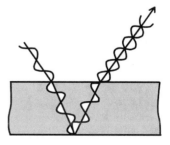

Insulators are those solids that are characterized because the range of prohibited energies, between the valence band and the driving band, is very large. There is no current conduction, since the electrons would need a very large supply of energy to be able to excite them and to move from the valence band to the conduction band.

Semiconductors are those solids that are characterized because the range of prohibited energies is small and, at a given temperature, some electrons can become thermally excited and pass to the conduction band. They could be electric current conductors, as conductivity increases with temperature. An example would be the diamond that under normal conditions is insulating, but with certain doping agents is conductive (wide band gap semiconductor).

Metals are those solids in which the range of forbidden energies is so small that there is practically no separation between the valence and driving bands, so they are very good conductors. Conductivity reduces with temperature.

Examples

– **Color in metals**

The color in metals is due to the variation of the reflectance with the wavelength, i.e., the dispersion of the reflectance.

All metals intensely absorb the light that hits them, so that it can only penetrate them a distance less than a wavelength. Being conductive materials, the absorbed light induces alternating electric currents on the surface of the metal, which dissipate rapidly by emitting the absorbed energy in the range of the visible. This is what gives them strong reflection.

The mineral being "dyed" a color will depend on the most favorable electronic transitions on its surface, which will cause more of a given λ to be absorbed and, therefore, the reflection for the corresponding color is greater.

– **Color in the diamond**

The explanation of color in the diamond is based on the theory of bands, since the diamond, one of the forms of carbon, is an insulator with a very wide range of forbidden energies (5.4 eV) but, in some cases, it is a semiconductor.

Examples of color in diamond: yellow diamond, blue diamond.

– Yellow diamond

When nitrogen (N^{5+}) enters as impurity into the diamond, replacing in the carbon structure at a ratio of 1 nitrogen atom per 100,000 carbons, the diamond is yellow.

The explanation for this is that the N^{5+} has one more electron than the C^{4+}, and each nitrogen atom provides an electron that has no place in the valence band and will be placed at a level of the forbidden energy range.

From this level, it is easy to pass to the conduction band when it is excited, because the energy needed to overcome is 2.2 eV and, as this value falls within the energy range of the visible spectrum, specifically corresponding to blue and violet, it means that these colors are absorbed and the transmitted color is yellow.

Since transition from the electron to the conduction band is due to the donation of a dopant (nitrogen) with excess electrons, the level of the range of forbidden energies in which nitrogen is located is called the donor level.

When the nitrogen concentration is 1 per 1000 carbons, the resulting color is green.

If the concentration is higher, it can absorb the entire visible spectrum and the resulting color is black.

– Blue diamond

When, instead of nitrogen it is boron (B^{3+}) that is introduced into the crystal structure of the diamond, replacing the C^{4+} carbon, there is a lack of an electron.

It is necessary to understand that when there is a lack of an electron, in the forbidden energy band there is what is called a hole or electronic hole.

This level that causes the presence of boron is called the acceptor level and the energy is 0.4 eV, this being the energy absorbed, and the color transmitted is blue.

The energy required to reach the acceptor level from the valence band is very small and that the electrons can be easily excited by thermal energy and pass to the accepting level, thus leaving gaps in the valence band and generating electron movement, so blue diamonds are conductors of electricity.

8.4.3 Point Defects and Chemical Composition

– *Solid solution*

Solid solution is the solid-state dissolution of one mineral phase in another. It originates as a result of chemical variability in minerals due to the existence of specific defects. On an atomic level, this is demonstrated by mixed crystals and can be caused by the following mechanisms: replacement, omission, interstitial formation.

– *Replacement*

It consists of the substitution of one ion with another in the same atomic position of the crystalline structure. The valence of the ions that are replaced must be the same.

Ion radii cannot differ by more than 15% in ion compounds. They may be

- *Complete*: When substitution can be in any proportion.
- *Incomplete*: When substitution cannot be in any proportion.

- *Omission*

Omission mechanism consists of the existence of a vacancy as a result of the absence of an ion in the structure (Fig. 8.7b).

Example

Pyrrhotite is a non-stoichiometric iron monosulfide in the compositional range between FeS and $Fe_{0.875}S$. Pyrrhotite occurs abundantly in nature, and it is a close second, after pyrite, among the most abundant iron sulfides in ore deposits and crustal rocks. It is a common mineral in igneous and metamorphic rocks and, to a minor extent, in extraterrestrial rock bodies and dust particles [3–5].

Pyrrhotite is the main carrier of rock magnetism in many rock types; therefore, it is important to understand how the distinct structure types are distributed in rocks and how this distribution affects the overall magnetic properties of rocks.

The basic modules in pyrrhotite are layers of iron atoms octahedrally coordinated by sulfur (Figure). There are two principal types of layers: An A layer, with completely filled octahedral Fe positions and a B layer, with some Fe positions being vacant (Fig. 8.23).

Fig. 8.23 Structure of pyrrhotite showing the vacancies of iron, one of four layers bearing iron-site vacancies. Here, the B position is shown vacant

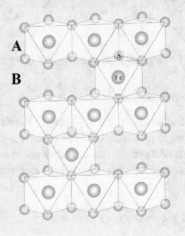

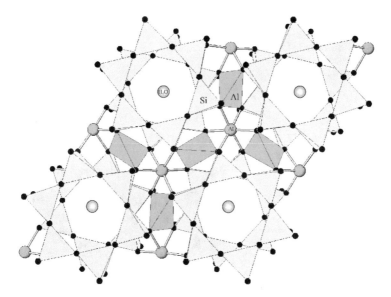

Fig. 8.24 Beryl structure, projection perpendicular to the crystallographic axis *c* showing the large structural gaps (parallel to *c*) occupied by water molecules, forming an isomorphic series

– *Interstitial formation*

Interstitial formation is the presence of an ion in a structural spacing, such as in Fig. 8.24.

The solid solution represents a disordered state and depends on the temperature. At high temperature, atomic substitution is higher because atomic vibrations are higher and atomic positions dilate. The solid solution is homogeneous and formed by one phase. At low temperature, vibrations are lower and atoms are more static. The extent of the solid solution is smaller. The decomposition, unmixing, or exsolution appears with more than one component or mineral phase by lowering the temperature. It manifests as small acicular inclusions that are crystallographically oriented (Fig. 8.25).

The curve, *solvus* or *spinodal*, separates the solid solution from the components A and B, which are immiscible at low temperature (Fig. 8.26).

Isomorphism is the phenomenon that causes isomorphic crystals or minerals to exist. Isomorphic crystals or minerals have the same crystal structure and stoichiometry but different chemical compositions.

Fig. 8.25 Transmission polarizing microscope image of a
pyroxene showing oriented
acicular exsolution

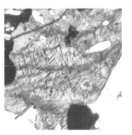

Fig. 8.26 Temperature—
composition phase diagram of
two components, A and B,
which are immiscible at low
temperature (RT)

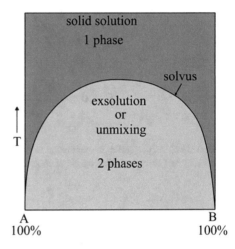

The Chemical Formula of a Mineral

Most minerals are composed of two or more elements, and their formulas,
recalculated from the results of quantitative chemical analyses, indicate the
atomic proportions = stoichiometry of the present elements.

A quantitative chemical analysis provides basic information regarding the
atomic formula of a mineral but not as to its position in the structure. The sum
of the percentages in the analysis should be 100%.

Chemical analyses of minerals are obtained by various techniques: wet
way, (dissolving the mineral), optical emission, X-ray fluorescence, atomic
absorption spectroscopy, electron microprobe.

The analyses provide basic information regarding the atomic formula of a
mineral, based on the percentage by weight of the elements or oxides of the
elements that form it.

The analyses do not give information about the structural position of these
elements.

The steps to obtain a chemical formula are presented below, with the
example of an olivine.

Example

Given the percentages by weight of the oxides of the constituent elements of
an olivine, column (1) of Table E1, and knowing that the molecular weights
of said oxides are in column (2), the procedure for obtaining the chemical
formula and the percentage of the pure end members of the olivine series,
which is the way it is usually expressed, is described below

Table E1 Procedure for obtaining the chemical formula, knowing the weight (%) of the oxides
and their molecular weights

	Oxides % (1)	Molecular weight (g/ mol) (2)	Proportion molecular (3)	Proportion cationic (4)	Proportion Oxygen (5)	Cations based on four oxygen (6)
SiO_2	39.30	60.09	0.654	0.654	1.308	1.004
FeO	19.30	71.85	0.269	0.269	0.269	0.413
MgO	41.40	40.31	1.027	1.027	1.027	1.578

1. Molecular proportions

The molecular proportions, column (3), are obtained by dividing the percentage by weight of each oxide (analysis of column (1) of Table E1) by its
molecular weight. If the analyses were elementary, the percentage by weight
of each element would be divided by its atomic weight. The results can be
given in %, for which the quotient must be multiplied by 100.

2. Cation proportions

The cation proportions, column (4), are obtained as the product of the
molecular ratio and the subscript of the corresponding cation.

3. Oxygen proportions

The oxygen proportions, column (5), are obtained by multiplying the subscript of oxygen in each oxide by the molecular ratio.

4. Cation proportions based on a given number of oxygen (6)

The generic formula of the mineral must be taken into account, in this case, of
the olivine, $(Fe,Mg)_2 SiO_4$, which has four oxygen.
   The cation proportions, column (6), based on a given number of oxygen,
column (5), are obtained by multiplying the cation ratio (column 4) by the
oxygen factor (quotient between the base number of oxygen and the sum of
the calculated oxygen).

5. Atomic relations

The atomic relations are obtained by considering column (6) of Table E1 and the generic formula of the mineral, in this case, olivine:

  $(Mg_{0.4}Fe_{1.6})\ SiO_4$ or $(Mg,Fe)_2SiO_4$

It should be noted that the sum of silicon must be 1 and the sum of the rest of cations must be 2, which is approximately true.

The formula is also often expressed in terms of the compositions of the end members, when it comes to solid solutions. Olivine is expressed according to the percentage of forsterite (fo), the magnesium-rich member, and fayalite (fa), the iron-rich member. To do this:

1. The molecular proportions of iron and magnesium, 0.269 + 1.027–1.296, are added (column 3 of Table E1)
2. The molecular proportion of the above-mentioned cations is divided by 1.296
3. The quotient is multiplied by 100.

  Finally, the result obtained is: $Fa_{20.76}Fo_{79.24}$ (20.76% of fayalite and 79.24% forsterite).

Variations in composition, relative to ideal, can be represented by different diagrams. Among them are bar diagrams, and triangular diagrams, both of which represent solid solutions.

– Bar diagram

The bar diagram is used to observe the variation of two components. Each end of the diagram represents 100% of one of the two phases of the solid solution series. The other end represents 100% of the other phase. The intermediate values correspond to the composition of the intermediate phases.

Example
Representation of the composition of an olivine

The extreme components, forsterite (Mg_2SiO_4), Fo, and fayalite (Fe_2SiO_4), Fa, are located at the ends of the diagram (Fig. E1).

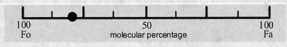

Fig. E1 Representation of the composition of the olivine with 20.76% fayalite and 79.24% forsterite ($Fa_{20.76}Fo_{79.24}$) in a bar diagram

– Triangular diagram

The triangular diagram represents the variation of three components instead of two, as in bar diagrams. Each side of the triangular diagram would represent a bar diagram, so that the variation of two of the three components can be observed in the same way as in a bar diagram.

Example

Representation of the composition of a pyroxene, based on three components, wollastonite ($CaSiO_3$), enstatite ($MgSiO_3$), and ferrosilite ($MgSiO_3$)).

These components are located at the vertices of the triangular diagram.

Any pyroxene composition that includes only two of the three components can be represented along one edge of the triangle.

The En-Wo edge represents 0% of Fs, the En-Fs edge represents 0% of Wo, and the Fs-Wo edge represents 0% of En.

The Wo vertex represents 100%, the vertex En represents 100%, and the Fs vertex represents 100%. For example, the Wo vertex represents 100% $CaSiO_3$ and the horizontal lines between this vertex and the base of the triangle indicate variations from 100% to 0% of $CaSiO_3$ (Fig. E2).

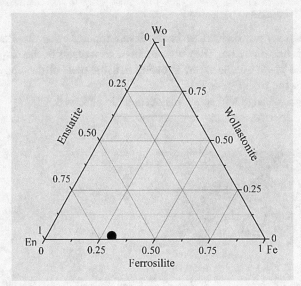

Fig. E2 Triangular diagram of the composition of pyroxene, with 1.66% wollastonite, 68.72% enstatite, and 29.62% ferrosilite ($Wo_{1.66}En_{68.72}Fe_{29.62}$)

8.5 Linear Defects

Linear defects are discontinuities in the crystalline structure that affect a lattice row and are called dislocations. An example of a linear defect is shown in Fig. 8.27.

They affect the structure in a deeper way than point defects and have higher energy.

They are not in thermodynamic balance with the crystal, unlike point defects.

The Burgers vector *b* defines the value and direction of the displacement of the atoms from their ideal lattice position.

Dislocation glide explains the plastic deformation in the materials in the crystalline state.

There are two basic types of dislocations: edge dislocations, screw dislocations.

– *Edge dislocations*

Edge dislocations are defined as linear discontinuities with the Burgers vector perpendicular to the line direction. They are produced as a result of the movement of one crystalline plane over another, and they do not affect all cells equally (Fig. 8.28).

– *Screw dislocations*

Screw dislocations are defined as linear discontinuities with the Burgers vector parallel to the line direction. They are produced as a result of the displacement of one crystalline plane over another, affecting half the cells in the upper and lower planes but not the rest (Fig. 8.29).

Often, a mix of edge and screw dislocations is observed.

(a)   (b)

Fig. 8.27 a Transmission electron microscope image of dislocations in vermiculite **b** scheme of dislocations of (**a**)

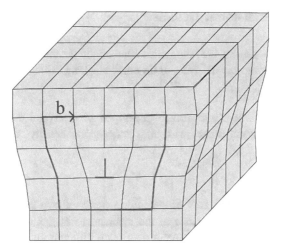

Fig. 8.28   Edge dislocation is shown by the inverted T and b is the Burgers vector

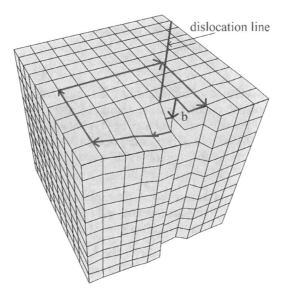

Fig. 8.29   Screw dislocation scheme, b is the Burgers vector

8.6   Two-Dimensional Defects

Two-dimensional defects are anomalies that affect the crystalline planes. They include the following: crystal face, grain edge, stacking faults, and polytypism.

– *Crystalline face*

A crystalline face is a plane surface developed on a crystal during its growth. Crystal faces are crystal planes with high lattice-point density and mainly simple Miller indices.

– *Grain boundary*

The grain boundary is the area of separation between two crystals of the same species. It arises as a consequence of the mechanism of grain growth, or crystallization, when two crystals that have grown from different nuclei "meet".

– *Stacking faults*

Stacking faults are irregularities in the sequence of the crystalline planes in the structure. They affect the long-range order.

Types of stacking faults in compact packaging include the following: twins, intrinsic defects, extrinsic defects.

– *Twin*

Twinning is the association of individuals of the same crystalline species, with different crystallographic orientations and related by some element of symmetry called twin law. The symmetry element may be (1) rotation axis of second or fourth order or inversion-rotation axis of first order; or (2) mirror plane.

Twins are referred to by their twin laws. It is also commonly referred to by a name that refers to

– The town where it was first discovered. Example: Carlsbad twin of the orthose.
– The shape of the twin. Example: Swallow-tail twin in gypsum.
– The mineral or minerals that most often present it. Example: Spinel twin.

Twins can be recognized because they have incoming angles, stretch marks, brightness differences, etc., that indicate contact between individuals.

The contact surface between the individuals of the twin can be a plane.

The twin elements are twin plane and twin law.

The twin types are contact twins and interpenetration twins. In the contact twins, the individuals of twins are joined by flat surface. In the interpenetration twins, the twin individuals are joined by an irregular surface and are interpenetrated.

Example

Contact twins		
Spinel: octahedral face (111) is the twinning plane	Gypsum: the (010) is the twinning plane	Cassiterite: the plane (101) is the twinning plane
Rutile: the plane (101) is the twinning plane		Albite: twinning is most commonly of repeated lamellar type involving very thin lamellae of the mineral

(continued)

(continued)

Interpenetration twins		
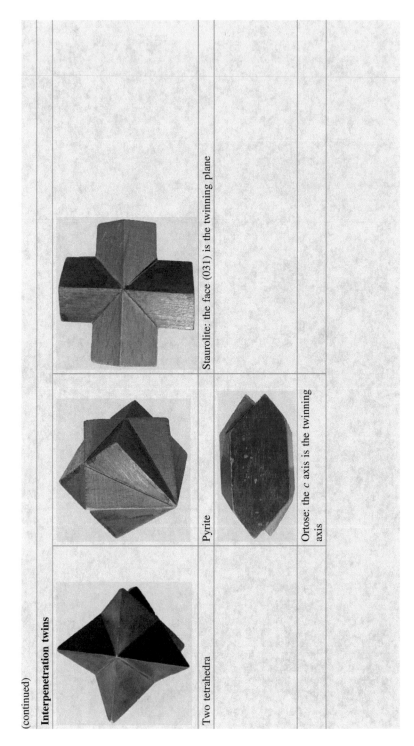		
Two tetrahedra	Pyrite	Staurolite: the face (031) is the twinning plane
	Ortose: the c axis is the twinning axis	

To understand them, you must consider the normal sequence of layers in compact packaging.

- Normal sequence in compact cubic packaging is ABC ABC...
- Normal sequence in compact hexagonal packaging is AB AB AB AB ...

Twins appear when an inversion occurs in the normal succession of layers of the corresponding packaging, or a twin plane (intrinsic defect) or polysynthetic twins (extrinsic defect).

- *Intrinsic defect*

An intrinsic defect is the result of adding a plane in a position that does not belong to it.

- *Extrinsic defect*

An extrinsic defect is the result of extracting a particular plane in the stacking sequence.

- *Polytypism*

Polytypism is a phenomenon that causes polytypes to exist. Polytypes are crystals and minerals that differ in the stacking periods/sequences of identical layers. The polytypism affects only one dimension, unlike in polymorphism.

Example
Mica (Fig. 8.30), ZnS, SiC.

Fig. 8.30 Example of polytype in mica

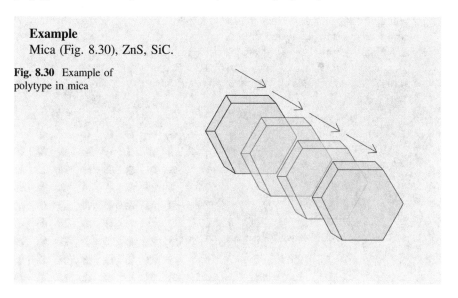

8.7   Three-Dimensional Defects

This type of defect implies that the three-dimensional nature of the crystal is broken by the presence of the defect.

Inclusions or voids within the crystalline mass (Fig. 8.31), which are a consequence of the growth process, are considered three-dimensional defects. They can be formed before (protogenetics), during (syngenetic), or after (epigenetic) the host.

Fig. 8.31 Example of inclusion of zircon in feldspar

Exercises

1. In Fig. E1, draw lines between perfectly ordered zones following the scheme in Fig. 8.1 and areas that do not follow that scheme, to achieve a structure in domains due to disorder.
2. Calculate the number of vacancies per cubic meter in NaCl at 1000 K. The energy for vacancy formation is 2.02 eV/atom. The atomic weight of Na is 22.99 g/mol and of Cl is 35.453 g/mol, and the density of NaCl is 2.16 g/cm^3.

Fig. E1 Scheme for drawing ordered and disordered domains of a crystalline structure

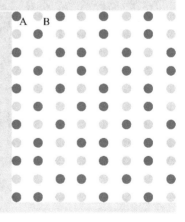

Questions

1. Pair each defect with its type

Crystalline faces	Linear
Displacement	Point
Impurity	Bidimensional

2. When an atom moves to another position, it is because

    ◯ a. potential energy is higher
    ◯ b. potential energy is lower
    ◯ c. kinetic energy is higher
    ◯ d. kinetic energy is lower

3. The movement of atoms in solid-state diffusion may be by

    ◯ a. vacancy, impurity, exchange between pairs of atoms and circular exchange between atoms
    ◯ b. vacancy, interstitial atom and pairs of atoms
    ◯ c. vacancy and exchange between pairs of atoms
    ◯ d. vacancy and interstitial atom

4. From the moment a crystal or mineral is formed, it is subject to changes in its physical and chemical environment. Such changes may be

    ◯ a. subtle changes in link length or major structural transformations and chemical changes on an atomic scale or reactions that cause new mineral species
    ◯ b. subtle changes in link length or major structural transformations and/or chemical changes on an atomic scale or reactions that cause new mineral species
    ◯ c. subtle changes in link length or major structural transformations or chemical changes on an atomic scale or reactions that cause new mineral species
    ◯ d. subtle changes in link length or major structural transformations

5. The solid solution represents a messy state and depends on temperature

    ◯ a. an orderly state and depends on the pressure
    ◯ b. an orderly state and depends on the temperature
    ◯ c. a messy state and depends on the pressure
    ◯ d. a messy state and it depends on the temperature

6. The defects that allow solid-state diffusion are

    ◯ a. dislocations
    ◯ b. two-dimensional

    ◯ c. inclusions

    ◯ d. point

7. Enter the name of defects in which there is association of two vacancies of different signs.

    Response: [   ]

8. Write the type of disorder caused by the central atom in Fig. Q1 that acquires each time one of the different positions drawn.

Fig. Q1 Scheme of type of disorder to be indicated

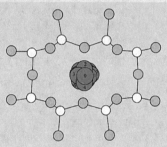

    Response: [   ]

9. Write the number of components represented in a bar diagram.

    Response: [   ]

10. Write the name of the defect that causes an empty space of the structure (structural vacuum) to be occupied by an atom.

    Response: [   ]

11. Write the name of the line separating the solid solution area from the exolution area into a composition-temperature diagram.

    Response: [   ]

12. Point defects are not in thermodynamic balance with crystal, unlike two-dimensional defects

    ◯ True

    ◯ False

13. Twins are called by their twin law, which is some element of symmetry that relates the individuals that make up the twin.

    Select one:

    ◯ True

    ◯ False

14. Polytypes are crystals and minerals that differ in the stacking of any type of structural layers.

   ⭕ True
   ⭕ False

15. Write the name of the point defect shown in Fig. Q2.

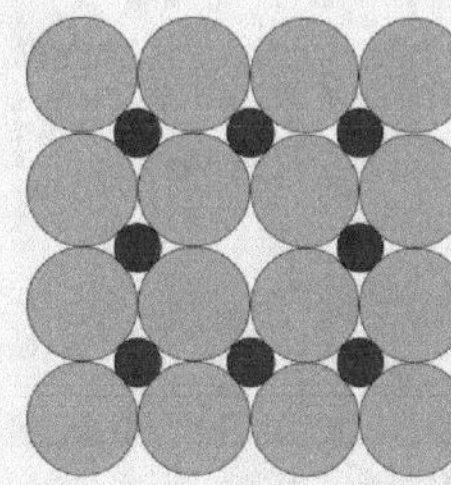

Fig. Q2 Scheme of point defect to be indicated

Response: _____

References

1. Amorós JL (1975) El, cristal. Urania, Barcelona
2. Hahn T (Ed) (1995) International tables of crystallography Vol. A Space-Group Symmetry (4th ed.). the international union of crystallography, Kluwer Academic Publishers, Dordrecht/Boston/London
3. Rochette P, Lorand J-P, Fillion G, Sautter V (2001) Pyrrhotite and remanent magnetization of SNC meteorites: a changing perspective on Martian magnetism, earth planet. Sci Lett 190:1–12
4. Keller LP et al (2002) Identification of iron sulphide grains in protoplanetary disks. Nature 417:148–150
5. Louzada KL, Stewart ST, Weiss BJ (2007) Effect of shock on the magnetic properties of pyrrhotite, the Martian crust, and meteorites. Geophys Res Lett 34:L05204. https://doi.org/10.1029/2006GL027685

Chapter 9
Polymorphism and Polymorphic Transformations Transformation Order—Disorder

Abstract This chapter shows that the crystal is a dynamic entity since, at temperatures different from absolute zero, the atoms, ions or molecules that form the crystals suffer thermal vibrations, causing them to move from their positions of equilibrium. The phase transitions are an example of this dynamism. The interpretation of the polymorphic changes, from the thermodynamic point of view, is explained using Gibbs free energy. The structural aspect of these changes is also presented. Finally, the polymorphic transformations are classified, using illustrative examples.

9.1 Introduction

At temperatures other than absolute zero, atoms, ions or molecules that make up crystals undergo thermal vibrations, causing them to move from their equilibrium positions.

The amplitude of such vibrations depends, among other factors, on the nature of the bond, and the size of the atom in relation to the atomic position of the crystal structure.

These vibrations can change the symmetry of the crystal or mineral and explain the phenomenon of polymorphism.

The crystal, in this case, ceases to be a static entity to become a dynamic entity, the phase transitions constituting an example of this dynamism. The behavior of these transitions can be explained mathematically using the principles of thermodynamics.

Thermodynamics is a logical consequence of two elementary physical axioms—the law of conservation and the law of energy degradation. Thermodynamics allows us to deduce and/or predict results in all those processes in which energy exchanges occur, such as in minerals transformation.

The principles of thermodynamics describe the behavior of three fundamental physical quantities—temperature, energy and entropy—which characterize thermodynamic systems.

© The Author(s), under exclusive license to Springer Nature Switzerland AG 2022
C. Marcos, *Crystallography*, Springer Textbooks in Earth Sciences,
Geography and Environment, https://doi.org/10.1007/978-3-030-96783-3_9

The *first law of thermodynamics* or law on energy conservation is energy cannot be created or destroyed, only transformed. It is expressed by Eq. 9.1:

$$\Delta U = Q - W \tag{9.1}$$

where

U   is the internal energy of the system
Q   is the amount of heat supplied
W   the work done by the system

In this way, when supplying a certain amount of heat (Q) to a physical system, its total amount of energy may be calculated as the heat supplied minus the work (W) performed by the system on its surroundings.

The *second law of thermodynamics* or law of entropy implies that the degree of disorder of systems increases to a break-even point, which is the state of greatest disorder of the system. This law introduces the concept of entropy (S), which represents the degree of disorder. In every physical process in which there is an energy transformation, a certain amount of energy is heat, and it cannot perform work. That heat released by the system increases the system's disorder; that is, its entropy. The change in entropy (ΔS) will be:

$$\Delta S \geq \Delta Q/T \tag{9.2}$$

The *third law of thermodynamics* states that when a system is carried to absolute zero (K), the process of that system stops and entropy has a constant minimum value.

The *fourth law of thermodynamics* is known as the zero law or the law of thermal balance and states that if two systems are in thermal balance independently with a third system, they must also be in thermal balance with each other. It can be expressed logically as follows: if A-C and B-C, then A-B. It means that by contacting two bodies with different temperatures, they exchange heat until their temperatures are equal. Examples of this law include when we get into hot or cold water. We will notice the temperature difference only during the first few minutes, as our body will then enter into thermal balance with the water and we will no longer notice the difference. The same is true when we enter a hot or cold room. We will notice the temperature at first, but then we will stop perceiving the difference because we will enter into thermal balance with it.

9.2 Stability and Equilibrium

At elevated temperatures, a certain atomic configuration may be thermodynamically favorable; that is, it may have the highest thermodynamic probability, while at lower temperatures, the most probable thermodynamic configuration may be different.

This thermodynamic probability is related to the free energy G (Gibbs) or F (Helmholtz) which it is minimum, when the probability is maximum. Therefore, in any mineral transformation, the free energy will tend to a minimum value.

Thus, the most stable state at any given temperature, T, will have the lowest free energy. However, it must be taken into account that many natural minerals are not thermodynamically stable, since changes are possible that will decrease free energy, but the kinetics of such changes (speed of transformation) can be slow and maintain unstable structures.

There are two types of possible instability: Instability and metastability. The meaning of these terms can be understood using the example in Fig. 9.1, which illustrates a hillside and a sphere.

Example

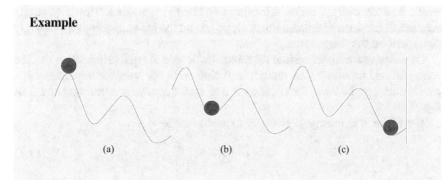

(a)   (b)   (c)

Fig. 9.1 Equilibrium: **a** Unstable, **b** metastable, **c** stable

At position 1, the sphere is in an unstable situation and a small change in its position will reduce its free energy (in this case potential energy).

In position 2, the sphere has acquired a minimum in free energy, but still has a higher energy than if it were in position 3, which is the stable position. In situation 2 it is said to be in a metastable state.

To reach position 3, the sphere must first go through an intermediate, less stable and higher-energy situation, which acts as a barrier to change. This can be overcome if extra energy is supplied, the so-called activation energy.

The change in free energy of a mineral when it takes part in a transformation can be represented by Fig. 9.2.

The reaction coordinate can be any variable that defines the progress of the transformation, for example, the transformation from a high-temperature state to a low-temperature state, below the transformation temperature. The difference in energy from a highly metastable state with F1 free energy to another final state with F2 free energy is negative and is called the driving force for the transformation. To

Fig. 9.2 Change in free
energy of a mineral when it
undergoes a transformation

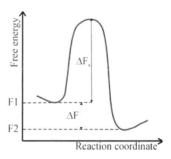

go from state F1 to state F2, it is necessary to overcome the energy barrier F_a, for
which an extra energy, called activation free energy, is needed. This is generally
supplied in the form of thermal fluctuations, so that the transformations are strongly
dependent on the temperature.

Considering a transformation, the mineral will tend to equilibrium when the free
energy of said transformation tends to a minimum. In this way, the term metastable
equilibrium can be used for situation 2 and stable equilibrium for situation 3 of
Fig. 9.1.

The Gibbs free energy is given by expression 9.3:

$$G = U + PV - TS \tag{9.3}$$

where

U is the internal energy
P is the pressure
V is the volume
T is the temperature
S is the entropy (defined in Chap. 8, item 8.1)

Equation 9.3 can be put in the form:

$$G = H - TS \tag{9.4}$$

where

H is the enthalpy:

$$H = U + PV \tag{9.5}$$

If we consider a system composed of a single mineral, and we apply energy in
the form of heat to it, the increase in energy ΔU will be proportional to the heat
supplied but not the same because part of it will be used in the thermal expansion of
the mineral.

Fig. 9.3 Evolution of the
G-T diagram

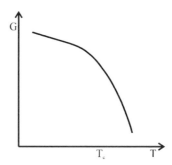

Because the absolute values of many thermodynamic quantities cannot be obtained, the differences between different states are used. Therefore, it is advisable to define a standard state for which the value of a particular quantity is specified at a reference temperature and pressure, which are 25 °C (298.15 K) and 1 atm. When a mineral changes its structure in a polymorphic transformation, there will be a change in enthalpy. If it decreases, ΔH is negative and the process is exothermic, and heat is released. When ΔH increases, the process is endothermic because it absorbs heat.

In the stability of a transformation, entropy must be considered, in addition to enthalpy, as it occurs in the aragonite-calcite transformation. The standard state enthalpy of aragonite is -1207.74 kJ mol^{-1} and of calcite is -1207.37 kJ mol^{-1}. The aragonite-calcite transformation at 25 °C and 1 atm involves an increase in enthalpy of 0.37 kJ. However, at that temperature, calcite is the stable polymorph of $CaCO_3$, which means that entropy plays an important role.

The free energy that is generally used when there is no change in volume of the phase is the Helmholtz free energy:

$$F = E - TS \tag{9.6}$$

The most stable phase at atmospheric pressure (the PV term is negligible) is the phase with the lowest internal energy.

The variation in Gibb's free energy G with temperature, for a single phase, is a curve like that shown in Fig. 9.3.

9.3 Polymorphism and Polymorphic Transformations of Crystals and Minerals

Polymorphism is the phenomenon that causes polymorphic crystals or minerals to exist through polymorphic transformations.

A polymorphic transformation is the phase change that a crystal or mineral undergoes due to the change in the physical conditions (P and T) of the mineral environment.

Polymorphic crystals and minerals are those that have the same chemical composition but different crystal structures. Examples are graphite-diamond (C), aragonite-calcite ($CaCO_3$), silica (SiO_2) transformations, kyanite—andalusite—sillimanite (Al_2SiO_5).

Examples

Figures E1, E2, E3 and E4

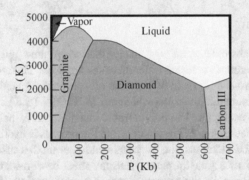

Fig. E1 Phase diagram diamond (cubic)—graphite (hexagonal)

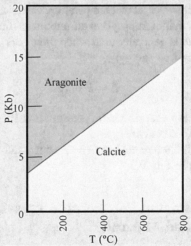

Fig. E2 Phase diagram calcite (rhombohedral)—aragonite (orthorhombic) $CaCO_3$

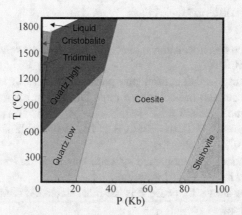

Fig. E3 Phase diagram low quartz (rhombohedral)—high quartz (hexagonal) SiO_2

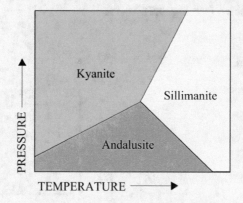

Fig. E4 Diagram of Al_2SiO_5 phases: kyanite-andalusite-sillimanite

9.4 Thermodynamic Aspect of Polymorphic Transformations

The simplest phase transformation is that which occurs between two polymorphic forms of a mineral.

In order for transformation from one phase to another to take place, the free energy curves (G curves) of each phase must intersect at a temperature called the transformation temperature, T_c.

At temperatures above T_c the phase B is the most stable and, at temperatures below T_c, the most stable phase is A.

The transformation is accompanied by a sudden change in free energy (Fig. 9.4), and it is the latent heat of transformation.

At temperatures above T_c, phase B is the most stable. At temperatures below T_c, the most stable phase is A.

When the temperature increases, for one curve to cross the other it is required that the high-temperature phase has higher entropy, and the change in entropy is discontinuous.

At the transformation temperature, as the temperature increases, the change in free energy is positive (see Fig. 9.5).

Fig. 9.4 T-G diagram showing a reversible transformation at temperature T_c

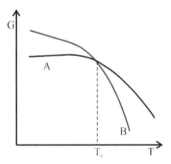

Fig. 9.5 T-G and T-H diagrams showing the evolution of a reversible transformation

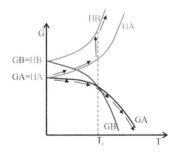

Fig. 9.6 T-G diagram
showing the evolution of an
irreversible transformation

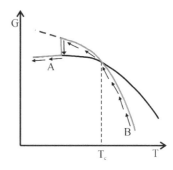

In the reversible transformation, the transformation temperature is the same in heating as in cooling (see Fig. 9.4).

In the irreversible transformation, the transformation temperature is different during heating and cooling (see Fig. 9.6).

The irreversible transformation is the most frequent in minerals since in these, atomic mobility is too slow for ideal thermodynamics to be fulfilled, so there are deviations from ideal behavior. Some degree of super-cooling or super-heating is generally required for the transformation to take place. In this case, the transformation is said to occur under non-equilibrium conditions.

9.5   Mechanisms of Polymorphism

– **Displacement**

Transformation by displacement implies the internal adjustment that allows passing from one substance to another is very small and requires little energy. The structure remains practically intact, and the bonds are not broken. Only a small displacement of the atoms and a readjustment of the bond angles are necessary. The transformation is reversible.

> **Example**
> **Transformation of quartz at 573 °C**
>   High Quartz $P6_222$.
>   Low Quartz $P3_121$.
>   Above the transformation temperature, the stable form of quartz is called high quartz or high temperature quartz. Below 573 °C, the stable form is called low quartz or low temperature quartz.
>   The difference between the two lies in the symmetry, revealed by the space group.

The loss of symmetry occurs when the temperature drops, due to the distortion of the bond angles, as a consequence of the distortion that can be carried out in two opposite directions (right and left, related by a 180° turn). In the transition, a twin may appear: dolphin twin (only in low quartz).

– Reconstruction

Transformation by reconstruction implies internal readjustment of the atoms leads to a breakdown of the bonds and a new distribution of the atoms. It requires a lot of energy. It is very slow and irreversible.

Example
Transformation of tridimite to low quartz
    See Fig. E3.

– Order–Disorder

Order–disorder transformation implies a change from a higher symmetric disordered structure of high-temperature to an ordered, low-temperature structure at lower symmetry.

Example
In potassium feldspar ($KAlSi_3O_8$) Al occupies an identical position to Si, which it replaces in a mineral. The high-temperature form of this feldspar, sanidine, shows disordered distribution of Al in the SiO_2 lattice (Fig. E5). Microcline, the low-temperature potassium feldspar, has an ordered distribution of Al in the SiO_2 lattice (Fig. E6).

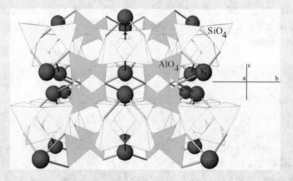

Fig. E5 Sanidine structure monoclinic

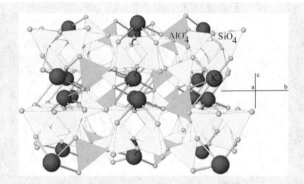

Fig. E6 Microcline structure triclinic

To characterize alkali feldspars, in addition to knowing their chemical composition, it is also necessary to determine the distribution of Al and Si in the tetrahedral positions as a function of crystallization temperature and thermal history (Appendix III).

Thus, alkali feldspars can exhibit a disordered Al-Si distribution if they cool rapidly after crystallization at high temperatures and adopt a structure corresponding to the temperature of maximum symmetry, high temperature feldspars. This is the case for volcanic rocks. If they cool slowly from high temperatures, or if they crystallize at low temperatures, they will show an ordered Al–Si distribution, as in the case of plutonic rocks. There are also feldspars whose structural state is intermediate.

The X-ray diffraction technique, crystalline powder method (Chap. 17), provides the data to calculate the triclinicity index, the Al, Si ordering in the tetrahedral sites of the mineral structure and the deformation index.

- The interplanar distances d_{131} and $d_{1\bar{3}1}$ provide the symmetry through the triclinicity index, using the Goldsmith and Laves (1954)[1] Eq. (E1):

$$\Delta = 12.25(d_{131} - d_{1\bar{3}1}) \tag{E1}$$

- The interplanar distances d_{110}, $d_{1\bar{1}0}$, d_{060} and $d_{\bar{2}04}$ provide the direct and reciprocal cell parameters to calculate the distribution of Al, Si in tetrahedral sites, using the method of Kroll (1971,[2] 1973,[3] 1980[4]).
- The deformation index can be calculated from the Kroll and Ribbe (1987)[5] plot.

---

[1] Goldsmith and Laves [1].

[2] Kroll [2].

[3] Kroll [3].

[4] Kroll [4].

[5] Kroll [5].

Application Example

Structural characterization of a potassium feldspar from a pegmatite in the vicinity of La Cocha, Villa Praga-Las Lagunas group, Tilisarao-Renca subgroup and Concarán subgroup, San Luis, Argentina (Wul et al. 2016, with permission).[6]
    The starting data are in Table E1.

Table E1 Reflection data of a potassium feldspar from a pegmatite

Reflexions	(131)	(1$\bar{3}$1)	(110)	(060)	($\bar{2}$04)
d (Å)	3.0321	2.9575	6.7218	2.1599	1.8064

The results of dimensions and angular values of unit cells of the potassium feldspar from pegmatite sample are in Table E2.

Table E2 Results of dimensions and angular values of unit cells of the potassium feldspar from pegmatite sample (* = reciprocal values)

a	b	c	α	β	γ	
8.5718	12.9539	7.2174	90.59	115.99	87.68	
$a*$	$b*$	$c*$	$\alpha*$	$\beta*$	$\gamma*$	V
0.1299	0.0860	0.1543	90.4748	64.0122	92.2934	719.7

The results of the triclinicity index and distribution (Al, Si) of potassium feldspar in the pegmatite are presented in Table E3.

Table E3 Results of the triclinicity index and distribution (Al, Si) of potassium feldspar in the pegmatite

Triclinicity	Distribution (Al, Si) in terms of b-c* and α*-γ*				
Δ	$\sum t_1$	Δt_1	$t_1 o$	$t_1 m$	$2t_2$
0.9321	0.9627	0.0044	0.9785	−0.0158	0.0186
	Distribution (Al, Si) in terms of $2\theta_{(060)}$–$2\theta_{(\bar{2}04)}$ and $\Delta\theta_{(130)}$-$2\theta_{(\bar{2}01)}$				
	$\sum t_1$	Δt_1	$t_1 o$	$t_1 m$	$2t_2$
	1.0103	0.9723	0.9913	−0.0095	−0.0052

---

[6] Wul et al. [6].

From this information, the triclinicity index (Δ) was calculated using the E.1 equation. The result reflects the degree of deviation, as a function of 2θ, of interplanar spacing between (131) and (13$\bar{1}$). The difference in spacing between (131) and (13$\bar{1}$) as a function of 2θ is 0.075. The triclinicity value is 0.9321, and values close to 1 indicate maximum ordering, slow crystallization and correspond to the triclinic system, while values close to 0 indicate monoclinic symmetry and fast crystallization.

The structural state, expressed as the Al content in specific tetrahedral sites, indicates that this feldspar of triclinic symmetry corresponds to low microcline, with $t_1 0$ close to unity, indicating an extreme (Al, Si) ordering, with Al predominantly in the $t_1 0$ position.

Exercises

1
  (a) **Indicate** the stable phase(s) in the calcite and aragonite stability diagram (Fig. 9.1) at the following temperatures and pressures: (1) 400 °C and 10 Kb, (2) 195 °C and 5 Kb, (3) 10 Kb.
  (b) **Mark**, with a cross, the place where kyanite, andalusite, and sillimanite coexist (Fig. 9.2). **Indicate** the range of pressures in which andalusite is stable. **Indicate** the temperature range in which kyanite is stable. **Mark** the place where sillimanite and andalusite coexist.
  (c) **Mark,** with a cross, the place(s) where three phases coexist. **Indicate** the stable phase(s) at the pressure of 80 kb. **Indicate** the stable phase (s) at the temperature of 600 °C. **Indicate** the stable phase(s) at the temperature of 1200 °C and at the pressure of 20 kb.

Questions

1. Polymorph crystals or minerals are those that
   - a. have the same chemical composition
   - b. have the same chemical composition but different crystal structures
   - c. have different chemical composition and the same crystal structure
   - d. have different crystal structures
2. The types of mechanisms of polymorphic transformations are
   - a. displacement, order-disorder, reconstruction
   - b. displacement, vacancy, reconstruction, impurity
   - c. displacement, order-disorder, reconstruction, position
   - d. order-disorder, reconstruction

3. What is the name of the abrupt change in internal energy that accompanies a phase transformation?

   Response:

4. Do the free energy curves of the phases involved in a polymorphic transformation intersect? Yes/No

   Response:

5. Do andalusite and sillimanite coexist sat high pressures? Yes/No

   Response:

6. Write the name of the mechanism type of a reversible polymorphic transformation with slight displacement of atoms and slight readjustment of the bond angles.

   Response:

7. Which type of transformation requires superheating? Reversible/Irreversible.

   Response:

8. At low temperatures the stable phase(s) in the phase diagram of aluminosilicates (andalusite, sillimanite, and kyanite) is andalusite.
   ○ True
   ○ False

9. As the processing temperature increases, the change in internal energy is negative.
   ○ True
   ○ False

References

1. Goldsmith JR, Laves F (1954) The microcline sanidine stability relations. Geochimica et Cosmochimica. Acta 5:1–19
2. Kroll H (1971) Determination of Al, Si distribution in alkali feldspar from X-ray powder data. Neues Jahrb. Mineral Monastsh 2:91–94
3. Kroll H (1973) Estimation of the Al, Si distribution of feldspar from the lattice translation tr [110] in Alkali feldspars. Contrib Mineral Petrol 39:141–156
4. Kroll H (1980) Estimation of the Al, Si distribution of alkali feldspar from laticce translations tr [110] and tr[110]. Revised diagrams. A Jb Miner Mh HI, 31–36
5. Kroll H, Ribbe PH (1987) Determining (Al, Si) distribution and strain in alkali feldspars using lattice parameters and diffraction-peaks positions: a review. Am Mineral 72:491–506
6. Wul J, Montenegro T1, López de Luchi MG (2016) Caracterización estructural defeldespatos potásicos depegmatitas de los alrededores de La Cocha, Grupo Villa Praga-Las Lagunas, Subgrupo Tilisarao-Rencay Subgrupo Concarán, San Luis, Argentina. Acta geológica lilloana 28(1):332–337

Part III
Crystallophysics

In this part, Chaps. 10 to 18, the physical properties of crystals are outlined, relating them to chemical composition and structure. Important properties to consider in this part are those derived from the interaction of electromagnetic radiation with matter. Optical properties, resulting from the interaction of visible light with crystals, are important in the routine identification of minerals. There are also those that allow the disposition of atoms in the structure, to identify crystalline phases, etc., as a result of the interaction of X-rays with crystals.

Chapter 10
Relationship Between Symmetry and Physical Properties

Abstract The physical properties of crystals are defined by the relationships between measurable quantities. These quantities may or may not depend on the direction in which they are measured. In certain cases, the value of a certain property is constant for all directions of the crystal; therefore, isotropic and anisotropic crystals will be differentiated. It will be shown that the properties can be represented by tensors, scalars being those of range 0. Depending on the symmetry of the crystal and the tensor range of the properties, they can be represented geometrically, which is especially useful for optical properties. The Curie–Neuman principle, which shows the relationship between the point symmetry of a crystal before the external influence and the symmetry of its physical properties, is explained. Finally, scalar properties, specific heat, and density are explained. In Chaps. 11 to 15, the optical properties of crystals and minerals are described, as a result of the interaction of light with them. The phenomena and properties that are observed with transmission and reflection polarization microscopes, instruments of routine use in the identification of minerals and rocks, are explained.

10.1 Physical Property

A physical property is any measurable or observable response from a mineral crystal to some external cause.

In addition to the shape, which may be completely missing from mineral fragments, physical properties are very useful for recognition. Some of these properties can be seen by simple observation, others require simple measurements and, finally, there are parameters that require more complex and costly instrumentation.

The physical properties of crystalline solids are divided into two categories:

– Non-directional
– Directional

© The Author(s), under exclusive license to Springer Nature Switzerland AG 2022   251
C. Marcos, *Crystallography*, Springer Textbooks in Earth Sciences,
Geography and Environment, https://doi.org/10.1007/978-3-030-96783-3_10

The non-directional properties include scalar properties, which do not depend on the direction in which they are measured. Examples are specific heat and density, and they are expressed by real numbers.

The directional properties are tensor properties, which depend on the direction in which they are measured.

10.2   Scalar Properties

10.2.1   *Density*

The density of a body represents the value of its mass per unit volume (g/cm^3). Its numerical value is equal to the *specific weight* which, in turn, indicates how many times the body in question weighs more than an identical volume of distilled water.

Density is directly related to the densification of atoms in the reticular cell and is, therefore, high in compounds with a high coordination number (metals) and low in compounds with lower coordination (compounds with residual or covalent bonds).

In general, there is no density but specific weight measures, based on the well-known Archimedes principle and by a very simple instrument, such as pycnometer, hydrostatic balance, and liquids with specific weight previously determined.

It is also convenient to assign an estimated value compared to a substance as a standard.

Accurate density determination is not always a safe diagnosis. Very few mineral samples lack darkness, point defects, impurities, voids, or fracture that so alter density values that it makes it only qualitative.

Density is given by the expression (10.1):

$$\rho = M/V \tag{10.1}$$

It is also expressed as

$$\rho = \frac{Z \cdot M}{N \cdot V} \tag{10.2}$$

where

Z is the number of formulas contained in the elementary cell
M is molecular weight in (g/mol)
N is the Avogadro number $6.02338 \cdot 10^{23}$ mol^{-1}
V is the volume of the elementary cell. It is obtained from the cell parameters using the expression (10.3):

$$V = a \cdot b \cdot c \sqrt{1 - \cos^2 \alpha - \cos^2 \beta - \cos^2 \gamma + 2 \cos \alpha \cos \beta \cos \gamma} \tag{10.3}$$

Table 10.1 Specific weight and atomic weight of cation in orthorhombic carbonates

Mineral	Formula	Atomic weight of cation (U)[1]	Specific gravity
Aragonite	$CaCO_3$	40.08	2.95
Stroncianite	$SrCO_3$	87.62	3.76
Witherite	$BaCO_3$	137.34	4.29
Cerussite	$PbCO_3$	207.19	6.55

10.2.2 Specific Gravity

The specific gravity is a number that expresses the relationship between its weight and the weight of an equal volume of water at 4 °C. The specific weight of a crystalline substance depends on two factors: Type of atoms and Packaging of atoms.

In isostructural compounds, the packaging is constant, and the elements with higher atomic weight usually have greater specific weight, as in orthorhombic carbonates (Table 10.1).

In polymorphic compounds, the chemical composition remains constant but the packaging varies. The C polymorphs are an example. Diamond has a specific weight of 3.5 and a structure with compact packaging, while graphite has a specific weight of 2.23 and the packaging is less dense than in the diamond.

The variation of the specific weight with the composition is clearly manifested in the solid solution series. Example: the specific weight in the olivine series varies between 3.3 in forsterite and 4.4 kg/m^3 in fayalite.

Box 10.1. Determination of Specific Gravity Using Hydrostatic Weight

Material:

- Balance
- Wooden bridge (placed on left saucer)
- Beaker of about 100 ml
- Wire to suspend the ore in the water

Procedure:

- Weigh the mineral in air. The mineral is placed in one of the dishes of the scale and the other dish is weighted to establish balance between the weights of both. The mineral weight in air is M_a (Fig. 10.1).

---

[1] U = Amu = mass equal to one-twelfth the mass of an atom of carbon.

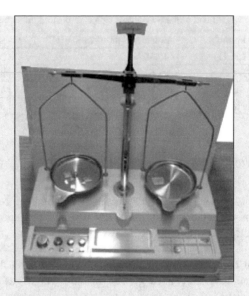

Fig. 10.1 Weighing mineral in air

- Place the wooden bridge over left saucer.
- Fill the beaker with 3/4 parts of deionized water.
- Place the beaker over the wooden bridge.
- Suspend the wire on the left side of the arm and immerse it in the liquid.
- Annotate the weight of the submerged wire, A_s (Fig. 10.2).

---

[1] U = Amu = mass equal to one-twelfth the mass of an atom of carbon.

Fig. 10.2 Water-immersed wire

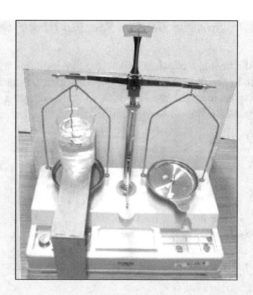

- Carefully remove the beaker and the wire and place the mineral on the wire spiral, which is then put back into the water and suspended from the arm (Fig. 10.3).

Fig. 10.3 Weighing the immersed mineral in water

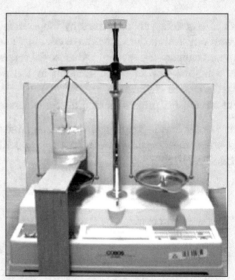

- Annotate the weight of the wire with the immersed mineral, AM_s
- Weight of the immersed mineral, $M_s = AM_s - A_s$
- Mineral specific gravity $= \frac{M_a}{M_a - M_s}$

10.2.3   Specific Heat or Specific Heat Capacity

Specific heat is a physical magnitude that indicates the ability of a material to store internal energy in the form of heat. Formally, it can be said that it is the energy needed to increase by one unit of temperature a quantity of substance. It is expressed by Eq. 10.4.

$$C_p = \left(\frac{\Delta Q}{m \cdot \Delta T} \right)_p \tag{10.4}$$

where

ΔQ is the amount of heat supplied
ΔT is the temperature increase
m is the mass
p the constant pressure at which the experiment is carried out

10.3   Tensor Properties

A tensor is a quantity characterized by the existence of several numbers (components) with physical meaning that have certain values for a given reference system. This amount cannot vary, even if the tensorial reference system and its components do. Knowing the law of transformation to the new reference system, the new components can be calculated.

There are tensors properties of different orders:

– Zeroth-order tensor properties: scalar magnitudes.
– First-order tensors properties: Vector properties. They are represented by expression (10.5)

$$A_i = a_i B \tag{10.5}$$

and they are characterized by three components according to the reference axes.

$$\begin{bmatrix} a_1 \\ a_2 \\ a_3 \end{bmatrix} .$$

This type is the same as pyroelectricity and pyromagnetism. They do not appear in center-symmetric crystals.

- Second-order tensors properties shall be represented by expression (10.6):

$$A_i = a_{ij}B_j \tag{10.6}$$

They are characterized by nine coefficients, each associated with a pair of axes taken in a certain order.

$$\begin{bmatrix} a_{11} & a_{12} & a_{13} \\ a_{21} & a_{22} & a_{23} \\ a_{31} & a_{32} & a_{33} \end{bmatrix}$$

Thermoelectricity belongs to this tensor type.

There are other physical properties represented by a second-order tensor which are characterized because the components $a_{ij} = a_{ji}$, represent a symmetrical tensor, in which case the coefficients are reduced to six.

$$\begin{bmatrix} a_{11} & a_{12} & a_{13} \\ & a_{22} & a_{23} \\ & & a_{33} \end{bmatrix}$$

Thermal expansion, compressibility, electrical conductivity, heat conductivity, electrical induction, magnetic induction, and optical properties belong to this group. When a second-order symmetrical tensor refers to its main axes, the coefficients are reduced to three.

$$\begin{bmatrix} a_{11} & & \\ & a_{22} & \\ & & a_{33} \end{bmatrix}.$$

- Third-order tensors are given by expression (10.7).

$$A_i = a_{ijp}B_{jp} \tag{10.7}$$

The representative matrix consists of 27 coefficients:

$$\begin{bmatrix} a_{11} & a_{12} & a_{13} \\ a_{21} & a_{22} & a_{23} \\ a_{31} & a_{32} & a_{33} \end{bmatrix} \begin{bmatrix} a_{112} & a_{122} & a_{132} \\ a_{212} & a_{222} & a_{232} \\ a_{312} & a_{322} & a_{332} \end{bmatrix} \begin{bmatrix} a_{113} & a_{123} & a_{133} \\ a_{213} & a_{223} & a_{233} \\ a_{313} & a_{323} & a_{333} \end{bmatrix}$$

Piezoelectricity and piezomagnetism belong to this group.

– Fourth-order tensors properties are characterized by 81 coefficients and given by expression (10.8).

$$A_{ij} = a_{ijmp}B_{mp} \tag{10.8}$$

Because of its intrinsic symmetry, the number of coefficients of the Fourth-order tensor is reduced to 21. The crystal symmetry reduces the number of independent coefficients again.

This group includes elasticity, damping of sound waves, photoelasticity, and photomagnetism.

The tensor properties are grouped according to the shape of the geometric surface it can represent, which is called the *representation quadric*. The 2nd degree surface is an ellipsoid (Fig. 10.4). A particular case is a sphere.

Properties that are represented by surfaces of higher order to the ellipsoid not only require a numerical value (module) for their correct expression but also must specify the direction in which the measurement was performed, such as cohesion, toughness, piezoelectricity, and pyroelectricity (Fig. 10.5).

Fig. 10.4 Ellipsoid

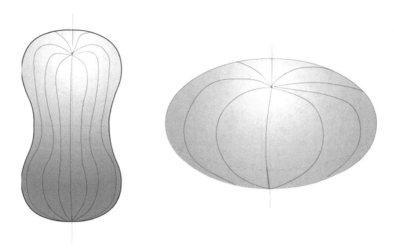

Fig. 10.5 Representation surfaces of higher order than the ellipsoid

10.4 Curie–Neumann Principle

The Curie–Neumann principle relates the symmetry of an observable effect to the symmetry of the cause and symmetry of the crystal.

The minimum symmetry of an effect is equal to the combined symmetry that exists in both the cause and the crystal.

Therefore, the symmetry of a physical property must include the symmetry of the point group of the crystal.

The symmetry of a physical property corresponds to the symmetry of an ellipsoid, the geometric place at the ends of the vectors that represents the value of the property in each direction of the crystal.

The geometric idea of this principle can be seen in Table 10.2. by the example of crystal optics (refraction of light—birefringence).

Table 10.2 Geometric idea of the Curie–Neumann principle, by the example of crystal optics

Symmetry		
Cause isotropic	+ Crystal point group	= Effect Indicatrix

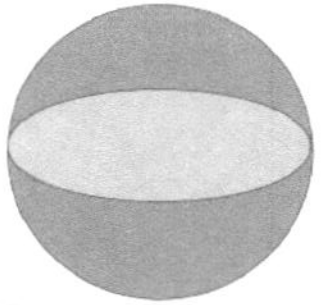

$\infty/m \ \infty/m \ \infty/m$

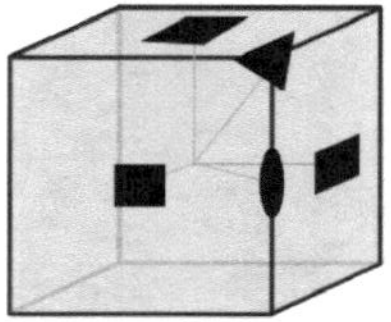

$m\bar{3}m$

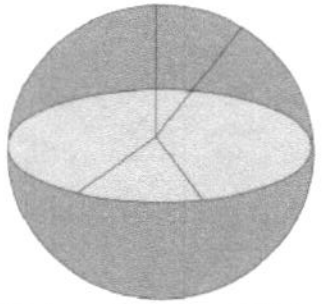

$\infty/m \ \infty/m \ \infty/m$

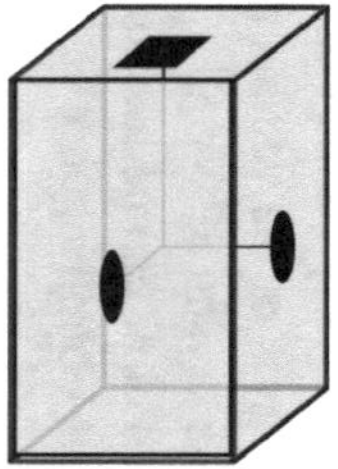

$4/m \ 2/m \ 2/m$

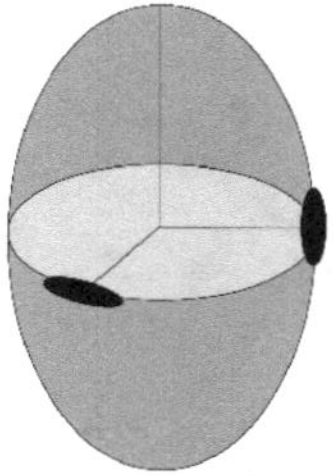

$\infty/m \ 2/m \ 2/m$
Rotational ellipsoid (two axes of different lengths)

(continued)

Table 10.2 (continued)

Symmetry		
Cause isotropic	+ Crystal point group	= Effect Indicatrix

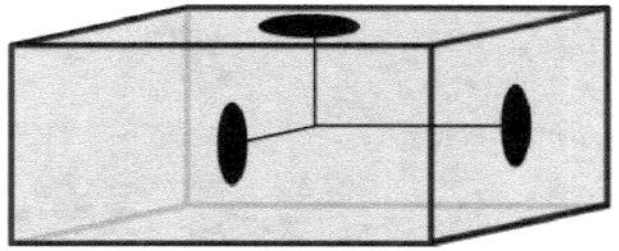

$2/m\ 2/m\ 2/m$

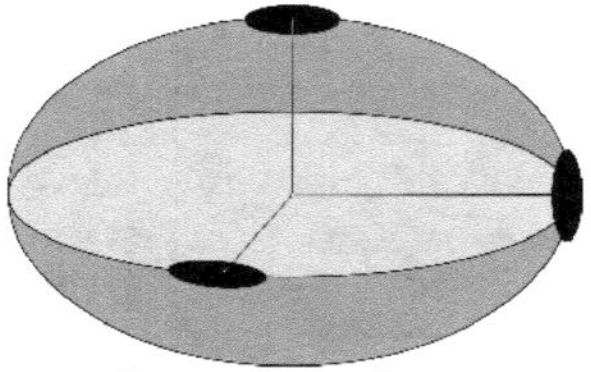

$2/m\ 2/m\ 2/m$
General ellipsoid (three axes of different lengths)

Questions

1   Using a number, indicate the order of the tensor of electrical conductivity
    Response: |

2   According to the Curie–Newmann principle, "The symmetry of the crystal
    plus the symmetry of the effect is equal to the symmetry of the cause."
    ○ True
    ○ False

3   Scalar properties depend on the direction in which they are measured.
    ○ True
    ○ False

4   Pair each property with property type and classEnd of form

Elasticity	Scalar
Pyroelectricity	Second-order symmetrical tensor, electrical
Specific heat	Second-order tensor, electric
Thermoelectricity	First-order tensor, magnetic
Pyromagnetism	Second-order symmetrical tensor, mechanical
Electrical conductivity	First-order tensor, electric

5   According to the Neumann–Curie principle:
    ○ a. The maximum symmetry of an effect can never be less than the sum
    of the symmetry of the cause and that of the crystal
    ○ b. The minimum symmetry of an effect can never exceed the sum of the
    symmetry of the cause and that of the crystal
    ○ c. The maximum symmetry of an effect can never be less than the sum
    of the symmetry of the cause and that of the crystal
    ○ d. The minimum symmetry of an effect can never be less than the sum
    of the symmetry of the cause and that of the crystal

6   Can specific heat be represented on a surface? Yes/No. Explain your
    answer.
    Response: |

Chapter 11
Interaction of Electromagnetic Waves with Crystals and Minerals

Abstract The characteristics of the electromagnetic waves and the electromagnetic spectrum are recalled, emphasizing the visible spectrum, in which the relationship between the different units used for their description is shown. The characteristics of visible light propagation as a wave movement are established. From one of the solutions of this equation, that of a plane sine wave and the temporal evolution of the electric field in the plane perpendicular to the direction of propagation, the concept of polarization is introduced. This is called optical polarization to distinguish it from electrical polarization. The solution of the wave equation and the velocity of the waves are presented to introduce the concepts of refractive index and absorption coefficient, which allow the distinction between isotropic and anisotropic, and transparent and opaque crystals and minerals. Within the section of light incidence in transparent crystals, the concepts of refraction and reflection are explained, in addition to the laws of refraction and reflection, critical angle and total internal reflection, and diffraction. In Chap. 10, the importance of the representation surfaces is emphasized. They take special relevance in the case of transparent crystals and minerals, since the ellipsoid of the indexes is the key to understanding not only variation of the refraction index according to the direction, but also of propagation of light in the crystals and minerals. In the case of opaque crystals and minerals, the concept of reflectance or reflectivity is introduced as a property that allows the understanding of these materials.

11.1 Electromagnetic Waves and Maxwell Equations

A *wave* is a physical disturbance that spreads in a certain medium.

An *electromagnetic wave* is a progressing electromagnetic field. Electromagnetic waves carry energy and quantity of motion.

The wave can be represented by sinusoidal function. The *intensity*, I, of a wave is proportional to the square of its amplitude. The *amplitude*, A, is a measure of the maximum variation of displacement that varies periodically in time. It is the

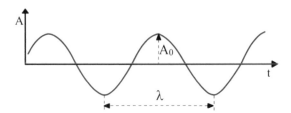

Fig. 11.1 Electromagnetic wave. A = amplitude, t = time, λ = wavelength

distance halfway between the ridge and the valley in a wave. The *wavelength*, λ, is the measure of the distance between two points in phase that the wave travels in a given time, t (Fig. 11.1).

The *frequency* (*f*) is the number of vibrations of the wave in the unit of time.

$$f = \frac{1}{t} \tag{11.1}$$

$$v = f\lambda. \tag{11.2}$$

An electromagnetic field is the excited state that is established in space by the presence of electrical charges. It is represented by two vectors:

- *electric field* $\vec{E}$
- *magnetic induction* $\vec{B}$

The vectors:

- *Electrical current density* $\vec{j}$
- *Dielectrical displacement density* $\vec{D}$ (or electrical flux density)
- *Magnetic vector* $\vec{H}$

are used to describe the effect of the field on an object and the behavior of the object under its influence.

Maxwell's equations—originally four partial differential equations, describe the propagation of any electromagnetic wave in any medium. They correlate charge and current density (sources) with electric and magnetic fields (results).

Maxwell equations represent those derived from the above five vectors of time and space.

The following (2nd rank) tensors are involved:

- *electrical conductivity* (σ)
- *dielectric constant* (ε)
- *magnetic permeability* (μ)

The simplified Maxwell's equations are

$$\nabla \mathrm{x} \vec{E} = -\frac{1}{c}\frac{\partial \vec{B}}{\partial t}$$

$$\nabla \mathrm{x} \vec{H} = -\frac{1}{c}\frac{\partial \vec{D}}{\partial t} \qquad (11.3)$$

$$\vec{D} = \varepsilon_r \varepsilon_0 \vec{E} = (1 - \chi)\varepsilon_0 \vec{E}$$

$$\vec{P} = \chi \varepsilon_0 \vec{E}$$

Maxwell equations can be combined, and the solution is a *wave equation* (Bloch wave) that must satisfy the electric field and magnetic field vectors.

11.2   Electromagnetic Wave Propagation

Electromagnetic wave propagation is the way in which electromagnetic waves propagate and it follows the general equation of wave motion. The temporal evolution of $\overrightarrow{E}$ or $\overrightarrow{H}$ in one plane perpendicular to the direction of propagation originates *polarization,* which can be called *optical polarization.* Depending on the shape, the said polarization is called *elliptical, circular,* or *linear.*

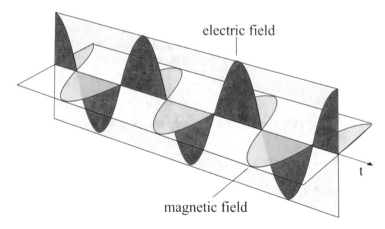

Fig. 11.2 Plane polarized wave (electric and magnetic fields perpendicular to each other and to the direction of propagation) (t = time)

One of the solutions of the wave equation corresponds to that of a *sinusoidal plane* or *linear wave* (Fig. 11.2). In this type of wave, the electric field and the magnetic field are always perpendicular to each other. The electric field and the magnetic field are perpendicular to the direction of propagation.

Therefore, these waves are transverse.

The wave equation for the electric field of a crystal without magnetic properties, with permeability in vacuum, $\mu_0 = 0$, is given by the expression:

$$E = E_0 e^{i(\tilde{k}s - \omega t)} \tag{11.4}$$

where

ω is the frequency
s is the distance traveled (thickness of the crystal crossed by the wave)
t is the time
$\tilde{k}s - \omega t$ difference is the phase

$\tilde{k} \cdot s$ product corresponds to the scalar product of the vectors k and s

$$\tilde{k} = \overrightarrow{k} \overrightarrow{s} \tag{11.5}$$

k is the wave vector whose modulus is

$$\tilde{k} = \frac{\omega}{c} \sqrt{\varepsilon_r + i \frac{\sigma \mu_0}{\omega}} \tag{11.6}$$

where
dielectrical constant/permittivity $\varepsilon = \varepsilon_0 \varepsilon_r$ and ε_0 = electrical field constant/vacuum permittivity, ε_r = relative dielectrical constant/permittivity.
c (equal to $2.998 \cdot 10^8$ m/s) is the light velocity in the vacuum

$$c = \frac{1}{\sqrt{\mu_0 \varepsilon_0}} \tag{11.7}$$

in vacuum, with $\sigma = 0$ and $\varepsilon = \varepsilon_0$ and $\varepsilon/\varepsilon_0 = \varepsilon_r = 1$

$$\tilde{k} = \frac{\omega}{c} = \frac{2\pi}{\lambda}. \tag{11.8}$$

The velocity in a crystal is modified:

$$v = \frac{c}{\tilde{n}} \tag{11.9}$$

being

$\tilde{n}$ complex refractive index

$$\tilde{n} = n + ik \tag{11.10}$$

so that

$$\tilde{k} = \frac{\omega}{c}\tilde{n} \tag{11.11}$$

therefore:

$$\tilde{k} = \frac{\omega n(1+ik)}{c}. \tag{11.12}$$

Equation 11.4 can be put in the following form:

$$E = E_0 e^{-\frac{\omega}{c}nks} e^{i\omega\left(\frac{ns}{c}-t\right)}. \tag{11.13}$$

This plane wave loses energy as it passes through the crystal. The decrease in energy is represented by the exponential term of the real part, $-\frac{\omega}{c}nks$.

The attenuation of the wave is related to the absorption of electromagnetic energy. The absorption coefficient, μ, is defined as

$$\mu = \frac{\omega}{c}\tilde{n}. \tag{11.14}$$

In the absence of absorption, the complex part of the complex refractive index is zero and k (Eq. 11.12) is a real number

$$k = \frac{\omega}{c}n \tag{11.15}$$

with n being a real number.

11.3   Propagation of Light in a Transparent Crystal

1. When the light passes from a less dense medium, the air ($n_1 = 1$), to a denser medium, the crystal ($n_2 > n_1$), some of the light is transmitted through the crystal and some are reflected from the crystal.

   The transmission of light through crystal (Fig. 11.3) is governed by Snell's law

$$\frac{\sin i}{\sin r} = \frac{n_2}{n_1} = n \tag{11.16}$$

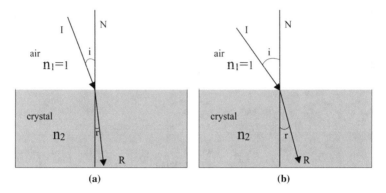

Fig. 11.3 Light incidence when passing from a less dense medium (air) to a denser medium (crystal). In (**a**) the angle of incidence, i, and therefore the angle of refraction, r, are smaller than in (**b**)

where

i is the angle of incidence: the angle between the incident beam I and the normal or perpendicular to the surface of separation of the two media N.

r is the angle of refraction: the angle between the refracted ray R and the normal to the surface of separation of the two media N. It increases as i angle increases.

The reflexion of light from the crystal (Fig. 11.4) is governed by the laws of reflexion:

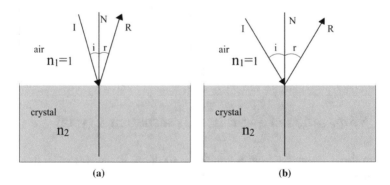

Fig. 11.4 Light reflexion when passing from a less dense medium (air) to a denser medium (crystal). In (**a**) the angle of incidence i and, therefore, the angle of reflexion r is smaller than in (**b**)

Fig. 11.5 Incidence of light when passing from a denser medium (crystal) to a less dense medium (air). Ray A with $i < i_c$ is transmitted to denser medium (air); ray B, with $i = i_c$, travels along the surface of separation of the two media; and C, with $i > i_c$, undergoes total internal reflexion

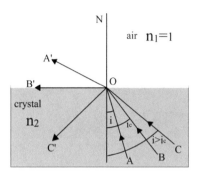

- The incident ray I, the reflected ray R, the perpendicular N to the surface of separation of the two media, as well as the angles of incidence i and reflexion r lie in the same plane. It is a plane perpendicular to the surface of separation of the two media.
- The angle of incidence i and the angle of reflexion r have the same value.

2. When the wave passes from a denser medium, the crystal ($n_2 > n_1$), to a less dense medium, the air ($n_1 = 1$), part of the light (the one incident at an angle below the critical angle i_c) is transmitted through the air, and another part is reflected internally (the one incident at an angle $> i_c$) (Fig. 11.5).
3. When light passes from one crystal to another medium, the air, perpendicular to the surface of the separation of the two media, the light slows down but is not deflected.

When the light passes from the air, with an angle of incidence different from 90°, to a crystal whose surfaces are parallel, and comes out again into the air, it deflects into the crystal.

When the light comes out of the crystal, it doesn't deviate from its initial trajectory (Fig. 11.6).

Fig. 11.6 When the light passes from one crystal to another medium, the air, perpendicular to the surface of separation of the two media the light slows down but is not deflected

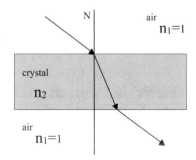

Fig. 11.7 Refraction of light
when it passes through a
prism

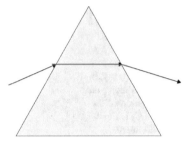

Fig. 11.8 Different refraction
of the component colors of
white light when passing
through a prism

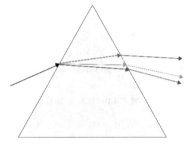

When white light passes through a prism, it is refracted or deflected within the prism and, when it comes out, it is deflected again following a different path than the incident light (Fig. 11.7).

There are two reasons for this:

1. The faces of the prism form an angle.
2. The difference in refractive index between air and prism material.

When white light passes through a prism, its component colors are refracted or deflected differently. This is because the frequency of the light of each color (monochromatic light) does not vary when it passes from one medium to another, but it varies with λ.

When Snell's law is applied to each component color of white light when it passes through a prism, blue color is the most deviated, as its associated refractive index is higher since the wavelength is shorter, according to normal dispersion (Fig. 11.8).

11.4   Electrical Polarization, Local Electric Field, and Velocity of Light in a Crystal

Ions, atoms, or molecules in a crystal or mineral interact and produce an electric field.

Fig. 11.9 Electrical dipole

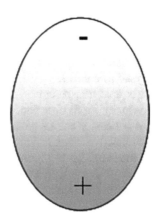

This field is modified by the application of any external electric field (electro-magnetic waves, visible light).

The electrical charges, positives and negatives, are displaced and each ion is converted into an *electrical dipole* (Fig. 11.9).

Electrical polarization is defined as the number of dipoles per unit of volume.

To determine the velocity of light in a particular direction through a crystal or mineral, the total *polarization* must be evaluated.

The polarization depends on the following:

- Number of dipoles
- Force that those dipoles exert on the other dipoles
- Local electric field
- Field generated in the crystal due to the interaction of the electrons
- The effect of dipoles on the polarization of a particular ion depends on the translation symmetry.

The dipoles become sources of new secondary waves that combine with each other and with the incident field and form the total field. Thus, the wave velocity in a crystal and in a given direction can be written using the following expression:

$$v = c\left(\frac{E_L}{P + E_L}\right)^2 \qquad (11.17)$$

where

v is the velocity in the crystal or mineral
c is the velocity of light in a vacuum
P is the polarization
E_L the local electric field

In this expression, it is observed that the velocity of an electromagnetic wave in a crystal or mineral varies according to the polarization and the local electric field.

It follows that the polarization is higher and the velocity is lower in a direction with high electronic density. The polarization is lower and the velocity higher in a direction with low electronic density.

These relationships allow us to predict how the velocity of light will differ from one crystal to another.

In isotropic transparent crystals, allowing visible light to pass through with the same velocity in all directions, refractive index will vary with density. Example: halite $n = 1.544$ y $= 2.17$ g/cm^3; garnet $n = 1.80$ y $= 4.32$ g/cm^3.

The refractive index is quite sensitive to variations in chemical composition and crystal structure.

In general, crystals that have atoms with high atomic numbers will have relatively high refractive indices (transparent crystals have refractive index in the range of ~ 1.3–2.1).

11.5   Electromagnetic Spectrum

Electromagnetic waves are classified according to their λ in the following:

- γ-rays
- X-Rays
- Ultraviolet rays (UVA)
- Visible light
- Infrared rays
- Radio waves

All these electromagnetic waves together are called the *electromagnetic spectrum* (Fig. 11.10).

Monochromatic light is light that includes a very small range of wavelengths and reaches the eye as a single color. Examples include red light and green light.

Polychromatic light is light that includes a wide range of wavelengths. An example is white light.

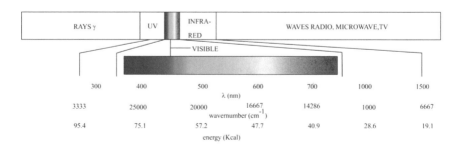

Fig. 11.10 Electromagnetic spectrum

11.6 Isotropic Crystals and Minerals

Isotropic crystals and minerals can be transparent and absorbent (opaque). They crystallize in the cubic system. Light travels with the same velocity, v, in any direction.

– **Transparent isotropic crystals and minerals**

The index of refraction, n, has the same value in any direction. Light travels in all directions with the same velocity (Fig. 11.11a).

Any section of an isotropic crystal observed with a polarizer appears bright in a complete rotation of it (Fig. 11.11b). That is because the light vibrates on a single plane, as the polarizer.

Any section of an isotropic crystal, when viewed between crossed polarizers (polarization planes or vibration directions forming a 90° angle), appears dark (extinguished) (Fig. 11.11c) in a complete turn. This is because in any turning position of the section, the light vibrates in the same direction as the polarizer, which is perpendicular to that of the analyzer that does not allow the passage of light.

The reflexion of linearly polarized monochromatic light at perpendicular incidence from the surface of a crystal or isotropic mineral does not change its polarization state, but its intensity decreases with respect to the incident in a proportion given by the reflectance.

$$R = \frac{I_R}{I_I} \tag{11.18}$$

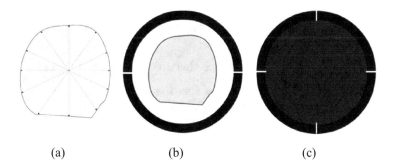

(a)   (b)   (c)

Fig. 11.11 Isotropic section of a transparent isotropic crystal: **a** Light vibrating in all directions in any section (all isotropic); **b** any isotropic section observed with a polarizer appears bright; **c** any isotropic section observed with crossed polarizers appears extinguished in a complete turn because, in any turn of the sample, its vibration direction is perpendicular to that of the polarizer and it is cancelled out

where

I_R is the intensity of the reflected light
I_I is the intensity of the incident light

$$R\% = \frac{(n-1)^2}{(n+1)^2} \tag{11.19}$$

– **Absorbent or opaque isotropic crystals and minerals**

Linearly polarized incident light is reflected without changing the polarization state; however, the reflectance on these absorbing isotropic surfaces depends on both the refractive index and the absorption coefficient. Its value in air is given by the expression:

$$R\% = \frac{(n-1)^2 + k^2}{(n+1)^2 + k^2}. \tag{11.20}$$

11.7   Anisotropic Crystals and Minerals

Anisotropic crystals and minerals can be transparent and absorbent/opaque. They are divided into

– uniaxial crystals and minerals, both transparent and absorbent, and belong to the crystal systems—tetragonal, hexagonal, and trigonal/rhombohedral.
– biaxial crystals and minerals, only transparent, and belong to the crystal systems —orthorhombic, monoclinic, and triclinic. The absorbent crystals and minerals belonging to these crystal systems cannot be called absorbents for the reasons explained below.

Anisotropic crystals and minerals are characterized because in them:

– The velocity of light varies with the direction.
– The value of the refractive index varies with the direction.
– Light is generally split into two rays or components. Each ray has its plane of polarization or direction of vibration perpendicular to the other. The faster ray is associated with the lower refractive index, the so-called ordinary ray. The slower ray is associated with the higher refractive index, the so-called extraordinary ray.
– There are one or two directions in which the light does not split. Each of these directions is called an *optical axis*. Crystals and minerals with one *optical axis* are called *uniaxial*. Crystals and minerals with two *optical axes* are called *biaxial*.

Any section perpendicular to an optical axis behaves like any section of an isotropic crystal or mineral, i.e., that section is isotropic.

When there are two optical axes, the angle between them is called 2V. The plane containing the two optical axes is called the *optic plane*.

An anisotropic crystal and any of its anisotropic sections that are observed with polarized light will appear clear in one full rotation of the crystal. That is because light is split into two rays vibrating in mutually perpendicular planes (Fig. 11.12a). If one of its directions of vibration is perpendicular to that of the polarizer and is canceled out, the other passes through (Fig. 11.12b).

If neither direction of vibration coincides with that of the polarizer, the light corresponding to the two rays passes through it (Fig. 11.12c).

An anisotropic crystal and any of its anisotropic sections that are observed between crossed polarizers, in a complete turn, presents four positions of darkness (*extinction positions,* every 90°) and four clarity positions (every 90°). The extinction positions occur when the directions of vibration of the crystal coincide with those of the polarizers (Fig. 11.13a). The clarity positions occur when the directions of vibration of the crystal and the polarizers do not coincide (Fig. 11.13b).

When one of the directions of vibration of the crystal is parallel to that of a polarizer and, therefore, perpendicular to that of the other since these are crossed, it is canceled (extinction). The same happens with the other direction of vibration of the crystal or mineral or anisotropic section.

To move from a light position to a dark or extinguishing position, turn the crystal or mineral or the anisotropic section by 45°.

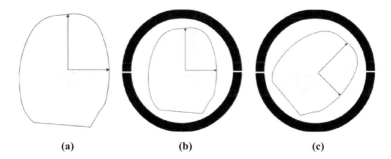

(a)   (b)   (c)

Fig. 11.12 Anisotropic section of a transparent anisotropic crystal observed with polarized light: **a** Light vibrating in two directions in any section except the isotropic section; **b** anisotropic section observed with a polarizer, with one of its vibration directions (corresponding to the rays into which the polarized light has split) parallel to that of the polarizer appearing bright because the light leaves the crystal vibrating in the direction of the polarizer; **c** anisotropic section observed with a polarizer with any of the two directions of vibration coinciding with the polarizer appearing bright because the light corresponding to the two rays leaves the crystal

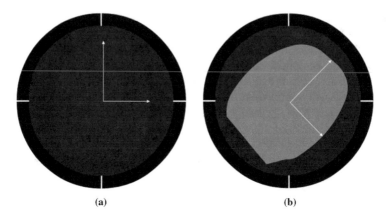

Fig. 11.13 Anisotropic section of a transparent anisotropic crystal between crossed polarizers: **a** extinction position due to the directions of vibration of the crystal coinciding with those of the polarizers; **b** bright position due to the directions of vibration of the crystal not coinciding with those of the polarizers

The reflected light consists of two mutually perpendicular and linearly polarized vibrations. Each has a value of R, depending on the value of n. Each vibration can be isolated by rotating the section on the microscope stage until the direction of vibration matches that of the polarizer. Thus, two values of R, maximum (R_2) and minimum (R_1), can be obtained separately for each section.

$$R_1\% = \frac{(n_1 - 1)^2}{(n_1 + 1)^2} \tag{11.21}$$

$$R_2\% = \frac{(n_2 - 1)^2}{(n_2 + 1)^2}. \tag{11.22}$$

The *bireflectance of the* section is given by the difference ($R_2 - R_1$) and is a function of the birefringence.

11.7.1   Uniaxial Crystals and Minerals

– **Transparent uniaxial crystals and minerals**

In these crystals and minerals, the two extreme refractive indices are ordinary refractive index, n_ω, associated with ordinary ray (follows Snell's law) and extraordinary refractive index, n_ε, associated with the extraordinary ray (does not follow Snell's law). These indices can be measured in the section parallel to c axis.

There are other indices, $n_{\varepsilon'}$, with an intermediate value between n_ω and n_ε.

- *Positive uniaxial* crystals and minerals: $n_\varepsilon > n_\omega$.
- *Negative uniaxial* crystals and minerals: $n_\omega > n_\varepsilon$.

The birefringence is the difference between the indices, n_ε and n_ω.

The section that provides the ordinary and extraordinary reflectance, R_ω and R_ε, is any section parallel to the optical axis. Both values represent the extreme values of the section and the crystal, and they are what characterize it. The difference between both provides the maximum bireflectance and is what characterizes the crystal.

$$R_\omega\% = \frac{(n_\omega - 1)^2}{(n_\omega + 1)^2} \tag{11.23}$$

$$R_\varepsilon\% = \frac{(n_\varepsilon - 1)^2}{(n_\varepsilon + 1)^2}. \tag{11.24}$$

- **Absorbent/Opaque uniaxial crystals and minerals**

Linearly polarized incident light is split into two linearly and mutually perpendicular polarized vibrations. Each of the vibrations has its own values of n and k (absorption coefficient) and, therefore, of the reflectance, whose extreme values are given by the following expressions:

$$R_\omega\% = \frac{(n_\omega - 1)^2 + k_\omega^2}{(n_\omega + 1)^2 + k_\omega^2} \tag{11.25}$$

$$R_\varepsilon\% = \frac{(n_\varepsilon - 1)^2 + k_\varepsilon^2}{(n_\varepsilon + 1)^2 + k_\varepsilon^2}. \tag{11.26}$$

R_ω and R_ε being the ordinary and extraordinary reflectance, respectively, and
k_ω and k_ε being the ordinary and extraordinary absorption coefficients.

The bireflectance is given by the difference between R_ω and R_ε.
The birefringence is given by the difference between n_ω and n_ε.
The biabsorbance is given by the difference between k_ω and k_ε.

11.7.2 Orthorhombic, Monoclinic and Triclinic Crystals and Minerals

– **Orthorhombic, monoclinic and triclinic transparent crystals and minerals**

They are characterized by possessing two optical axes and are, therefore, called biaxial.

In biaxial crystals and minerals, the sections perpendicular to any of the optical axes behave like the isotropic sections of isotropic crystals and minerals.

They are characterized by three principal refractive indices $n_\gamma > n_\beta > n_\alpha$.

In orthorhombic crystals and minerals, each principal refractive index is associated with a crystallographic axis, with nine possibilities.

In monoclinic crystals, only an extreme refractive index is associated with a crystallographic axis.

In triclinic crystals and minerals, there is generally no alignment with symmetry. There are other indices $n_{\gamma'} > n_{\beta'} > n_{\alpha'}$ with intermediate values between the principal n_γ, n_β, n_α values.

– *Positive biaxial* crystals and minerals: n_β closer to n_α than to n_γ.
– *Negative biaxial* crystals and minerals: n_β closer to n_γ than to n_α.

Birefringence is the difference between the principal indices n_γ and n_α.

Biaxial crystals are characterized by three values of *reflectance*: maximum (R_g), minimum (R_p) and intermediate (R_m), depending on the major, minor and intermediate refractive indices, which are now symbolized by (n_g), (n_p) and (n_m), respectively.

$$R_g\% = \frac{(n_g - 1)^2}{(n_g + 1)^2} \tag{11.27}$$

$$R_m\% = \frac{(n_m - 1)^2}{(n_m + 1)^2} \tag{11.28}$$

$$R_p\% = \frac{(n_p - 1)^2}{(n_p + 1)^2}. \tag{11.29}$$

Obtaining these values from two sections of the crystal is necessary The bireflectance is given by the difference $(R_g) - (R_p)$.

– **Orthorhombic, monoclinic and triclinic absorbent/opaque crystals and minerals**

In this case, they cannot be called biaxial because they do not have optical axes but circularly polarized axes (Fig. 11.14).

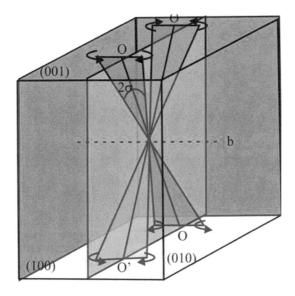

Fig. 11.14 Orthorhombic crystal showing the axes of circular polarization, O–O and O'–O' (Figure taken from Cameron,[1] after Berek[2])

The angle between each pair of these axes, symbolized as 2σ, increases as the absorption of light increases.

When the angle (2σ) between two of these circular polarization axes with opposite direction of rotation is large, the crystal is very absorbent. When the angle decreases, the absorption is very small. In the extreme case, when this angle is zero, both axes are fused into one and the absorption is zero, so these axes would now correspond to the optical axes of a transparent crystal.

There are three values of reflectance that characterize the crystal, as well as three values of the refractive index and three values of the absorption coefficient. One aspect to consider is that these values do not have to coincide in the same direction as the crystal.

$$R_g\% = \frac{\left(n_g - 1\right)^2 + k_g^2}{\left(n_g + 1\right)^2 + k_g^2} \tag{11.30}$$

[1] Cameron [1].

[2] Berek [2].

$$R_m\% = \frac{(n_m - 1)^2 + k_m^2}{(n_m + 1)^2 + k_m^2} \qquad (11.31)$$

$$R_p\% = \frac{(n_p - 1)^2 + k_p^2}{(n_p + 1)^2 + k_p^2}. \qquad (11.32)$$

To obtain these values, two sections of the crystal are necessary. The bireflectance is given by the difference $(R_g) - (R_m)$.

11.8   Dispersion

Dispersion is the variation of the refractive index n with the wavelength λ, with the temperature T. In general, the variation of n with T in crystals can be considered negligible.

An adequate treatment of dispersion implies deepening the atomic theory of matter, although it can be simplified considering that a crystal consists of ions, atoms or molecules arranged periodically and orderly in space. These ions, atoms or molecules can behave like dipoles and the sum of all is polarization, addressed in the previous section.

For a transparent crystal that absorbs very few λ into the visible or near, the visible refractive index decreases nonlinearly as the wavelength increases, and the resulting curve is called a *dispersion curve*. These dispersion curves typically have a steeper slope in materials with a high refractive index compared to those with low refractive, and solids containing transition elements (Fe, Ti), compared to those that lack them. Dispersion is normal when the refractive index decreases as λ increases (Fig. 11.15).

The refractive indices of a given crystal for specific λ can be symbolized using the λ as a subscript. In Fig. 11.15, it is observed that the refractive index n_g for $\lambda = 687$ nm is 1.5840, which can be expressed as $n_{\gamma_{687}} = 1.5840$.

The dispersion capacity of a crystal can be expressed by its *dispersion coefficient*, which is defined as

$$dispersion\ coefficient = n_F - n_C \qquad (11.33)$$

and by its *dispersion power,* given by

$$dispersion\ power = \frac{n_F - n_C}{n_D - 1}. \qquad (11.34)$$

The latter expression shows the ability of a transparent crystal to separate white light into its different components (colors). The subscripts correspond to the Fraunhofer lines, associated with λ individuals. In Table 11.1, a list of them is presented.

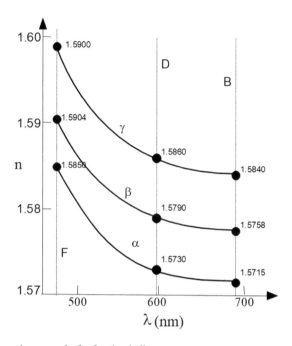

Fig. 11.15 Dispersion normal of refractive indices

Table 11.1 Fraunhofer lines

Fraunhofer lines	λ(nm)
A	759.4
B	687
C	656.3
D	589.3
E	526.9
F	486.1
G	430.8

Dispersion is anomalous when the refractive index increases as λ decreases (Fig. 11.16). It is produced in strongly colored transparent crystals as a result of the absorption of certain λ from the visible region of the spectrum (selective absorption of light).

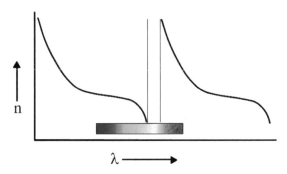

Fig. 11.16 Anomalous dispersion

Exercises

1. (a) Calculate the velocity of light considering the data provided (second to fifth columns) in the following isotropic minerals:

	Formula	Molecular weight	ρ (g/cm^3)	n	v
Fluorite	CaF$_2$	78.08	3.18	1.433	
Halite	NaCl	58.45	2.16	1.544	
Almandine	Fe$_3$Al$_2$Si$_3$O$_{12}$	476.82	4.32	1.83	
Diamond	C	12.011	3.51	2.42	

   (b) Establish the relationship between molecular weight, density, refractive index, and velocity of light in these minerals.

2. Indicate the optical characteristics (second to fifth columns) of the minerals that crystallize in the following crystal systems:

	Isotropic or anisotropic	Principal refractive indices	More birefringent section	Number of optical axes
Cubic				
Monoclinic				
Hexagonal				
Orthorhombic				
Rhombohedral				
Triclinic				
Tetragonal				

3. Indicate the birefringence, if it is an isotropic or anisotropic section, optical sign, Miller indices of the most birefringent section and the corresponding section of the ellipsoid of the indices of the following minerals, whose refractive indices are given:

	Indices	Birefringence	Isotropic or anisotropic	Optical sign	(hkl) of the most birefringent section	Section of the indicatrix
Andalucite	$n_\alpha = 1.632$ $n_\beta = 1.638$ $n_\gamma = 1.643$					
Olivine	$n_\alpha = 1.674$ $n_\beta = 1.692$ $n_\gamma = 1.712$					
Almandine	$n = 1.830$					
Rhodochrosite	$n_\omega = 1.816$ $n_\varepsilon = 1.597$					
Diamond	$n = 2.42$					

4. Indicate the Miller indices of the following sections:

   (a) isotropic section of a cubic mineral
   (b) isotropic section of a hexagonal mineral
   (c) isotropic section of a rhombohedral mineral
   (d) isotropic section of a tetragonal mineral
   (e) anisotropic section of a hexagonal mineral
   (f) anisotropic section of a rhombohedral mineral
   (g) anisotropic section of a tetragonal mineral

5. From each of the following crystal sections, indicate whether it is isotropic or anisotropic:

   (a) section perpendicular to one of the optical axes of an orthorhombic mineral
   (b) Sect. (210) of a tetragonal mineral
   (c) Sect. (101) of a cubic mineral
   (d) Sect. (0001) of a rhombohedral mineral
   (e) Sect. (010) of a tetragonal mineral
   (f) section perpendicular to the acute bisector of a triclinic mineral
   (g) Sect. (1010) of a hexagonal mineral

6. Draw the optical axis on the following crystal sections of a uniaxial mineral: (001), (110), (111), (101), (011), (100).

7. (a) What is the optical sign of a mineral with $n_\omega = 1.68$ and $n_\varepsilon = 1.67$.
   (b) What crystal system(s) does it belong to?
   (c) Is mineral birefringent?

(d) Write the Miller indices of the most birefringent section of that mineral. Draw this section indicating the optical axis (axes) and refractive indices that characterize it.

8. The values of refractive indices, measured in five different sections of a mineral belonging to the hexagonal system, are the following:

	(00.1)	(11.1)	(11.0)	(10.1)	(12.1)
n_{max}	1.566	1.566	1.566	1.566	1.566
n_{min}	1.566	1.564	1.562	1.565	1.563

Indicate the following:

(a) for each of the sections, whether it is isotropic or anisotropic
(b) the birefringence of each section and that of the mineral
(c) the principal refractive indices of the mineral
(d) the section that provides the principal refractive indices of the mineral

9. Apatite is a mineral that crystallizes in the hexagonal system. The maximum and minimum values of the refractive indices measured in several crystal sections were $n_{max} = 1.667$ and $n_{min} = 1.666$. The maximum value was constant in all sections. The thin section shows blue color. Indicate

(a) whether the apatite is isotropic, uniaxial, or biaxial.
(b) if the apatite is anisotropic, birefringence, and the most birefringent section.
(c) the mineral indices with which the measured maximum and minimum indices would correspond.
(d) if the mineral is anisotropic, the optical sign.
(e) if the mineral is anisotropic, whether the interference colors of the more birefringent section are of high or low order.
(f) if the apatite is uniaxial, the Miller indices of the section parallel to the optical axis.
(g) if the apatite is uniaxial, whether the section perpendicular to the optical axis is isotropic or anisotropic and the corresponding Miller indices.
(h) which refractive indices could have been measured in the section perpendicular to the optical axis.

10. The epidote ($Ca_2Fe^{3+}Al_2(OH)Si_3O_{12}$) crystallizes in the monoclinic system, with the Y-axis of the ellipsoid of the indices coinciding with the crystallographic axis b. It is colorless to pale yellowish green, the optical sign is negative, and the refractive indices are $n_\gamma = 1.734$, $n_\beta = 1.725$, $n_\alpha = 1,715$. Indicate

(a) if this mineral prepared in thin sheet (cemented with Canada balsam, $n \cong 1.54$) and observed with the transmission polarizing microscope, in orthoscopic arrangement and without analyzer, will show relief change and, if shown, if it will be highly appreciable, unappreciable or moderately appreciable.

(b) the relationship between acute and obtuse bisectors and optical normal with crystallographic axes.

(c) the most birefringent section.

(d) the optical axial plane in relation to any crystallographic direction or crystal plane.

Questions

1. Pair each section with its corresponding optical characteristics

Section (210) of a tetragonal crystal	Section with n_β and n_α
Section (101) of a cubic crystal	Section with n
Section perpendicular to the acute bisector of a triclinic crystal	Section with n_β and n_α

2. An anisotropic mineral crystallizes in which of these systems:
   - ◯ a. hexagonal and cubic
   - ◯ b. tetragonal, hexagonal and rhombohedral
   - ◯ c. tetragonal and cubic
   - ◯ d. tetragonal, hexagonal and cubic

3. Crystals in which the light is not split are
   - ◯ a. tetragonal, hexagonal, rhombohedral, orthorhombic, monoclinic, cubic
   - ◯ b. tetragonal, hexagonal, rhombohedral, orthorhombic, cubic, triclinic
   - ◯ c. tetragonal, hexagonal, rhombohedral, triclinic, monoclinic, cubic
   - ◯ d. cubic

4. What is the name of the optical constant that relates the velocity of light in the vacuum and in a crystal?

   Response: [   ]

5. What is the name of the direction of a crystal along which the light does not split?

   Response: [   ]

6. How many extreme refractive indices characterize the minerals crystallizing in the tetragonal system?

Response: [   ]

7. Write the name of the direction of the transparent crystal perpendicular to the plane containing the optical axes.

Response: [   ]

8. Write the number of refractive indices of a crystal belonging to the orthorhombic system.

Response: [   ]

9. Write the number of refractive indices of crystals belonging to the corresponding crystal system:

Crystal system	Number of refractive indices
Cubic	
Monoclinic	
Hexagonal	
Orthorhombic	
Rhombohedral	
Triclinic	

10. Will the crystal section perpendicular to the normal optics of a monoclinic crystal be isotropic or anisotropic?

Response: [   ]

References

1. Cameron EN (1961) Ore microscopy. John Wiley and Sons Inc., New York
2. Berek M (1931) Die Schnittkurven der komplexen Indikatrix absorbierender Kristalle mit ihren reellen Symmetrieebenen. Sonder-Abdruck aus dem Neuen Jahrbuch f. Mineralogie etc., Beilage-Band 64. Abt. A., 123–136

Chapter 12
Representation Surfaces of Optical Properties of Crystals

Abstract The importance of the representation surfaces explained in Chap. 10 will be emphasized. They are especially relevant in the case of transparent crystals and minerals, since the ellipsoid of the indexes will be key in understanding not only variation of the refraction index according to the direction but also propagation of light in the crystals. In the case of opaque crystals and minerals, the indicatrix surface of reflectance is described; reflectance or reflectivity is the property that best allows the understanding of these materials.

12.1 Introduction–Representation Surfaces of Optical Properties of Crystals

It is useful to be able to represent geometrically the variation of a crystal optical property with the direction in the crystal.

In transparent crystals (minerals), when a linearly polarized light beam interacts with a transparent crystal, each direction of light vibration corresponds to a single value of the crystal refractive index. This property can be represented by a radius vector, whose length is proportional to the value of the property, and the direction is that of the crystal on which the property has been measured. The set of radius vectors, all with the same origin, give rise to a surface whose representation is possible if the magnitude of the property is a real number.

The *ellipsoid of the indices* is a surface that represents the variation of the refractive index with the direction of a transparent crystal or mineral and, therefore, the propagation of light in it. To draw it, an origin is chosen and radius vectors are drawn.

In absorbent crystals (opaque minerals) some optical properties require a complex number to define completely. In this case, the property cannot be represented geometrically by a three-dimensional surface. However, the complex number can be split into two parts and represent separately the variation of each with the direction. If the vibration in the crystal is elliptical and not linear, a simple radius vector cannot be used to represent it. Thus, even if the property is a real number, it

Table 12.1 Types of sections as a function of symmetry in crystals

Section	Line of optical symmetry	Cubic crystals	Uniaxial crystals	Orthorhombic crystals	Monoclinic crystals	Triclinic crystals
Uniradial	∞	All sections	Circular sections			
Symmetrical	1		Vertical sections	(*010*) (*010*) (*001*)		
	2		General sections	(*h0l*) (*0kl*) (*hk0*)	(*h0l*)	
Asymmetrical				General sections	Other sections	All sections

cannot be represented geometrically by a surface. Reflectance is the most commonly used optical property to characterize them.

The *indicatrix surface of reflectance* [1][1] is a surface that represents the variation of the reflectance with the direction. If the reflectance is a real number, it can be represented by a radius vector for a given direction of vibration.

As a function of optical symmetry, several sections can be distinguished in the crystals (Table 12.1).

In a sphere, there are an infinite number of planes of symmetry that pass through the center of the sphere, and they are called planes of optical symmetry. When a radius vector is perpendicular to a plane of optical symmetry, it implies that the corresponding vibration in the crystal is linear. Therefore, the use of the spherical surface is a way of verifying that in an isotropic crystal each vibration is linear and corresponds to the same refractive index or reflectance.

Each section of the sphere represents a uniradial section of the crystal, and each diameter is a line of optical symmetry.

On a surface of revolution, the basal plane (perpendicular to the axis of revolution) and all the main planes (parallel to the axis of revolution) are planes of symmetry. Each of the main planes is associated with crystal sections that have Miller indices (*hk0*), (*h00*) or (*0k0*) which are characterized by two lines of optical symmetry.

The basal sections (uniradial) are associated with crystal sections (*00l*), and they are characterized by infinite lines of optical symmetry.

The existence of a single line of optical symmetry implies that the two mutually perpendicular vibrations are linearly polarized, while the basal section is indistinguishable from any section of an isotropic crystal.

---

[1] Hallimond [1].

In the ellipsoid of the orthorhombic crystals, there are three mutually perpendicular planes of symmetry. All sections perpendicular to each of these planes have at least one line of optical symmetry.

- The pinacoidal sections $(h00)$, $(0k0)$, $(00l)$ have two lines of optical symmetry.
- The sections of type $(h0l)$, $(0kl)$ and $(hk0)$ have an optical symmetry line. In these sections, the two vibrations are linearly polarized and fixed for all wavelengths of light.
- The general sections (hkl) do not have lines of optical symmetry.

In the ellipsoid of the monoclinic crystals, there is only one line of optical symmetry.

In the ellipsoid of the triclinic crystals, there are no lines of optical symmetry.

12.2 Ellipsoid of the Indices or Optical Indicatrix of the Transparent Crystals

The indices ellipsoid is a sphere in the isotropic crystals or minerals, and the module of the radius vectors is n.

The indices ellipsoid is an ellipsoid in the anisotropic crystals or minerals and the module of the radius vectors is different as n varies between two or three extreme values:

- n_ω and n_ε, *ordinary* and *extraordinary refractive indexes*, respectively, in the uniaxial ellipsoid coinciding with its two semi-axes.
- n_γ, n_β, and n_α, with $n_\gamma > n_\beta > n_\alpha$, in the biaxial ellipsoid coinciding with its three semi-axes.

12.2.1 Ellipsoid of the Indices and Indicatrix Surface of Reflectance of the Isotropic Crystals and Minerals

When a linearly polarized vibration is transmitted or reflected from a transparent isotropic crystal or mineral, it continues to be linearly polarized, regardless of the orientation of the vibration.

As the refractive index is a real number and a unique value for any direction, the optical indicatrix is a sphere.

In this surface, any section is circular–isotropic (Fig. 12.1). It corresponds with any section of the crystal or mineral whose generic Miller indices are $(h00)$, $(0k0)$, $(00l)$, $(hk0)$, $(h0l)$, $(0kl)$, (hkl).

Reflectance is also a real number, so the indicatrix surface of reflectance is a sphere.

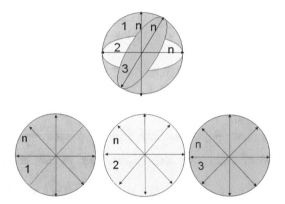

Fig. 12.1 Sections of the isotropic indicatrix

12.2.2 Ellipsoid of the Indices and Indicatrix Surface of Reflectance of the Anisotropic Crystals and Minerals

In the anisotropic crystals and minerals, linearly polarized light is split into two linearly polarized rays and vibrates in mutually perpendicular planes. For a normal wave[2] there are only two characteristic waves that the crystal will transmit or reflect.

In reflection and at perpendicular incidence, the perpendicular to the surface defines the normal wave. One can speak of two directions of vibration of a given section if the normal wave is defined. If each of the directions of vibration of the section is parallel to the direction of vibration of the polarizer, it is reflected as a linearly polarized wave, i.e., its state of polarization does not change.

– **Uniaxial crystals and minerals**

In transparent uniaxial crystals, the ellipsoid of the indices is an ellipsoid of revolution (Fig. 12.2).

In this ellipsoid, there are three special sections:

1. Circular section

   Is an isotropic section, it is the optical axis section. It corresponds to sections of the crystal or mineral with Miller indices (00l). Its refractive index n_ω can be measured in any direction. It has two elliptical sections.

---

[2] A unit vector which is perpendicular to an equiphase surface of a wave and has its positive direction on the same side of the surface as the direction of propagation.

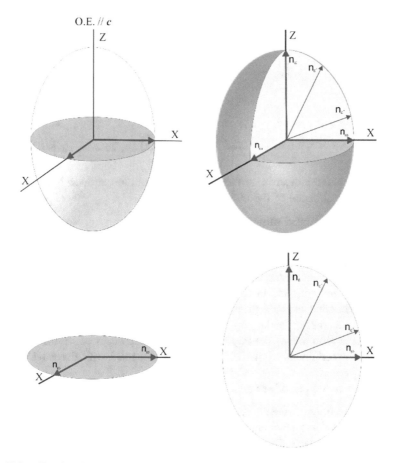

Fig. 12.2 Ellipsoid of the indices of uniaxial transparent crystals

2. Section parallel to optical axis (parallel to *c* axis), it is an anisotropic section. It corresponds to the sections of the crystal or mineral with Miller indices (*h00*), (*0k0*), (*hk0*). The refractive index n_ε is parallel to optical axis, and the refractive index n_ω is perpendicular to optical axis. It is the most birefringent section. It is the most pleochroic section (if the crystal or mineral is pleochroic). It is the section showing the higher order interference colors.

3. Section inclined with respect to the optical axis, anisotropic section. It corresponds to sections of the crystal or mineral with Miller indices (*hkl*), (*0kl*), (*h0l*). It has two refractive indices, $n_{\varepsilon'}$ (between n_ω and n_ε) and n_ω. It can be positive or negative (Fig. 12.3).

Fig. 12.3 Ellipsoid of the
indices, positive **a** and
negative **b**, of uniaxial
transparent crystals and
minerals

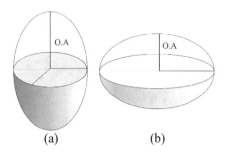

(a)   (b)

The reflectance indicatrix surface is also a revolution surface but of a higher range than the ellipsoid of the indices.

– **Biaxial crystals and minerals**

The optical indicatrix is a three-axis ellipsoid with two circular sections, equally inclined with respect to the major and minor axes of the ellipsoid. Each of the circular sections is perpendicular to one of the optical axes. Along the optical axes, the light does not split and the polarization state is maintained.

In orthorhombic crystals or minerals, the optical indicatrix has six possible orientations, according to the choice of the crystallographic axes, since each one must coincide with one of the three binary axes of the ellipsoid. In these sections, the vibrations are linearly polarized but can be dispersed with the wavelength of light.

There are two types of sections in the optical indicatrix (Fig. 12.4):

1. Circular sections. There are two circular sections. They are optical axis sections, isotropic sections. They correspond to sections of the crystal or mineral whose Miller indices depend on it. The refractive index n_β can be measured in any direction.

2. Elliptical sections. They are sections containing two axes of the ellipsoid and anisotropic sections:

   – Section Z–X (optical axial plane). This section contains the two optical axes. It coincides with the crystal sections $c–a$, $c–b$, $a–c$, $b–c$, $a–b$, $b–a$. The Z-axis is associated with n_γ. The X-axis is associated with n_α. It is the more birefringent section. It is the more pleochroic section (if the crystal or mineral is pleochroic). This section shows the higher order interference colors.

   – Section Z–Y. It coincides with $c–a$, $c–b$, $a–c$, $b –c$, $a–b$, $b–a$. The Z-axis is associated with n_γ. The Y-axis is associated with n_α. It contains the optic normal which coincides with Y-axis.

   – Section X–Y. It coincides with $c–a$, $c– b$, $a–c$, $b–c$, $a–b$, $b– a$. The X-axis is associated with n_α. The Y-axis is associated with n_β. It contains the optic normal which coincides with Y-axis.

   – Sections containing one of the three axes of the ellipsoid and sections containing none of the axes of the ellipsoid.

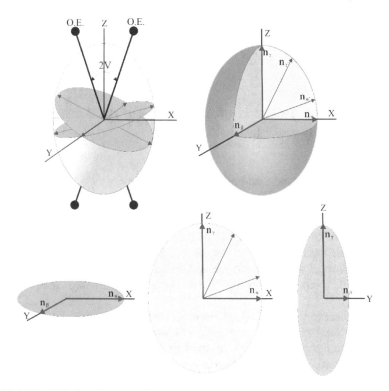

Fig. 12.4 Biaxial indicatrix and section types

12.3 Surface Indicatrix of Reflectance of the Absorbent (Opaque) Crystals and Minerals

12.3.1 Surface Indicatrix of Reflectance of the Isotropic Crystals and Minerals

In absorbent (opaque) isotropic crystals and minerals, the indicatrix equation contains a complex number instead of a real number for the refractive index of transparent crystals and minerals. For this reason, a three-dimensional surface cannot be represented. However, the variation of the refractive index and absorption coefficient with direction can be represented separately. Each of these representations gives rise to a sphere.

Reflectance is a real number and can also be represented by a sphere. The infinite number of planes of optical symmetry shown by this surface indicates that, for each direction, the reflected vibration is linearly polarized.

12.3.2   Surface Indicatrix of Reflectance of the Anisotropic Crystals and Minerals

– Uniaxial crystals and minerals

In these crystals, the indicatrix surfaces for n and k are revolution surfaces of the 8th range, and the surface indicatrix of reflectance is of the 24th range (Fig. 12.5).

On a surface of revolution, the basal plane (perpendicular to the axis of revolution) and all the main planes (parallel to the axis of revolution) are planes of symmetry. Each of the main planes is associated with crystal sections with Miller indices $(hk0)$, $(h00)$, or $(0k0)$, which are characterized by two lines of optical symmetry.

The basal sections (uniradial) are associated with crystal sections $(00l)$, and they are characterized by infinite lines of optical symmetry. The existence of a single line of optical symmetry implies that the two mutually perpendicular vibrations are linearly polarized, while the basal section is indistinguishable from any section of an isotropic crystal.

– Orthorhombic, monoclinic, and triclinic crystals and minerals

The indicatrix surfaces for n and k are, in general, surfaces of the 8th range, and the surface indicatrix of reflectance is of the 24th range. Each of these surfaces has three mutually perpendicular planes of symmetry. All sections perpendicular to one of the planes of symmetry have at least one line of optical symmetry.

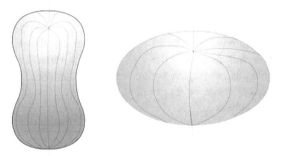

Fig. 12.5 Examples of revolution surfaces in uniaxial crystals

– The pinacoidal sections (*h00*), (*0k0*), (*00l*) have two lines of optical symmetry.
– The sections (*h0l*), (*0kl*), and (*hk0*) have an optical symmetry line. In these sections, the two vibrations are linearly polarized and fixed for all wavelengths of light.
– The general sections (*hkl*) do not have lines of optical symmetry. In these sections, the vibrations can be elliptically polarized and can be dispersed with the wavelength of light. In this case, the direction of polarization and the elliptical relationship are the same in both vibrations.

The ellipticity ratio is zero for sections perpendicular to a plane of symmetry. For sections that are not perpendicular to a plane of symmetry, the ratio of ellipticity increases as a function of the inclination of these sections to the plane of symmetry, until the maximum value of the unit originating the circular polarization is reached.

There are four directions in the crystal in which light is circularly polarized, and they are called *axes of rotation* or *axes of circular polarization* (Fig. 11.16). Due to elliptical polarization, only the sections perpendicular to one of the three planes of optical symmetry are geometrically representable (Fig. 12.6).

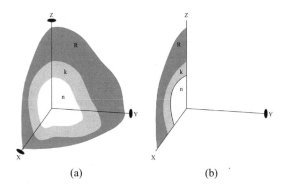

(a)   (b)

Fig. 12.6 Sections to be represented: **a** orthorhombic crystal; **b** monoclinic crystal

---

Questions

1 Indicate the characteristics of the sections of the indices ellipsoid to which the following crystal sections relate:

(010) of a cubic mineral
(010) of a tetragonal mineral
(010) of a monoclinic mineral
(213) of a cubic mineral
(111) of a hexagonal mineral
(321) of an orthorhombic mineral
(001) of a cubic mineral
(011) of a rhombohedral mineral

(102) of an orthorhombic mineral

Response: [   ]

2 Indicate the generic Miller indices or another crystallographic characteristic that allows us to perfectly identify the crystal sections corresponding
   to the following sections of the indices ellipsoid:

   (a) circular section of the uniaxial indices ellipsoid
   (b) circular section of the biaxial indices ellipsoid
   (c) circular section of the isotropic indices ellipsoid
   (d) elliptical section whose semi-axes are n_ω and $n_{\varepsilon'}$
   (e) elliptical section whose semi-axes are n_γ and n_β
   (f) elliptical section whose semi-axes are $n_{\alpha'}$ and $n_{\beta'}$

   Response: [   ]

3 Draw the optical axes on the crystal sections of a positive monoclinic
   mineral which coincide, respectively, with the corresponding sections of
   the indices ellipsoid Z–X, Z–Y, X–Y.

4 Write the refractive indices of a crystal section of a monoclinic mineral
   which coincides with the Z–X section of the indices ellipsoid. Indicate
   whether the section is isotropic or anisotropic.
   Response: [   ]

5 Write the refractive indices of a crystal section of a monoclinic mineral
   which coincides with the Z–X' section of the indices ellipsoid. Indicate
   whether the section is isotropic or anisotropic.
   Response: [   ]

6 Write the refractive indices of a crystal section of a monoclinic mineral
   which coincides with the section of the indices ellipsoid perpendicular to
   the normal optics. Indicate whether the section is isotropic or anisotropic.
   Response: [   ]

7 Write the refractive indices of a crystal section of a monoclinic mineral
   which coincides with the section Y–X of the indices ellipsoid. Indicate
   whether the section is isotropic or anisotropic.
   Response: [   ]

8 Write the name of the direction of the transparent crystal perpendicular to
   the plane containing the optical axes.
   Response: [   ]

Reference

1. Hallimond AF (1970) The polarizing microscope, 3th edn. Vickers Instruments York, England

Chapter 13
The Polarizing Microscope

Abstract Transmission and reflection polarization microscopes are described. Emphasis is placed on the differences in their observation and illumination systems, such as the absence of a condenser in the reflection microscope, since the objective acts as such in the illumination system. An illuminator makes the light coming from the lamp deviate towards the specimen located on the microscope stage and, from it, be reflected and pass through the objective to the eyepiece and, from it, to the eye. The orthoscopic and conoscopic arrangement of the polarization microscope is also described. The preparation of a thin section and a polished sample to observe the properties of transparent minerals and opaque minerals is described with the transmission and reflection polarization microscopes, respectively.

13.1 Polarizing Transmission Microscope

The polarizing transmission microscope is composed of (Fig. 13.1):

- Light source.
- Rotating stage (Fig. 13.2). The microscope turn stage is the surface where the object to be examined is placed. Allows you to make turns and angle measurements.
- Substage set. The substage set is the set of elements located under the stage. These elements are:

  - Polarizer (Fig. 13.3). The polarizer transmits light vibrating in one direction, generally east–west. The polarizer is below the stage and above the light source.
  - Condenser lens. Condenser lenses direct a cone of light onto the object being examined. There are two condenser lenses–Inferior condenser lens and superior condenser lens.
  - Iris diaphragm. The iris diaphragm reduces the illuminated area on the object to be examined.

© The Author(s), under exclusive license to Springer Nature Switzerland AG 2022   299
C. Marcos, *Crystallography*, Springer Textbooks in Earth Sciences,
Geography and Environment, https://doi.org/10.1007/978-3-030-96783-3_13

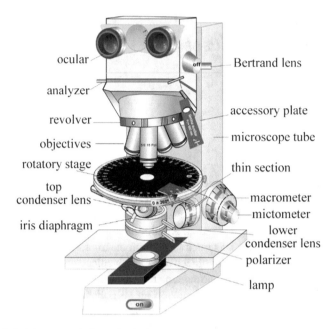

ocular

analyzer

revolver

objectives

rotatory stage

top
condenser lens

iris diaphragm

Bertrand lens

accessory plate

microscope tube

thin section

macrometer

mictometer

lower
condenser lens

polarizer

lamp

Fig. 13.1 Polarizing transmission microscope

Fig. 13.2 Rotating stage

Fig. 13.3 Polarizer

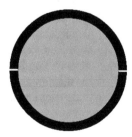

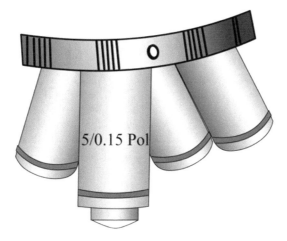

Fig. 13.4 Revolver with objectives

- Analyzer. The analyzer is the polarizer located above the objective. It vibrates in a direction perpendicular to that of the lower polarizer.
- Objectives (Fig. 13.4). Objective lens provide a real and magnified image of the object on the stage. The objectives are characterized by:

  - Magnification

    Low magnification –2.5x, 5x. Increases the image 2.5 or 5 times, respectively. Allows general observation of the object on the plate. Allows you to focus on an image before using another higher magnification lens.
    Medium magnification –10x, 20x.
    High magnification –40x, 50x, 100x.

  - *Angular aperture A.A.* Angle between the most divergent rays that can enter the lens from a focused point on the plate object.
  - *Numerical aperture NA or N_A:*

$$NA = n \cdot \sin \alpha \qquad (13.1)$$

  where

  n = index of refraction

  α = maximum half-angle (Fig. 13.5)

- The *resolving power* (δ) is the minimum distance at which two points can be discriminated. This limit is determined by the wavelength of the illumination

Fig. 13.5 Maximum
half-angle

source, in this case visible light (400–700 nm). Optical microscopes have a
maximum resolution limit of 200 nm.

The resolving power is given by expression (13.2)

$$\delta = \frac{\lambda}{2N_A} \qquad (13.2)$$

N_A being the numerical aperture.

Working distance. This is the distance between the lowest part of the lens and
the highest part of the object on the plate. The greater the increase of magnification and N_A, the shorter the working distance. Caution should be used when
focusing.

Depth of focus. This is the power to focus in depth. It is a reverse function of the
numerical aperture.

– Accessory plates (Fig. 13.6). The accessory plates or the so-called auxiliary
  plates are (a) quartz wedge (retard variable with thickness); (b) gypsum (retard
  ∼ 546 nm); and (c) mica sheet (retard of 1/4 λ ∼ 140 nm).
– Bertrand lenses. When inserted, they are placed on top of the analyzer (upper
  polarizer). They allow observation of the interference figure in conoscopy, with
  condenser lens.
– Eyepiece or ocular (Fig. 13.7). The eyepiece or ocular provides a virtual and
  magnified image, usually 10x, of the lens image.
  The oculars include a reticle (Fig. 13.8) that marks the directions of vibration of
  the polarizers.

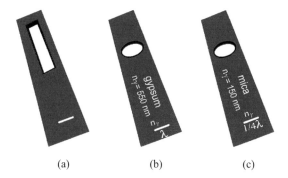

(a) (b) (c)

Fig. 13.6 Accessory plates **a** quartz wedge, **b** gypsum, **c** mica sheet

Fig. 13.7 Eyepiece

3

Fig. 13.8 Reticle wires

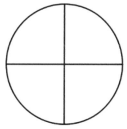

– Focusing screws (Fig. 13.9). The focusing screws allow focus on the object on the stage by varying the distance between the object and the objective. There are two screws–a macrometer with coarse adjustment and a micrometer with fine adjustment.

Fig. 13.9 Focusing screws

13.2   Reflection Polarizing Microscope

This microscope is used to observe the properties of opaque minerals (they do not let light through with thicknesses of up to 30 μ).

Their components (Fig. 13.10) are basically the same as those of a polarizing transmitted light microscope. The difference is that this microscope requires the following:

–   Illuminator. The illuminator is above the objectives. Its mission is twofold:

   –   To deviate the light from the light source through the objective to the polished sample on the microscope stage.
   –   To deviate the light reflected by the sample through the lens to the eyepiece.

–   Objective. The objective acts as such in the observation system. It acts as a condenser in the lighting system.

13.3   Illumination Types

The stage object may be illuminated with different arrangements of optics but, in this chapter, only the so-called *Köhler illumination* and the bright field technique are used.

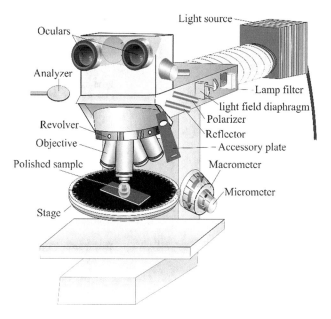

Fig. 13.10   Reflection polarizing microscope

The features of Köhler illumination are as follows:

-   The luminous surface of the lamp is imaged into the condenser-aperture diaphragm plane.
-   The luminous field diaphragm is placed in a plane where a back image of the stage object is formed.
-   The stage object outside the illuminated area does not scatter light or suffer possible harmful effects through heating, as a consequence of the previous features. Bright-field technique consists of illuminating the object on the microscope stage from below and to observe from above. The typical appearance of a bright-field microscopy image is a dark object on a bright background, hence the name.

Two types of illumination can be used with the polarizing microscope: Orthoscopic and conoscopic.

-   *Orthoscopic illumination*

Orthoscopic illumination is caused by light rays travelling parallel to each other and perpendicular to the object on the microscope stage. In Fig. 13.11, a scheme of this illumination type is shown.

Divergent rays from the luminous surface of the lamp intersect, after passing through the collector, at the place where an image of the luminous surface is

Fig. 13.11 Scheme of
orthoscopic illumination

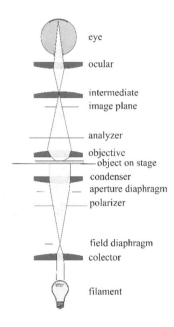

formed, and where the aperture diaphragm[1] of the condenser is located. These rays
are called *principal rays*. The condenser makes them parallel, and they pass through
the stage object as a cylindrical bundle, delimiting the illuminated area of the object.

The parallel rays coming from the luminous surface of the lamp are brought to
focus by the collector to its focal plane.[2] In this plane, the luminous field diaphragm
is placed, and the image of the stage object is formed. These rays are marginal to the
aperture area of the condenser and converge and intersect in front of the collector,
where the luminous field diaphragm is placed. These rays are marginal to the
aperture area of the condenser, which brings them back into focus, but on the stage
object. Half the width of the aperture area corresponds to the aperture angle.

---

[1] Diaphragm is a thin opaque structure with an opening (aperture hole through which light travels)
at its center. Also, it is called a stop (an aperture stop, if it limits the brightness of light reaching the
focal plane, or a field stop for other uses of diaphragms in lenses). If adjustable, the diaphragm is
known as an iris diaphragm.

[2] The front focal point of an optical system, by definition, has the property that any ray that passes
through it will emerge from the system parallel to the optical axis. The back focal point of the
system has the reverse property: rays that enter the system parallel to the optical axis are focused
such that they pass through the back focal point. The front and back focal planes are defined as the
planes, perpendicular to the optic axis, which pass through the front and back focal points. An
object infinitely far from the optical system forms an image at the back focal plane. For objects a
finite distance away, the image is formed at a different location, but rays that leave the object
parallel to one another cross at the back focal plane.

With reflected light, auxiliary lenses are required because the lens plays a double function; it acts as a condenser in the illuminating system and as a lens in the observation system. Due to the lack of an aperture in its focal plane, a system that includes auxiliary lenses must be adopted. In this way, the illuminating system includes an additional plane; thus, an additional image of the luminous surface of the lamp is formed in the plane where the aperture diaphragm of the condenser is placed. The reflector can be placed at any level in the tube, above the objective and below the eyepiece, if it is of plane-glass type. The part of the reflected light microscope that lies between the lamp and the objective acting as a condenser is called the reflected light illuminator and comprises a condenser aperture diaphragm, a luminous field diaphragm, a reflector, and an auxiliary lens.

With reflected light and Köhler illumination, the lenses are arranged to form an extra image both of the stage object and of the filament. This has the effect of producing the sequence of lamp-aperture diaphragm-field stop, which is the reverse of transmitted light.

– *Conoscopic illumination*

Conoscopic illumination is an optical technique to make observations of a transparent object on the microscope stage in a cone of converging rays of light. When a strongly convergent light cone is generated, parallel light rays with a wide range of directions will pass through the crystal or mineral section. The parallel light rays are focused on the back focal plane of the lens where the rays with different inclination relative to the axis of the microscope produce image points at different positions. When this image is observed with crossed polarizers, characteristic interference figures are generated, reflecting the symmetry and optical properties of the anisotropic crystals and minerals. This interference figure can also be observed directly by removing the eyepiece and placing a small aperture fixed diaphragm where it is best observed.

In Fig. 13.12, a scheme of this illumination type is shown.

13.4   Sample Preparation

– **Thin sections**

Thin sections (Fig. 13.13) are preparations for observations with transmitted light. Preparation of thin sections consists of:

– Cementing thin sheets of rock to a glass slide.
– Gentle grinding up to a standard thickness of 30 μm.
– Cementation of a coverslip on the rock sheet.

Fig. 13.12 Scheme of
conoscopic illumination

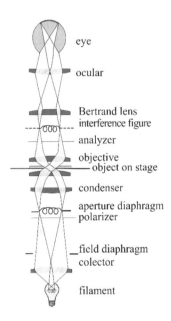

eye

ocular

Bertrand lens
interference figure

analyzer

objective
object on stage

condenser

aperture diaphragm
polarizer

field diaphragm
colector

filament

Fig. 13.13 Thin section

– **Polished samples**

Polished samples (Fig. 13.14) are preparations for observations with reflected light.
Preparation of polished samples consists of:

– Cutting a sample piece with a diamond saw.
– Cold-curing resin spilled on a 1.5 cm diameter by 1 cm high mould on which
  the sample has been placed.
– Sample roughing, once removed, to achieve a very smooth surface and a sample
  thickness of 25–30 μm.
– Polishing using diamond paste and a lubricant, to achieve a mirror polished
  surface.

Fig. 13.14 Polished sample

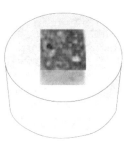

Questions

1. Thin section used with the transmission polarizing microscope has a standard thickness of
   - ○ a. 1.5 cm
   - ○ b. 30 micrometers
   - ○ c. 30 millimeters
   - ○ d. 15 cm
2. With the polarizing transmission microscope, an isotropic section can be distinguished from an anisotropic section using the
   - ○ a. Bertrand lens
   - ○ b. accessory plate
   - ○ c. iris diaphragm
   - ○ d. analyzer
3. What are the names of the lenses under the microscope stage that direct a cone of light over the object (mineral or rock preparation) to be examined?
   Response: [   ]
   True
4. An illuminator is a device characteristic of the polarizing transmission microscope.
   ○True
   ○False
5. Write the name of the device that allows light from the light source to be deviated through the lens to the polished sample on the stage of the polarizing reflection microscope.
   Response: [   ]
6. Does the polarizing microscope stage allow angle measurements? (Yes or No).
   Response: [   ]

7. Will when two polarizers with their perpendicular vibration directions are inserted into the light path of the polarizing microscope either clarity or darkness are observed?

Response: [   ]

8. The macrometer is a screw that allows centering of the object on the stage by varying the distance between object and objective.
   ○ True
   ○ False

9. The lens provides a real magnified image of the object on the microscope stage.
   ○ True
   ○ False

10. The eyepiece provides a virtual and magnified image of the lens image.
   ○ True
   ○ False

Chapter 14
Optical Properties of Transparent Crystals and Minerals

Abstract This chapter describes the properties of transparent crystals and minerals using the transmission polarization microscope, both in orthoscopic arrangement with polarized light and analyzed polarized light (cross-polarizers) and with conoscopic illumination. The concept of optical activity is defined, and the dispersion in crystals that present it is described through the colors and figures of interference.

14.1 Orthoscopic Arrangement of the Microscope

14.1.1 Observations with Plane Polarized Light

(a) **Requirements**

- Low (2.5x) or medium (10x, 20x) objectives.
- The upper condenser lens must be lowered.
- The iris diaphragm must be open.

(b) **Properties**

- **Color and pleochroism**

Color is the response of the eye to the visible range (approximately 350 nm to 700 nm) of the electromagnetic spectrum.

The color of a mineral depends on its composition, structure, presence of certain chromophore elements (Cr, Ti, Mn, Fe, Co, Ni, Cu) and small mixtures of other phases.

When visible light interacts with a crystal or mineral it can be transmitted, reflected, refracted, scattered or absorbed. If no component (wavelength) of light is absorbed by the crystal or mineral, it is colourless. When certain components of white light are absorbed by the crystal or mineral, it is coloured, and its colour results from the combination of wavelengths of white light that have not been absorbed.

Systematic study with the transmission polarizing microscope

© The Author(s), under exclusive license to Springer Nature Switzerland AG 2022   311
C. Marcos, *Crystallography*, Springer Textbooks in Earth Sciences,
Geography and Environment, https://doi.org/10.1007/978-3-030-96783-3_14

Table 14.1 Chromophores that cause color in some minerals

Mineral	Color	Chromophores
Almandine garnet	Red	Fe^{2+}
Emerald (beryl)	Green	Cr^{3+} and/or V^{3+} in octahedral coordination
Chrysoberyl	Yellow	Fe^{3+} in octahedral coordination
Zircon	Various	U^{4+}
Apatite	Green, yellow	Rare earths (neodymium, praseodymium)
Corundum (ruby)	Red	Cr^{3+} in octahedral coordination
Corundum (sapphire)	Blue	Fe^{2+} and Ti^{4+} in octahedral coordination
Sinhalite	Brown	Fe^{2+}
Olivine	Green	Fe^{2+} in octahedral coordination
Synthetic Spinel	Blue	Co^{2+}

One of the main causes of the colour of minerals is *selective absorption*. Such absorption involves an electronic transition of an electron from the fundamental state to the excited state (of higher energy). When the energy necessary for the electron to transit, the absorbed energy, is within the range of visible light, it is eliminated from the incident light and, as a consequence, the light observed by the human eye is a combination of wavelengths transmitted by the crystal or mineral.

The electronic transitions that most condition color are those that affect the elements of the first series of transitions, as well as some rare earths, including neodymium and praseodymium in lanthanides and uranium in actinides. These elements are called chromophores.

Table 14.1 shows the chromophores that cause color in some minerals.

Absorption can occur by crystal field transitions; charge transfer or the presence of color centers.

It must be taken into account that the colour of a crystal or mineral observed with a microscope is usually observed with polarized light and may vary slightly from its observation with non-polarized light.

Example
Colorless as fluorite or calcite (Fig. 14.1a). Green as chlorite (Fig. 14.1b). Brownish as augite (Fig. 14.1c).

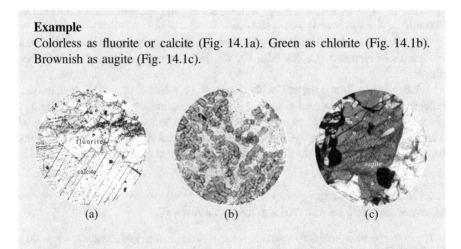

(a)   (b)   (c)

Fig. 14.1 Examples of mineral color

Pleochroism is the change in color of a crystal or mineral with observation direction.

It is a property that only anisotropic and coloured crystals can present, but not all of them do.

It is due to an uneven absorption of light (selective absorption) by the coloured crystals or minerals in different orientations; if it is large, it can be seen with the naked eye, although it is better seen with a polariser.

Example

Cordierite is a silicate that crystallizes in the orthorhombic system and has three extreme colors—violet yellow, lighter violet and dark blue violet—that can be seen with the naked eye or bright field or with a polarizer (Fig. 14.2).

Fig. 14.2 The three extreme colors that can be observed in cordierite according to orientation

Two of the colors can also be observed simultaneously with a dicroscope (Box 14.1) or a device composed of two polarizers (Fig. 14.4), each with its direction of vibration perpendicular to the other.

Box 14.1: Dicroscope

Dicroscope is an instrument that allows simultaneous observation of the pleochroic colors of a gem mineral.

It consists of a metal tube with a rectangular opening at one end and a lens at the other end. A scheme of the tube can be seen in Fig. 14.3.

Inside the tube is an elongated calcite cleavage sheet that allows observation of a double image of the rectangular opening through the microscope.

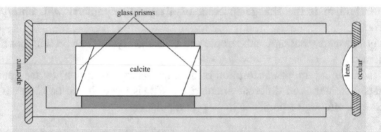

Fig. 14.3 Dicroscope scheme showing the window in front of which the gem mineral to be observed is located, the calcite prisms and the eyepiece

Procedure

The gem mineral to be observed is placed on a light source.

The dicroscope is brought close to the gem mineral and it is observed, through its eyepiece, whether or not the gem mineral has pleochroism.

If the gem mineral presents pleochroism, two images will be observed— one with each of the pleochroic colours for a certain position of the gem mineral. In another position, other colours or only one of them may be observed, depending on orientation and symmetry.

Absorption can vary in a similar way to refractive indices (Fig. 14.4).

– Pleochroism in uniaxial crystals

Uniaxial crystals can be selectively absorbed in:

– one direction, that of the extraordinary ray ω or that of the ordinary ray ε,
– two directions, that of the ordinary and extraordinary rays (see Fig. 14.5).

Both directions may not have the same amount of selective absorption. In this case, when the crystal or mineral is rotated, it transmits different colours depending on whether the direction of the ray ω or the ray ε is parallel to the direction of vibration of the polarizer.

– Pleochroism in biaxial crystals

Biaxial crystals can have different light absorption in three directions, depending on whether the light is vibrating parallel to n_γ (Z indicatrix axis associated), n_α (X indicatrix axis associated), or n_β (Y indicatrix axis associated).

Fig. 14.4 Crossed polarizers

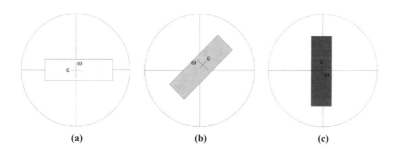

Fig. 14.5 Variation of pleochroism in a uniaxial crystal: **a** The color is light because it is the color associated with the ray ε is parallel to the direction of vibration of the polarizer. **b** The color is light gray because it is in an intermediate direction between the rays ω and ε is parallel to the vibration direction of the polarizer. **c** The color is dark gray because this is the color associated with the ray ω parallel to the direction of vibration of the polarizer

The directions of vibration corresponding to the intermediate refractive indices $n_{\alpha'}$, $n_{\beta'}$ and $n_{\gamma'}$ are generally associated with intermediate absorptions or, in other words, the same intermediate transmitted colours between those corresponding to n_{α} and n_{β} or n_{β} and n_{γ}, respectively.

Box 14.2: Observation procedure of pleochroism

1. Placing the microscope in an orthoscopic arrangement with polarized light and the thin film on the slide.
2. Observong the color of the grain or mineral section.
3. Turning the grain or mineral section 45° and 90° and observe if there has been a change in colour, i.e., if it has pleochroism.

One of the extreme colors of the section will be observed when the direction of vibration associated with that color is parallel to the direction of vibration of the polarizer. Turning 90° and placing the other direction of vibration of the crystal section parallel to the polarizer, the other extreme color will be seen. In positions intermediate to these, intermediate colors will be seen at the extremes.

– **Habit**

Habit is the most common way to present a crystal or mineral. It can be prismatic, acicular, tabular, and laminar, among others.

Box 14.3: Habit examples
Examples

See Fig. 14.6.

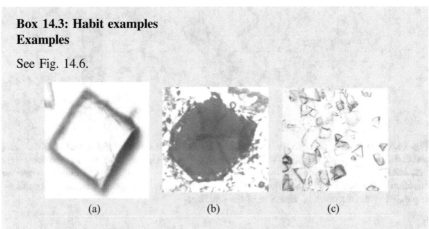

(a)   (b)   (c)

Fig. 14.6 Habit examples from cleavaged octahedrons: **a** Calcite rhombohedral, **b** garnet octagonal and **c** fluorite triangular faces

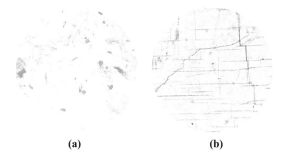

(a)   (b)

Fig. 14.7 Cleavage **a** One cleavage system in muscovite **b** Two cleavage systems in plagioclase

– **Cleavage and angle of cleavage**

Cleavage is the breaking of a crystal or mineral by particular crystal planes.

It is usually referred to by the Miller indices of the plane of cleavage, so the octahedral cleavage is the breakage according to planes {111}, the rhombohedral cleavage $\{10\bar{1}1\}$, etc.

The marks left by these cleavage planes when a mineral section is cut perpendicular to them are called cleavage traces.

The traces shown in section (001) of muscovite (Fig. 14.7a) are parallel and form one cleavage system, while the cleavage traces in section (001) of plagioclase (Fig. 14.7b) correspond to two cleavage systems with an angle about 90°.

The cleavage angle is the angle formed by two intersecting cleavage planes.

In the case of sections showing two cleavage systems, the angle between them, the cleavage angle, can be measured, and it can be useful in distinguishing one mineral from another.

Box 14.4: Measurement procedure of the cleavage angle

1. Placing the microscope in orthoscopic bright field arrangement and the thin section on the microscope stage.
2. Placong the trace of a cleavage plane of a focused grain or mineral section coinciding with one of the reticle wires, E-W, for example, and note the angle (Fig. 14.8a).
3. Turning the grain or mineral section until the other trace of cleavage coincides with the reticle wires, E-W, and note the angle (Fig. 14.8b).
4. Subtracting the values noted in sections (2) and (3) to obtain the cleavage angle.

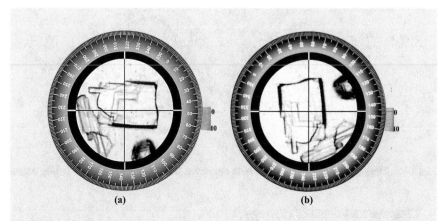

Fig. 14.8 Cleavage angle measurement **a** Trace of cleavage plane of a focused mineral grain coinciding with the reticle wire E-W 48° **b** the other trace of cleavage plane coinciding with the reticle wire E-W 146°

– Relief

The relief is related to greater or lesser appreciation of the contour of the crystal or mineral grain. It is expressed as low, moderate, high, or very high. It is a function of the refractive index, since the relief shows the difference between the refractive index of the crystal or mineral and the medium in which it is embedded. The greater the difference between the refractive indices, the greater the relief; while the lesser the difference, the lesser the relief. Table 14.2 shows the relief of some minerals according to their refractive index.

It cannot be determined whether the crystal or mineral or the surrounding medium has the highest refractive index. One approach is to use Becke's line test when the surrounding medium is known: Canada Balsam with $n = 1.54$ or a crystal or mineral whose refractive index is known. To embed grains, special liquids with well-defined n ~ 1.4–1.8 are used.

To observe the test, the following must be in place:

– Microscope in orthoscopic arrangement and bright field.
– Half-closed iris diaphragm.
– Upper condenser lens removed.

Table 14.2 Relief of some minerals according to their refractive index

Relief	Index of refraction	Example
Low	1.50 a1.58	gypsum (1.52–1.53)
Moderated	1.58–1.67	calcite (1.658–1.486)
High	1 67–1.76	corundum (1.76–1.77)
Very large	> 1.76	zircon (1.90–2.00)

Box 14.5: Observation procedure of the Becke test line

1. Placing the microscope in orthoscopic arrangement with polarized light, half-closed iris diaphragm, and upper condenser lens removed.
2. Focusing the mineral grain or section on the stage, with a crystallographic or optical direction coinciding with the direction of vibration of the polarizer (E-O).
3. Defocusing slightly the grain or mineral section.
4. If the defocusing is carried out by increasing the distance between the objective and the mineral in the thin section placed on the plate, the Becke line (bright line) will be introduced into the medium with the highest refractive index (Fig. 14.9).

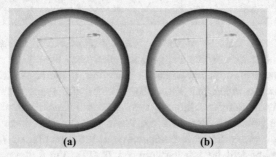

(a)   (b)

Fig. 14.9 Fluorite **a** Focused in thin section **b** Unfocused and with the Becke bright line outside the fluorite mineral unfocused

Example 1

Fluorite has a refractive index of 1.43, lower than the 1.54 of the Canada balsam (resin used to glue the mineral or rock and the coverslip onto the glass), and when unfocusing, the Becke bright line is outside the mineral (Fig. 14.9).

Example 2

Anhydrite has an average refractive index of 1.58, higher than the 1.54 of the Canada balsam and, when unfocusing, the Becke bright line is inside the mineral (Fig. 14.10).

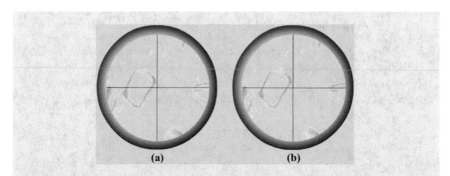

Fig. 14.10 Anhydrite: **a** Focused in thin section; **b** unfocused and with the Becke bright line inside the mineral

In anisotropic crystals or minerals with a pronounced birefringence, such as carbonates, it is easy to observe a change in relief when the crystal is rotated. Such relief will be more pronounced when the direction of the crystal or mineral with which the refractive index that shows the greatest difference with the refractive index of the surrounding medium is associated coincides with the direction of vibration of the polarizer (Fig. 14.11a). When the direction located at 90° with respect to the previous direction coincides with the direction of vibration of the polarizer, the relief will be less pronounced because the difference between the index of refraction associated with that direction and that of the medium that surrounds the crystal or mineral is less (Fig. 14.11b). The relief change indicates anisotropy.

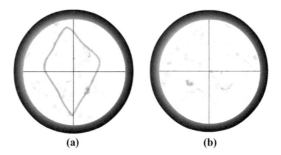

Fig. 14.11 Relief in calcite: **a** Relief very pronounced. Short diagonal, which coincides with the lowest refractive index, is parallel to the direction of vibration of the polarizer. **b** Relief low. Long diagonal, which coincides with the highest refractive index, is parallel to the direction of vibration of the polarizer

14.1.2 Observations with Polarized and Analyzed Light

(a) Requirements

- Low (2.5x) or medium (10x, 20x) magnification objectives.
- The upper condenser lens must be lowered.
- The iris diaphragm is open.
- Analyzer inserted and rotated 90° from the polarizer, which is known as "crossed polarizers"

(b) Properties

- **Retardation**

Retardation is the difference in trajectory between fast and slow rays. The advantage is that the fast ray takes over the slow ray during the time it takes for the slow ray to pass through the mineral or crystal section (Fig. 14.12). The light from the polarizer splits into two components when it enters the crystal or mineral section. Within this section, one of the components travels faster than the other, producing a

Fig. 14.12 Scheme showing the difference in the path of 1/2λ between the fast and the slow wave inside the crystal

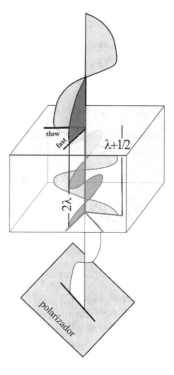

difference in trajectory, or retardation. When they leave the crystal, both components maintain their directions of vibration but return to the velocity and wavelength of light from the polarizer, keeping the retardation constant.

The retardation is symbolized by Δ and expressed by:

$$\Delta = c(t_N - t_n) \qquad (14.1)$$

where

c is the light velocity in a vacuum
t_N is the time required for the slow wave to pass through the crystal or mineral
t_n is the time it takes for the fast wave to pass through the crystal or mineral

Considering the following relationships:

$$v = e/t \qquad (14.2)$$

and

$$n = c/v \qquad (14.3)$$

where

v is the light velocity in the crystal or mineral
c is the light velocity in vacuum
e is the thickness
t is the time
n is the refractive index

Substituting in expression (14.1) of the retardation:

$$\Delta = e(N - n) \qquad (14.4)$$

where

e is the thickness
N is the highest refractive index associated with the slow wave
n is the lowest refractive index associated with the fast wave

Then retardation depends on the thickness and birefringence $(N–n)$.

– **Interference colors**
Interference colors are the colors that are observed as a consequence of the interference of the waves when they leave the anisotropic crystal or mineral. The waves are those that result from light splitting when it interacts with an anisotropic crystal or mineral. They are perpendicular polarized. The interference implies that the trajectory difference is

$$\frac{2n+1}{2}\lambda \tag{14.5}$$

The resulting wave vibrates in a plane that is perpendicular to that of the polarizer and parallel to that of the analyzer and, therefore, passes light through it.

The directions of vibration of the slow and fast wave of the mineral or crystal section do not coincide with the directions of vibration of the polarizers.

The analyser transmits the various components of white light, i.e., the various colours, in different ways. They depend on the retardation. A given retardation produces a certain interference color. These colors vary with the section of anisotropic crystal or mineral, because the retardation depends on the thickness and the birefringence.

> **Newton scale**
>
> The Newton scale is the grouping of interference colors. It is divided into orders—first, second, third, fourth, and higher. Each order groups a series of colors, and each color is associated with a retardation or path difference and, therefore, with a wavelength.

The higher the order of the interference color, the higher the birefringence.

Each section will show an interference color, which depends on the birefringence.

The section showing the highest order interference color will be the most birefringent or very close to it. The Miller indices of this section in crystals or uniaxial minerals are (*h00*), (*0k0*), (*hk0*). In crystals or biaxial minerals this section corresponds to the Z-X section of the biaxial indicatrix (Table 14.3).

The Michel Lévy chart consists of a large range of colors, corresponding to the interference colors, produced by the slow and fast rays when they leave the crystal. These colors depend on the wavelength of light that passes through the analyzer and the wavelength that is cancelled. Figure 14.13 shows the Michel Lèvy normal vision colour chart and that of colour-blind people. The retardation is represented in abscissa, the thickness of the crystal or mineral is represented in ordinates, and the birefringence in the lines starting from the origin of the system. In this way, it is possible to know approximately the retardation and, therefore, the interference colour, thickness or birefringence of the crystal or mineral, if two of the mentioned characteristics are known.

Table 14.3 Description of the birefringence and its relation to the order of the interference colour

Birefringence	Description	Color order
0.00–0.018	Low	First
0.018–0.036	Moderate	Second
0.036–0.055	High	Third
>0.055	Very large	Fourth and >

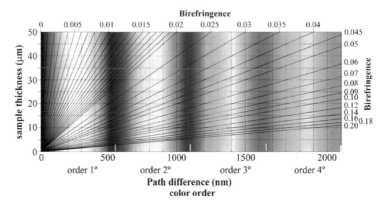

Fig. 14.13 Michel Lèvy colour chart[1]

– Grains and Twins

Grains are the association of one or more crystals or minerals with different crystallographic orientation. A twin is defined as two crystal individuals with a symmetrical relationship. Polycrystals or twins are easily recognized when it observed
in the microscope in orthoscopic arrangement and crossed polarizers, due to the
different crystallographic orientation of the individuals that form it. This is because,
in a certain position, they may show different interference colors or some may be in
extinction position and others not.

Example
See Fig. 14.14.

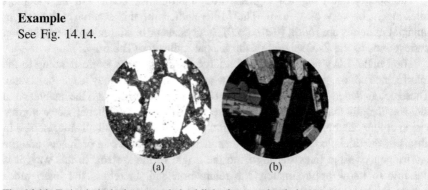

(a)   (b)

Fig. 14.14 Twined plagioclases **a** polarized light, **b** crossed polarizers

[1] Sørensen [1].

– **Zoning**

It is the compositional variation within a mineral. It often affects color and, between crossed polarizers, it is observed by changes in birefringence or extinction orientation.

Example
See Fig. 14.15.

Fig. 14.15 Tourmaline zoned

– **Alteration**

It appears as an area of turbidity on feldspars or as a dark edge on olivine grains and fractures (Fig. 14.16).

It is due to the reaction of some element of the original mineral with CO_2 or with water that is in contact with it, originating a new mineral phase.

Example

Fig. 14.16 Olivine with dark edge of alteration

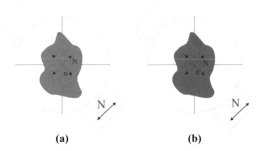

(a)   (b)

Fig. 14.17 Schemes showing the concept of addition. **a** Without accessory plate. **b** With accessory plate

- **Vibration directions**

It can be said that the directions of vibration in a section of a crystal or mineral are the directions in which slow and fast rays of light vibrate. These directions can be located considering the concepts of *addition* and *subtraction* of retardations. For this purpose, an accessory plate must be used, and the anisotropic section must be on the microscope stage at 45° from the extinction position.

The addition is the sum of the retardation of the mineral or crystal section and that of the accessory plate.

- The interference colour of the crystal or mineral section rises in order.
- It occurs when the directions of vibration of the slow and fast waves of the mineral or crystal section coincide with those of the accessory plate (Fig. 14.17).

Example

When a 30 μm thin section with a light blue interference color ▨ and a retardation of 713 nm changes to blue ▨ with retardation 1263 nm when inserting the gypsum accessory plate (retardation of 550 nm) in the microscope, it is because there has been addition of retardations, 713 + 550 = 1363 nm.

The substraction is the subtraction of the retardation of the mineral or crystal section and that of the accessory plate.

- The interference colour of the crystal or mineral section goes down in order.
- It occurs when the directions of vibration of the slow and fast waves of the mineral or crystal section are perpendicular to those of the accessory plate (Fig. 14.18).

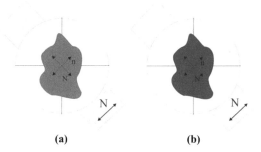

(a) (b)

Fig. 14.18 Schemes showing the concept of subtraction. **a** Without accessory plate. **b** With accessory plate

Example
When a 30 μm thin section with a light blue interference color ▮ and a retardation of 713 nm changes to gray ▮ with a retardation of 163 nm when inserting the gypsum accessory plate (retardation of 550 nm) in the microscope, it is because there has been a subtraction of the retardations, 713–550 = 163 nm.

– **Optical sign of elongation**
Elongation is the relationship between the long direction of a grain or elongated mineral section and the direction of vibration with which the highest refractive index of that grain is associated.

Box 14.6: Optical sign of elongation determination procedure
Determination of optical sign of elongation

The determination procedure consists of:

1. Determining the vibration directions.
2. Inserting the gypsum (550 nm) or mica (150 nm) accessory plates and observing the interference colours, which can be compared with those in the colour chart.

If the colours have increased in order, there has been addition of retardations because, in that case, the maximum and minimum refractive indices of the grain or mineral section coincide respectively with those of the accessory plate.

If the colours have decreased in order, there has been subtraction of retardations because, in that case, the maximum and minimum refractive indices of the grain or mineral section do not coincide respectively with those of the accessory plate.

- At 90° from the position where the addition of retardations occurs, there is now a subtraction of retardations.
- At 90° from the position where subtraction of retardations occurs, there is now addition of retardations.
- Optical sign of longation is *positive* if the higher refractive index coincides with the long direction of the grain (*slow length*).
- Optical sign of longation is *negative* if the lower refractive index coincides with the long direction of the grain (*fast length*).

Example

In tourmaline, subtraction of retardations occurs when the maximum refractive index (parallel to the long direction of the section) is perpendicular to that of the accessory plate (Fig. 14.19) and the elongation is positive or the mineral is slow length.

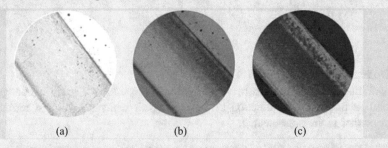

      (a)   (b)   (c)

Fig. 14.19 Images of a section of tourmaline parallel to the optical axis **a** with polarized light; **b** with crossed polarizers; **c** with crossed polarizers and gypsum accessory plate, where it can be seen that the interference colors have decreased in order because of the subtraction of retardations

– Extinction and extinction angle

The extinction is the darkness that is observed as a consequence of the constructive interference of the waves when they leave the crystal or anisotropic mineral.

The difference in the trajectory of the waves after leaving the crystal is $n\lambda$. The waves are in phase. The resulting wave vibrates in a plane that is parallel to that of the polarizer and perpendicular to that of the analyzer and, therefore, does not pass light through it. The directions of vibration of the slow and fast wave of the mineral or crystal section coincide with the directions of vibration of the polarizers.

An isotropic or amorphous section remains extinct in a full rotation of the stage.

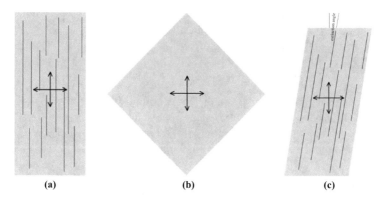

Fig. 14.20 Extinction types **a** straight, **b** symmetrical, and **c** oblique or inclined

In an anisotropic section, four extinction positions are observed, at 90° from each other and at 45° from the clarity positions.

The types of extinction are straight or parallel, symmetrical, oblique or inclined.

Straight extinction occurs when, in a crystal or mineral section between crossed polarizers, a crystallographic direction (cleavage trace or crystal plane trace) coincides with an optical direction (direction of vibration of the crystal or mineral section (Fig. 14.20a).

Straight extinction is observed when the crystallographic direction of the mineral or crystal section in extinction position coincides with one of the wires of the crosshair, which indicates the direction of vibration of one of the polarizers.

Symmetrical extinction is similar to parallel extinction but, in this case, extinction is observed when the crystallographic direction is diagonal to the vibration directions of the crystal or mineral section (Fig. 14.20c).

Straight and symmetrical extinction is presented by uniaxial crystals and the sections cut parallel to (*100*), (*010*), and (*001*) of orthorhombic crystals and the section cut parallel to (*010*) of monoclinic crystals.

Oblique or inclined extinction is observed when a crystallographic direction (cleavage trace or crystal plane) does not coincide with an optical direction (direction of vibration of the crystal or mineral section) (Fig. 14.20c). Oblique extinction is presented by biaxial crystals and minerals.

The extinction angle is the angle between a direction of vibration of the mineral section and a crystallographic direction (crystal or cleavage face trace).

Box 14.7: Extinction angle determination procedure
Determination of the extinction angle

The determination procedure consists of:

- Placing the mineral section in the extinction position (crossed polarizers) and noting the degrees of the microscope stage.
- Then placing the mineral section with a crystallographic direction parallel to the direction of vibration of the polarizer and noting the position on the stage.

Example

The mineral section of Fig. 14.21a has a crystallographic direction parallel to the lower polarizer E-W and, in this position (311.25°), the section is not in extinction (14.21b). The extinction occurs when the section is at 275° (14.21c). Therefore, the extinction angle is the difference between 311.25°–275° = 36.25°.

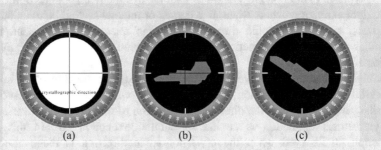

(a)   (b)   (c)

Fig 14.21 a Mineral section with a crystallographic direction parallel to the lower polarizer E-W (position at 311.25°); **b** the same mineral section as (**a**) with crossed polarizers; **c** mineral section placed at the extinction position (275°)

14.2   Conoscopic Microscope Arrangement

(a) **Requirements**

- High magnification objective (>50x).
- Superior condenser lenses.
- Open iris diaphragm open.
- Analyzer inserted and rotated 90° from the polarizer.
- Bertrand lenses inserted.

(b) **Properties**

  – **Interference figures**

An interference figure is formed in transparent anisotropic crystals or minerals by the interference of light.

The interference figure is very important in determining optical characteristics of a crystal or mineral—uniaxial or biaxial, optical sign or optical character (positive or negative), 2 V angle for biaxial—and even for estimating mineral chemistry.

The interference figure consists of melatope, isogyres, and isochromes (Fig. 14.22).

– *Melatope* is the point of emergence of the optic axis or optical axes.
– *Isogyres* are the extinguished zones that correspond to areas of the crystal or mineral where the directions of vibration coincide with those of the polarizer and analyzer.
– *Isochromes* are the zones of equal interference color (or retardation) that correspond to areas of the crystal or mineral where their directions of vibration do not coincide with those of the polarizer and analyzer.

Each section of an anisotropic crystal or mineral produces an interference figure with different aspect.

– **Interference figures of uniaxial crystals** (tetragonal, hexagonal, and rhombohedral)
– *Figures with centered optical axis*

This figure is presented by the sections perpendicular to the optical axis (parallel to the crystallographic axis *c*) and does not change in a complete turn of the stage (Fig. 14.23). The Miller indices of these sections are (*00l*).

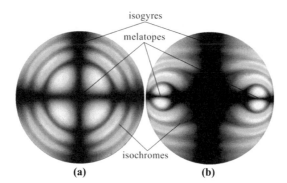

Fig. 14.22 Interference figures showing the melatopes, isogyres, and isochromes **a** uniaxial interference figure; **b** biaxial interference figure (adapted from Olaf Medenbach (https://homepage. ruhr-uni-bochum.de/olaf.medenbach/) with permission)

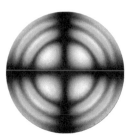

Fig. 14.23 Centered optical axis uniaxial interference figure (from Olaf Medenbach with permission)

– *Flash figure*

A flash figure, also known as an optic normal figure, is present in the sections parallel to the optical axis (Fig. 14.24). The Miller indices for these sections in uniaxial systems are $(h00)$, $(0k0)$, and $(hk0)$.

– *Off-center optical axis interference figure*

The off-center optical axis interference figure is present in the sections that form an angle different from 0° or 90° with the optic axis (Fig. 14.25). The Miller indices of these sections are (hkl).

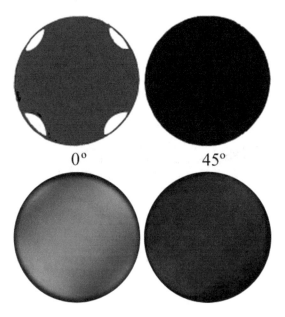

Fig. 14.24 Flash interference figure at 0° and 45°. The bottom shows a real flash figure

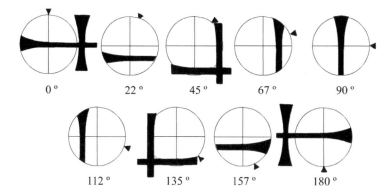

Fig. 14.25 Off-centre optical axis interference figure in uniaxial systems at different turning positions of the microscope stage

- **Interference figures of biaxial crystals or minerals** (orthorhombic, monoclinic and triclinic)

- *Centered acute bisectrix interference figure*

This figure is observed in sections perpendicular to the acute bisectrix—the Z axis if the crystal is (+) or the X axis if the crystal is (−). The two branches of isogyres are clearly different, one branch being broader than the other (Fig. 14.26a). When the microscope stage is turned, the cross is transformed into two hyperbolic curves that move away from each other and come together again (Fig. 14.26b).

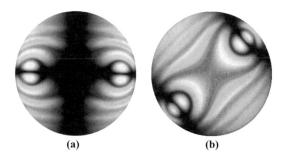

Fig. 14.26 Centered acute bisectrix interference figure: **a** Extinction position. **b** 45° off extinction position (from Olaf Medenbach with permission)

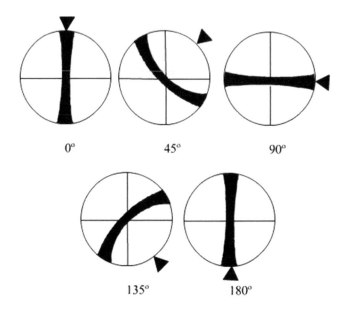

0°   45°   90°

135°   180°

Fig. 14.27 Optical axis biaxial interference figure centred on different rotation positions

– *Centered optical axis interference figure*

This figure is observed in sections that are perpendicular to one of the optical axes of the crystal or mineral (Fig. 14.27).

– *Flash figure*

The flash figure is observed in sections perpendicular to the normal optic (Y axis perpendicular to the optical plane). It is not easy to distinguish it from the uniaxial flash figure.

Box 14.8: Formation of interference figures

The shape of isogyres and isochromes can be understood by relating the directions of propagation of the convergent rays in the crystal to those in the indicatrix. In doing so, the following must be considered:

– In the anisotropic indicatrix, each section represents one of the crystals in which the light propagates perpendicularly to it.

- The components in which the light is split vibrate perpendicularly and are associated with the semi-axes of the corresponding section of the indicatrix.
- At each point in the interference figure, a convergent ray, perpendicular to a given section of the indicatrix, will strike.
- Considering the uniaxial indicatrix as an example, in each section the semi-axes are associated with the directions of the extreme refractive indices.
- In each section, two wave normals with different obliquity, WN1 and WN2, can also be considered (Fig. 14.28).

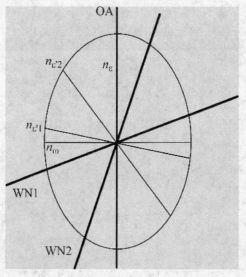

Fig. 14.28 Section parallel to the optical axis showing the wave normals, WN1 and WN2, with different obliquity

- A normal wave is perpendicular to a wave front (surface that joins wave points that are in phase).
- In an isotropic crystal, the normal wave and the direction of propagation coincide.
- In an anisotropic crystal, the normal wave and the direction of propogration do not coincide.
- Each normal is perpendicular to a section of the indicatrix and has a common direction of vibration, corresponding to the ordinary ray (Fig. 14.29).

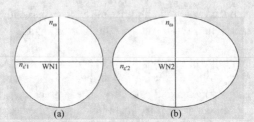

Fig. 14.29 Sections perpendicular to the normal wave WN1 (**a**) and WN2 (**b**), showing the associated refractive indices

- Its associated refractive index, n_ω, is perpendicular to the section.
- The other index of refraction, $n_{\varepsilon'}$, lies in the said section.
- The birefringence is higher in Fig. 14.29b and, therefore, the retardation is too, and the interference colour will be of higher order.
- The birefringence of each indicatrix section of Fig. 14.30 is different, increasing as it moves away from the center of the field of view.

Fig. 14.30 Scheme showing the rays (1, 2, 3, 4) coming from the light source striking the sample on the microscope stage with different obliquity. Each would be perpendicular to a section of the optical indicatrix

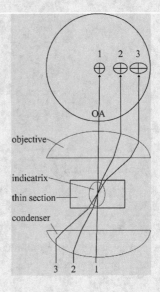

- The isogyres will correspond to the points where the sections have their semi-axes (associated with the extreme refractive indices of the section) coinciding with the directions of vibration of the polarizers (Figs. 14.29 and 14.30).
- The isochromes will correspond to the points where the birefringence is the same and the directions of the refractive indices do not coincide with the directions of vibration of the polarizers (Fig. 14.31).

Fig. 14.31 Scheme of
isogyres and isochromes
formation

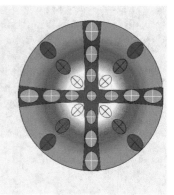

– **Optic sign**

The optic sign or optical character of a crystal or mineral can be positive (+) or
negative (−).

– Uniaxial crystals or minerals are

(+) when $n_\varepsilon > n_\omega$
(−) when $n_\omega > n_\varepsilon$

– Biaxial crystals or minerals are

(+) when $n_\gamma > n_\alpha$ and n_β approaches n_α
(−) when $n_\alpha > n_\gamma$ and n_β approaches n_γ

The optic sign can be determined by different methods. In Box 14.8, the
determination procedure of the refractive index from the interference figure is
exposed.

Box 14.9: Determination procedure of optic sign

The determination procedure consists of:

– Arranging the microscope to observe the interference figure (conoscopy,
  crossed polarizers, Betrand lens).
– Inserting the accessory plate (1st order red, gypsum) into the light path.
– Observing the quadrants where there has been addition or subtraction of
  retardation.

If, in quadrants 2 and 4, there has been subtraction of retardation, a yellow color will be observed when the thickness of the thin section on the microscope is 30 μm. In this case, the optical sign is positive (+), the imaginary line that joins the said quadrants is perpendicular to the direction indicated on the accessory plate for its maximum index, and the maximum refractive index of the mineral section is perpendicular to that of the accessory plate (Fig. 14.32).

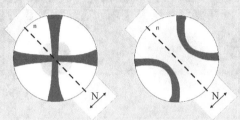

Fig. 14.32 Determination of the optic sign of a positive crystal with gypsum (1st order red) accessory plate

In the two opposite quadrants, SE-NW, there is addition of the retardation, and a blue color is in the NE-SW quadrants.

If, in the NE-SW quadrants, there has been subtraction of retardation, the sign is (−) (Fig. 14.33). The imaginary line that joins the said quadrants is parallel to the direction indicated on the accessory plate for its maximum index, and the maximum refractive index of the mineral section is perpendicular to that of the accessory plate.

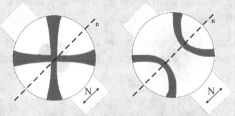

Fig. 14.33 Determination of the optic sign of a negative crystal with gypsum accessory plate

Example

Figure 14.34 shows the interference figure of a uniaxial positive crystal without accessory plate (a) and with gypsum accessory plate (b). Subtraction of retardation can be observed in the NE-SW quadrants, and the imaginary line that joins them is parallel to the direction corresponding to maximum refractive index of the accessory plate so that the maximum refractive index of the mineral section is perpendicular to that of the accessory plate.

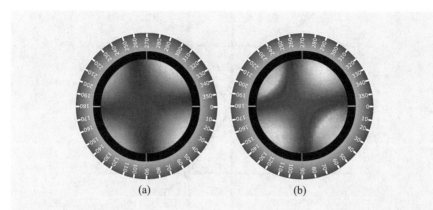

(a)   (b)

Fig. 14.34 Interference figure of a uniaxial positive crystal without accessory plate (**a**) and with gypsum accessory plate (**b**)

– **Optic angle**

The optic angle, 2 V, is the angle between the two optic axes in biaxial crystals or minerals.

It is necessary to distinguish between the angle 2 V and the apparent angle, 2E.

The apparent optic angle, 2E, is the distance separating the two melatopes in an acute bisectrix interference figure, at 45° from extinction position (Fig. 14.35).

This distance depends on 2 V and the refractive index n_β, the refractive index associated to light rays moving along the optic axes. These rays are refracted when they come out of the crystal or mineral at an angle, 2E, which is greater than the 2 V angle (Fig. 14.36).

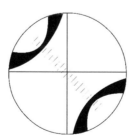

Fig. 14.35 2E measurement

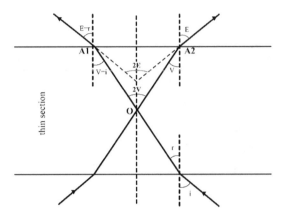

Fig. 14.36 Relation between the angles 2E and 2 V, representing the refraction of two rays that propagate along the optical axes OA1 and OA2 within the crystal or mineral section

14.3   Optical Activity

A section perpendicular to the optical axis of the quartz cut, thicker than the standard, shows a characteristic that cannot be explained through the theory of optical indicatrix. It is the ability to rotate the direction of vibration of the incident beam along certain directions during transmission.

This phenomenon can be seen in the quartz interference figure (Fig. 14.37), in which a cross appears more or less diffused in the centre of the figure, depending on the thickness of the section in which it is observed.

Fig. 14.37 Quartz interference figure (section thickness of 1 mm) (adapted from Olaf Medenbach with permission)

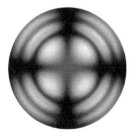

Any crystal belonging to one of the 11 enantiomorphic point groups (Table 14.4), such as quartz, will exhibit optical activity.

Table 14.4 Specific groups associated with optical activity

1	2	3	4	6	23	222	32	422	622	432

A right-hand crystal rotates the vibration counterclockwise. A left-hand crystal rotates the vibration clockwise.

The Airy spirals are the spirals of light visible when polarized light passes through two plates of left-handed and right-handed quartz between crossed polarizers. (Fig. 14.38).

Fig. 14.38 Airy spiral in a composite section with right and left handed quartz superimposed (adapted from Olaf Medenbach with permission)

14.4   Dispersion

The variation of refractive index with the λ can be observed in the colors and figures of interference.

– **Isotropic crystal dispersion**

The types of dispersion that can occur in isotropic crystals are normal and anomalous dispersion. It is usually studied by direct observations of n variation, using monochromatic light of different λ.

– **Uniaxial crystal dispersion**

The types of dispersion in uniaxial crystals are normal, anomalous, and birefringence dispersion.

The variation in the main refractive indices may be different for different λ. It implies possible variation in the shape and size of the indicatrix with the λ. The consequence is that it can change the optic sign in the visible spectrum.

For most crystals, the dispersion of the birefringence is usually small and is considered negligible. If this is significant, it can be detected by measuring refractive indices very accurately at different λ and also by the effects of interference colours observed in white light.

A uniaxial crystal that presents dispersion is characterized by a family of indicatrixes, each corresponding to a specific λ. The direction of the optical axis of each of the indicatrixes of this family is maintained for each λ, due to the symmetry of the crystal.

Although there is no dispersion of the orientation of the indicatrix in uniaxial crystals, the other two types of dispersion remain.

– **Biaxial crystal dispersion**

The types of dispersion that can occur are normal, anomalous, birefringence, optical axis, optical axial plane, and orientation of the optical indicatrix.

The size and shape of this indicatrix can vary with λ. The symmetry exerts control over its orientation.

The biaxial indicatrix is characterized by three binary axes, each perpendicular to a plane of optical symmetry, and it coincides with one of the main axes. In the orthorhombic system, there are three binary axes of symmetry, so there are six possible crystallographic orientations of the optical indicatrix, and each orthorhombic crystal has one of these orientations. In the monoclinic system, there is only one axis of binary symmetry and one of the main axes of the indicatrix must be parallel to it so that the indicatrix has the freedom to rotate around this fixed axis in any position and, for a given crystal, this position must be defined. In the triclinic system, there is no binary axis of symmetry, and the indicatrix can have any orientation.

The dispersion of optical properties with the λ is a general phenomenon, as it has been exposed; however, only in certain transparent crystals or minerals is the effect strong enough to be visible in the figure of interference, being able to appreciate bands of color in the isogyres.

The amount of dispersion can be sorted into the following:

– *noticeable*, if the isogyres show a slight coloring at the edges
– *weak*, if this coloration is more noticeable
– *strong*, if the coloration is stronger
– *extreme*, when the colored fringes cover a large part of the microscope field

– *Dispersion in orthorhombic crystals*

In the orthorhombic indicatrix, there is an independent variation of the three semi-axes *X*, *Y*, and *Z*, implying variation of the partial and total birefringences. Variation of the partial birefringence implies variation of the angle 2 V, *dispersion of the angle* 2 V, or *dispersion of the optical axes*. The dispersion of the optical axes in biaxial crystals is expressed by the *dispersion formula* r > v (Fig. 14.39) or *v > r, which states* whether the optical angle for red is greater or less than the optical angle for violet.

Example
Dispersion r > v (Fig. 14.39).

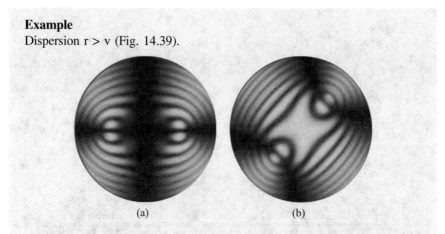

(a)   (b)

Fig. 14.39 Interference figure of orthorhombic crystal, cerussite (PbCO$_3$), with strong dispersión (r > v), **a** extinction position, **b** at 45° from the extinction position (adapted from Olaf Medenbach with permission)

The visible results of this kind of dispersion can be seen in the acute bisectrix interference figures. There may or may not be a change in the optical axial plane if the 2 V angle exceeds 0°. This means that there may be *dispersion of the axial optical plane* and change of optical sign.

Example
An example of axial plane dispersion is shown using brookite (Fig. 14.40).
    The refractive index associated with the direction of vibration parallel to the *c-axis is* n_α for λ less than 550 nm; n_β for λ less than 550 nm; and uniaxial for λ = 550 nm.

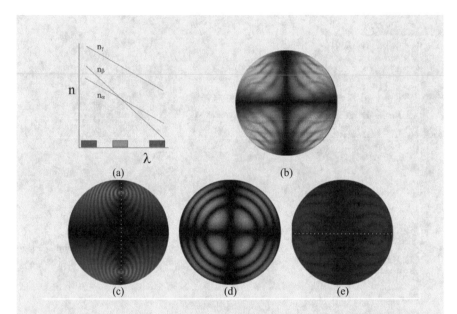

Fig. 14.40 Dispersion of the axial plane in brookite. **a** Dispersion of refractive indices; **b** the interference figure of brookite in white light; **c** interference figure for the blue light with the axial plane north–south direction, **d** interference figure of centered optic axis and **e** interference figure for red light with the axial plane east–west direction (adapted from Olaf Medenbach with permission)

– *Dispersion in monoclinic crystals*

In monoclinic crystals, the ellipsoid can rotate around the binary axis, and this implies that there may be dispersion of the orientation of the indicatrix, which is called *bisectrix dispersion*.

Only one of the main axes *X, Y,* or *Z* coincides with the only axis of symmetry for each; the other two axes must lie in the plane of crystallographic symmetry. There are three possibilities:

1. Acute bisectrix coincides with the crystallographic axis *b, and* the obtuse bisectrix and the normal optic lie in the plane of crystallographic symmetry.
2. Obtuse bisectrix coincides with crystallographic axis *b,* and the acute bisectrix and the normal optical lie in the plane of crystallographic symmetry.
3. Normal optics coincides with the crystallographic axis *b,* and the acute and obtuse bisectrixes lie in the plane of crystallographic symmetry.

With marked dispersion, in the interference figures, the blue fringes mark the output of the optical axes for red, and the red fringes mark the outputs of the optical axes for violet. The following types of dispersion can be observed:

Cross-dispersion when the crystallographic axis *b* coincides with the acute bisectrix.

Example

An example of axial plane dispersion is shown using heulandite (Fig. 14.41).

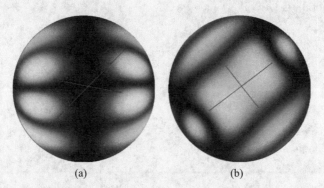

(a)   (b)

Fig. 14.41 Interference figure of orthorhombic crystal, heulandite (CaAl$_2$Si$_7$O$_{18}$), with cross-dispersión (r > v). The area of the isogyres where the optical axes come out for the red is coloured blue and the area of the isogyres where the optical axes come out for the blue is coloured red. **a** Extinction position. **b** At 45° from the extinction position (adapted from Olaf Medenbach with permission)

Parallel or horizontal dispersion when the crystallographic axis *b* coincides with the obtuse bisectrix.

Example

An example of axial plane dispersion is shown by low temperature sanidine (Fig. 14.42).

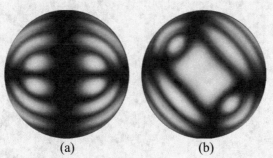

(a)   (b)

Fig. 14.42 Interference figure of monoclinic crystal, low temperature sanidine (KAlSi$_3$O$_8$), with horizontal dispersion. **a** Extinction position, **b** at 45° from the extinction position (adapted from Medenbach Olaf with permission)

Inclined dispersion when the crystallographic axis *b* coincides with the optical normal.

Example

An example of inclined dispersion is shown by high temperature sanidine (Fig. 14.43).

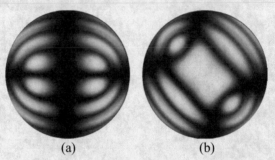

(a)   (b)

Fig. 14.43 Interference figure of a monoclinic crystal, high temperature sanidine (KAlSi$_3$O$_8$) with inclined dispersion. **a** Extinction position, **b** at 45° from the extinction position (adapted from Medenbach Olaf with permission)

Example

In Fig. 14.44, the interference figure of a monoclinic crystal with extreme inclined dispersion is shown.

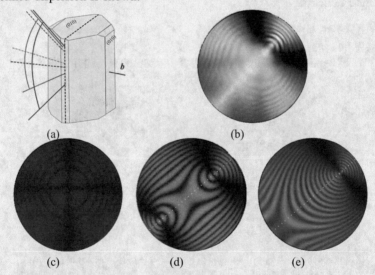

(a)   (b)

(c)   (d)   (e)

Fig. 14.44 Interference figures of a monoclinic crystal with extreme inclined dispersion. **a** 2 V angle for blue, green and red light. **b** Interference figure with white light. **c** Uniaxial interference figure with red light. **d** Biaxial interference figure with green light. **e** Biaxial interference figure with blue light (adapted from Olaf Medenbach with permission)

Dispersion in triclinic crystals

In the interference figure of a triclinic crystal, there is no correlation to symmetry. In these crystals, the crystallographic axes do not generally coincide with the optical directions.

The index ellipsoid can have any orientation with the wavelength so that no symmetry is observed in the interference figure.

These crystals may present dispersion of optical axes and dispersion of the principal vibration directions.

Example
Figure 14.45 presents the dispersion of axial plane for blue, green, and red light.

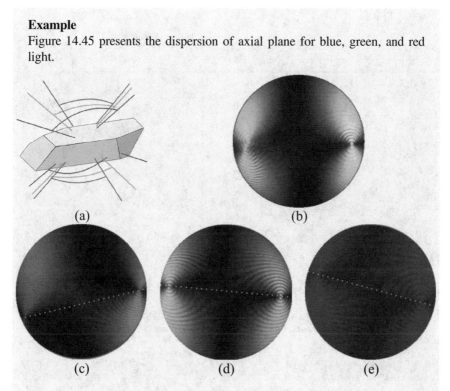

(a)   (b)

(c)   (d)   (e)

Fig. 14.45 Dispersion of a triclinic crystal. **a** 2 V angle for blue, green, and red light. **b** Interference figure in white light. **c** Interference figure for blue light showing the axial plane direction. **d** Interference figure of centered optic axis showing the axial plane direction. **e** Interference figure for red light showing the axial plane direction (adapted from Olaf Medenbach with permission)

Exercises

1. Staurolite

Figure E1 shows the relationship between the crystallographic and optical directions of a staurolite crystal. (a) Indicate the most birefringent section and the section containing the optical axes. (b) Indicate if there is addition or subtraction and annotate the retardation when the section with yellow interference color of first order and retardation of 325 nm is observed with crossed polarizers and the gypsum accessory plate and shows a yellow color of second order.

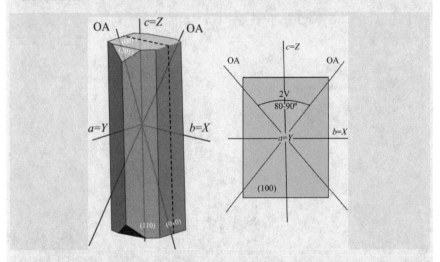

Fig. E1 Relationship between the crystallographic and optical directions of a staurolite crystal

2. Anisotropy, addition, and substraction of retardation

Materials

– Polarizing film
– Anisotropic transparent adhesive tape
– Glass slide

Procedure

(1) Paste a piece of tape in the glass slide.
(2) Place the glass slide between two polarizing films rotated 90° from each other and turn the slide or the upper polarizer to observe the extinction (Fig. E2).

Fig. E2 Glass slide between
two crossed polarizing films

(3) Turn the slide or the upper polarizing film to observe the interference
color (Fig. E3).

Fig. E3 Tape layer at 45°
extinction   position   and
between crossed polarizers

(4) Annotate the interference color and its retardation, using the Michel Lévy
chart (Fig. 14.12a).
(5) Place another tape layer over the previous one, with the same orientation.
Observe with the crossed polarizers the interference color and the retardation in the Michel Lévy chart (Fig. E4). Check that it is double of the
previous one. There has been addition (sum of retardation). The explanation is that both tape slides have the same thickness, and as they are
parallel, their directions of vibration coincide. The second sheet would
have acted as an accessory plate with respect to the first.

Fig. E4 Tape layer over the
previous one with the same
orientation, at 45° extinction
position and between crossed
polarizers

(6) Place another tape layer oriented perpendicularly over the previous one.
Observe with the crossed polarizers the color of interference (Fig. E5).
Observe the retardation in the Michel Lévy chart. Check that it is half of
the previous one. There has been subtraction (subtraction of retardation).
The explanation is that both adhesive tapes have the same thickness and,
as they are perpendicular, their vibration directions do not coincide. The
second tape would also have acted as an accessory plate with respect to
the first.

Fig. E5 Tape layer oriented perpendicularly over the previous one, at 45° extinction position and between crossed polarizers

(7)

    (a) Make schemes indicating the directions of vibration of the polarizers and the adhesive tape, and give the appropriate explanations of how they are related in the different events (2°, 3°, 4°, 5°, and 6°).

    (b) What is the retardation in tape?

Questions

1. In a uniaxial crystal or mineral, the section with higher-order interference colors is the one containing refractive indices
   - ○ a. maximum and intermediate of the crystal.
   - ○ b. maximum and minimum of the crystal.
   - ○ c. minimum and other intermediate between major and intermediate.
   - ○ d. intermediate and minimum of the crystal.
2. The image in Fig. 14.6 shows the observation of a transparent mineral section under the transmission polarizing microscope, with cross polarizers. The observed colors occur because the section is

Fig. E6 Transparent mineral section under the transmission polarizing microscope, with cross polar izers.

    ○ a. anisotropic of a cubic system crystal or mineral.

    ○ b. isotropic of a cubic system crystal or mineral.

    ○ c. isotropic of a crystal or mineral of any crystal system other than cubic.

    ○ d. anisotropic of a mineral of any crystal system other than cubic.

3. Ruby, the red variety of corundum, is a mineral that crystallizes in the rhombohedral system. The refractive indices have been measured in several crystal sections, and the maximum and minimum values obtained are $n_{max} = 1.77$ and $n_{min} = 1.76$. The maximum value has been constant in all sections. Select the answer you think is most correct to define that mineral.

   ○ a. birefringent with two extreme refractive indices
   ○ b. monorefringent
   ○ c. birefringent with three extreme refractive indices
   ○ d. birefringent with an extreme refractive index

4. Ruby, the red variety of corundum, is a mineral that crystallizes in the rhombohedral system. The refractive indices have been measured in several crystal sections, and the maximum and minimum values obtained are $n_{max} = 1.77$ and $n_{min} = 1.76$. The maximum value has been constant in all sections. The refractive index (indices) that could have been measured in the section perpendicular to the optical axis (axes) could have been

   ○ a. η_{max}
   ○ b. one intermediate index between n_{min} and n_{max}
   ○ c. n_{min}
   ○ d. n_{min} and n_{ma}

5. Can a red mineral be isotropic? (Yes or no).

   Response: [   ]

6. Can a red mineral with a single refractive index value be isotropic? (Yes or no).

   Response: [   ]

7. Indicate the optical character (isotropic or anisotropic) of a mineral that crystallizes in the hexagonal system.

   Response: [   ]

8. Of what order will the interference colors of a monorefringent crystal with very low birefringence be?
   Respond with one of the following terms: high, low, medium.

   Response: [   ]

9. What is the name of the type of extinction observed in a mineral section when some trace of cleavage or crystal plane of the same does not coincide with one of its directions of vibration?

    Response: [   ]

10. The epidote, $Ca_2Fe^{3+}Al_2(OH)Si_3O_{12}$, crystallizes in the monoclinic system with the Y axis of the indicatrix coinciding with the crystallographic axis b. It is colourless to pale yellowish green. The optical sign is negative and the refractive indices are $n_\gamma = 1.734$, n_β s 1.725, $n_\alpha = 1.715$.
    What optical property is related to the fact that the acute bisectrix coincides with the X axis of the optical indicatrix?

    Response: [   ]

11. What are Miller indices of the uniaxial section that displays the flash figure, and what section of the uniaxial indicatrix is it related to?

    Response: [   ]

12. Write the Miller indices of the most birefringent section of a uniaxial mineral or, in the case of biaxial minerals, by *means of the term axial* (since in these minerals the most birefringent section contains the two optical axes, i.e., that of the optical axial plane), of the following minerals: Fluorite (cubic), calcite (rhombohedral), beryl (hexagonal), sanidine (monoclinic), topaz (orthorhombic).

    Response: [   ]

13. The optical sign of a mineral can vary with the wavelength if refractive indices change.

    ○ True
    ○ False

14. Write the Miller indices for the more birefringent section of a tetragonal crystal.

    Response: [   ]

15. The centered optic axis interference figure is shown by the sections

    ○ a. perpendicular to normal optics
    ○ b. perpendicular to optical axis(s)
    ○ c. perpendicular to obtuse bisectrix
    ○ d. perpendicular to acute bisectrix

Reference

1. Sørensen BE (2013) A revised Michel-Lévy interference colour chart based on first-principles calculations. Eur J Mineral 25:5–10

Chapter 15
Optical Properties of Opaque Crystals

Systematic Study with a Polarizing Reflection Microscope

Abstract In this chapter, the properties of opaque crystals and minerals using the reflection polarization microscope are described, both in orthoscopic arrangement with polarized light and analyzed polarized light (cross-polarizers) and with conoscopic light. The dispersion of the reflectance and its relation with the color in reflection is described. In the case of anisotropic crystalline sections, the different dispersion of both reflectances and their relation with the reflection pleochroism is described.

15.1 Orthoscopic Arrangement of the Microscope

15.1.1 Observations with Polarized Light

(a) **Requirements**

- Low (2.5x) or medium (10x, 20x) magnification objectives.

(b) **Properties**

- Color

The color in reflection under the microscope is due to the scattering of the reflectance with the wavelength and can be expressed quantitatively by means of the scattering curve in the visible range of the electromagnetic spectrum. It varies between gray and white, with different shades. The color of a given mineral may appear different, depending on the minerals around it.

© The Author(s), under exclusive license to Springer Nature Switzerland AG 2022   355
C. Marcos, *Crystallography*, Springer Textbooks in Earth Sciences,
Geography and Environment, https://doi.org/10.1007/978-3-030-96783-3_15

Example
Blue (Fig. 15.1).

Fig. 15.1 Covellite (2.5x,
polarized light)

Yellow (Fig. 15.2).

Fig. 15.2 Chalcopyrite (2.5x,
polarized light)

Light yellow (Fig. 15.3).

Fig. 15.3 Pyrite (2.5x, polarized light)

Coppery red (Fig. 15.4).

Fig. 15.4 Copper (2.5x,
polarized light)

Orange-yellow (Fig. 15.5).

Fig. 15.5 Nickelite (2.5x,
polarized light)

– **Hardness and relief**

These are relative terms. When two minerals have similar hardness, they show very weak boundaries, indicating there is no difference in relief between them. However, when a mineral is harder than a neighboring one, polishing lowers the softer one and highlights the harder one.

At the junction of a hard and a soft mineral grain, there tends to be a kind of step that makes the bright line of Kalb visible.

Soft minerals, below 2 on the Mohs scale, acquire a smooth and bright polish. Examples are stibnite and covellite. However, in molybdenite or graphite, it is difficult to remove the grainy, matte appearance.

Minerals with medium hardness, 3–5 Mohs scale, polish well and quickly.

Hard minerals, above 5, as in magnetite, ilmenite, or nickel, acquire a soft polish quickly, while in arsenopyrite or wolframite, more polishing time is required to remove surface irregularities.

In general, soft minerals maintain a finely striped pattern until the end of the polishing. Medium-hard minerals generally show a homogeneous polished surface, and hard minerals show the marks of the coarser abrasive used in polishing.

Differences in the behavior of minerals when polishing often make it difficult to obtain a well-polished specimen.

The Kalb line is a bright line that appears at the junction between two soft minerals and a harder one. It allows us to determine which is harder and which is softer.

Box 15.1. Observation Procedure of the Kalb Line
The observation procedure consists of the following:

1. Focusing on the mineral section.
2. Defocusing slightly on the mineral section; increasing the distance between the objective and the section on the microscope stage.
   In this situation, the bright line of Kalb will be observed and will be introduced into the softer mineral.
   It will be introduced in the harder mineral by decreasing the distance between the objective and the mineral section.

– Cleavage and Parting

Example
Hematite (Fig. 15.6).

Fig. 15.6 Cleavage in hematites (2.5x, polarized light)

Galena
Galena displays a characteristic triangular parting that appears as a result of polishing (Fig. 15.7).

Fig. 15.7 Cleavage in galena (20x, polarized light)

– Exsolution

Exsolution is the process through which a solid solution separates into at least two different minerals. The component in smaller proportion is usually included in the one that is in greater proportion. Exsolution is best observed with crossed polarizers.

Example
Ilmenite (Fig. 15.8).

Fig. 15.8 Exsolution of
hematite in ilmenite (20x,
polarized light)

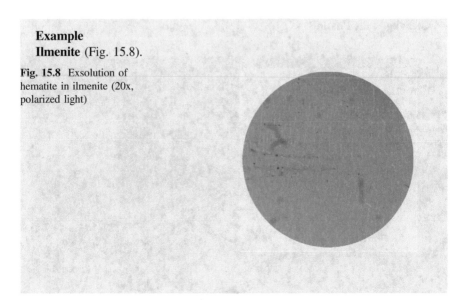

– **Form**

Form is a morphological character that can sometimes be useful in the study of opaque minerals. Many metallic minerals seem to have no defined shape (xenomorphic). Others are presented with well-defined shapes (idiomorphic). They are usually the hardest minerals. Those with high melting temperatures tend to develop crystal habits (idiomorphic). Certain minerals have a very marked tendency toward idiomorphism when isolated, such as pyrite and magnetite, or separate groups with rounded edges. There are minerals that are elongated or flattened, such as covellite, molybdenite, or hematites, for example. Others are irregular in shape and are called allotriomorphic, such as sphalerite, galena, nickelite, bornite, chalcopyrite, and chalcocite. Some minerals formed at low temperatures tend to show colloidal textures consisting of aggregates arranged in concentric, often convex or spherulitic layers, with an opal-like appearance. Example includes goethite and sphalerite.

This property is related to the origin of the mineral but is sometimes so constant that it is very useful to identify it.

– **Inclusions**

Inclusions can be solid, liquid, or gaseous phases trapped in the host mineral. They are formed before (protogenetic), during (syngenetic), or after (epigenetic) the host mineral.

– **Grains and Twins**

Twins and grains are best observed with crossed polarizers.

– **Reflection pleochroism**

Reflection pleochroism is a color variation or color intensity with direction, and it is due to the difference in dispersion with the wavelength of the reflectances of the mineral.

Example
Covellite
Very pronounced reflection pleochroism (Fig. 15.9).

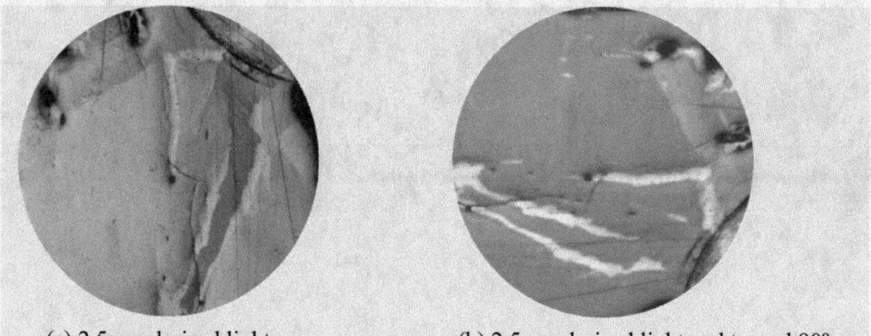

(a) 2.5x, polarized light.   (b) 2.5x, polarized light and turned 90°.

Fig. 15.9 Very pronounced reflection pleochroism in covellite

Ilmenite
Moderate reflection pleochroism (Fig. 15.10).

Fig. 15.10 Moderate reflection pleochroism in ilmenite (2.5x, polarized light)

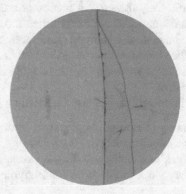

Marcasite
Weak reflection pleochroism (Fig. 15.11).

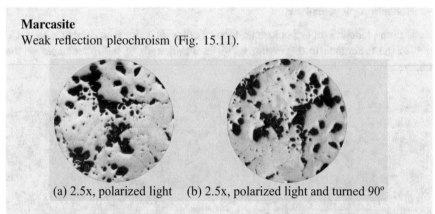

(a) 2.5x, polarized light   (b) 2.5x, polarized light and turned 90°

Fig. 15.11 Weak reflection pleochroism in marcasite

– **Bireflection**

Bireflection is the change in light intensity with direction. It is a manifestation of anisotropy. It depends on the difference between the two reflectances. The maximum bireflection of a section does not have to coincide with the maximum of the mineral. Only the vertical section of a uniaxial mineral and the section parallel to the maximum and minimum reflectances of an orthorhombic, monoclinic, or triclinic mineral will show the maximum bireflection for the mineral in question.

When bireflection is observed well defined in a mineral, it can be useful to relate the principal vibration directions with visible morphological characteristics such as exfoliation.

It is difficult to estimate bireflection, but three degrees can be distinguished:

1. Strong to medium bireflection: Strong, as in molybdenite, covellite, marcasite, and stibnite; medium, as in ilmenite, pyrrhotite, and nickel
2. Weak bireflection: Lollingite, arsenopyrite, and hematite.
3. Very weak bireflection: Chalcopyrite.

– **Reflectance**

Reflectance represents the amount of light that is absorbed as it passes through successive layers of constant thickness in the mineral. Reflectance is the ratio between reflected light and incident light, expressed as a percentage.

$$R\% = I_R/I_I. \tag{15.1}$$

Reflectance depends on the refractive index of the mineral, the refractive index of the medium in which it is found, the section orientation, and the absorption coefficient, in the case of opaque minerals.

Eye estimation can quickly establish an order of reflectance in ore minerals in a test tube, although it can be affected by color differences.

Reflectance in white light is categorized as follows:

Reflectance	Examples
Very high	Native elements such as platinium (71%), gold (72%), silver (85%), löllingite (53.5%), safflorite (53%), arsenopyrite and marcasite (52%), pyrite (51%)
Medium to high	Stibnite (47%), galena (43%), molybdenite (42%)
Medium to low	Hematite (30%), covellite (23%), digenite (22%)
Low	Ilmenite (19%), goethite (18%)

– **Zoning**

Zoning consists of compositional variation due to the segregation of the chemical components during crystal growth. Color and other optical properties may vary as a result. Zoning is best observed with crossed polarizers.

15.2   Observations with Crossed Polarizers (Polarized and Analyzed Light)

(a) **Requirements**

- Low (2.5x) or medium (10x) magnification objectives.
- Analyzer inserted and rotated 90° from the polarizer (crossed polarizers, like in transmission).

(b) **Properties**

Some properties such as exolution, twinning, or zoning are often better appreciated with cross-polarizers.

– **Internal reflections**

Internal reflections are reflections produced by light reflected from, for example, inclusions and fractures, of the mineral, when it is not totally opaque.

Example
See Fig. 15.12.

Fig. 15.12 Internal
reflections in cinnabar
(2.5x, cross-polarizers)

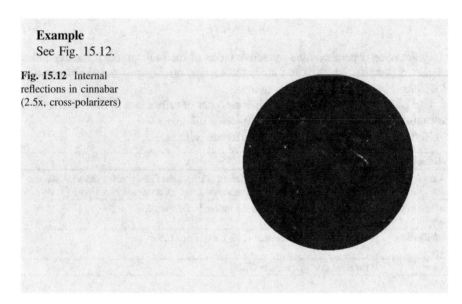

– **Isotropy-Anisotropy**

In optical isotropic minerals, the sections are extinct in a full stage rotation. Extinction in reflection does not imply total darkness, as occurs with crystals or transparent minerals in transmission. Some light is transmitted by the analyser, as a consequence of the weak ellipticity produced in the reflection of such minerals when the incident light is not totally perpendicular to the surface. However, the intensity of such light is the same in a complete turn of the microscope stage.

– Weakly anisotropic minerals. These minerals show a slight change with rotation, and it can be observed best if the polarisers are slightly uncrossed.
– Strongly anisotropic minerals. These minerals show a pronounced change in brightness and also a possible change in color with rotation.

In opaque anisotropic minerals, it is necessary to distinguish the behavior of symmetric and asymmetric sections between crossed polarizers.

Symmetrical sections are extinguished in white light. When a symmetrical section is rotated on the stage with respect to its extinction position, the resulting vibration reflected is rotated and is called *anisotropic rotation*. The rotation is always toward the higher amplitude vibration, and the maximum angle of rotation is below 45° of stage rotation. It is manifested by the polarization colors, which change when the section is rotated. The amplitude of the light transmitted by the analyser is proportional to the *anisotropy ratio* ($A = R_2/R_1$) since this ratio determines the angle of the anisotropic rotation.

Asymmetrical sections are not extinguished in the strict sense of the word, even in monochromatic light, due to the marked ellipticity of the vibrations.

– **Polarization colors**

Polarization colors are also known as *anisotropic tints*. Bright tints indicate strong dispersion of the anisotropy ratio.

Examples
Löllingite (Fig. 15.13).

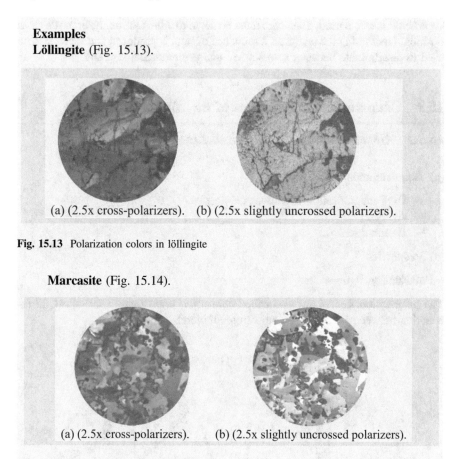

(a) (2.5x cross-polarizers).   (b) (2.5x slightly uncrossed polarizers).

Fig. 15.13 Polarization colors in löllingite

Marcasite (Fig. 15.14).

(a) (2.5x cross-polarizers).   (b) (2.5x slightly uncrossed polarizers).

Fig. 15.14 Polarization colors in marcasite

– *Polarization colors in symmetrical sections*

The tint observed for a given angle, by clockwise rotation of the stage, is the same as that observed for the same angle of counter-clockwise rotation.

A very sensitive test to distinguish symmetrical sections consists of slightly uncrossing the analyzer (1–5°) and observing the sequence of dyes produced that can be diagnostic criteria for certain minerals. Previously, a very precise adjustment of the analyzer is required.

– *Polarization colors in asymmetrical sections*

In asymmetrical sections, the tint observed for a given angle, by clockwise rotation of the stage, is not the same as that observed for the same angle of counter-clockwise rotation.

When this test is used, great care must be taken to adjust the analyzer to the cross position. Uncrossing the analyzer, a sequence of tints is produced and it can also be used to discriminate between symmetrical and asymmetrical sections.

15.3   Conoscopic Arrangement of the Microscope

15.3.1   *Observations with Polarized Light*

(a) **Requirements**

- High magnification objectives (> 40x).
- Analyzer.
- Bertrand lenses.

(b) **Properties**

– **Polarization figures**

The polarization figures are also called *convergent light figures* and allow differentiation between isotropic and anisotropic minerals (Fig. 15.15).

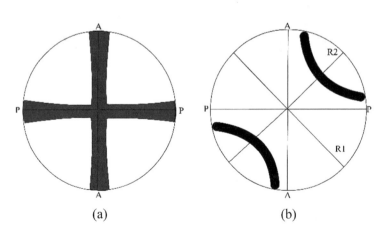

(a)   (b)

Fig. 15.15 Polarization figure, **a** isotropic and anisotropic sections in the extinction position; **b** anisotropic sections at 45° from the extinction position

– *Polarization figure in isotropic minerals*

The figure of isotropic minerals is a black cross. By crossing the polarisers slightly, the figure splits into two branches.

– *Polarization figure in anisotropic minerals*

The polarization figure of the anisotropic minerals is a cross that splits into two branches, when the microscope stage is rotated, which move in two opposite quadrants, associated with the vibration with the higher reflectance. Between crossed polarizers, with the plate in 45° position, separation of the isogyres indicates the amount of anisotropy of the section. Example: Chalcopyrite in a vertical section shows the isogyres barely open, and this indicates very little anisotropy.

Among other uniaxial minerals in a vertical section, molybdenite shows a great anisotropy without any kind of dispersion.

15.4 Dispersion

15.4.1 Color and Dispersion Effects

In many anisotropic minerals, the reflectance varies considerably according to the wavelength of the incident light, so the reflected beam has a color that is very characteristic. There is almost always a substantial mixture of white light, the result being a metallic tint like bronze or blue steel. Special caution is required in observing these effects because the appearance may be deceptively altered by contrast with a neighboring grain of another color or even by the nature of the illumination.

Dispersion of the optical constants produces dispersion of the reflectance and, if its curve has a maximum within the visible range, the substance appears colored (the type of dispersion curve and its position on the scale of reflectance can define the color of the reflection).

In opaque minerals, even with a small absorption, the dispersion of the absorption and, therefore, of the reflectance produces a marked color when the substance is observed with reflected light.

In anisotropic sections, the reflectance of one vibration can be dispersed independently of the reflectance of another. If there is a difference, it can be observed that the section changes color when the microscope stage is rotated, with a polarizer only. This phenomenon is the reflection pleochroism. In transmission, the observed color is the result of the combination of transmitted colors (not absorbed), whereas, in reflection, the dominant color is that for which absorption is greater, because this factor has a greater influence on the reflectance than on the refractive index.

Color and pleochroism are more evident in some minerals (e.g., covellite) when observed with oil immersion, due to the large difference in refractive indices. In the covellite and for a certain wavelength, the R in oil is higher than in air and the color changes sharply from blue to red.

15.4.2   Anisotropic Effects Between Crossed Polarizers

– Anisotropic effects in symmetrical sections

If the anisotropy ratio is dispersed, the amplitude transmitted by the analyzer varies for different wavelengths and the polarization colors are observed.

– Anisotropic effects in asymmetric sections

The asymmetrical sections are not extinguished in the strict sense of the word, even in monochromatic light, due to the marked ellipticity of the vibrations. But where the relation of ellipticity is nil, such sections will be extinguished in monochromatic light and may not be extinguished in white light. This is due to the dispersion of the direction of vibration.

15.4.3   Dispersion in the Polarization Figures

– *Dispersion in the polarization figures of isotropic sections*

The coloring of the isogyres of an isotropic section polarization figure reveals reflection-rotation dispersion.

The intensity of the color fringes is a measure of the amount of dispersion.

Minerals that have weak dispersion show black isogyres when observed with white light, while those with strong dispersion show isogyres with bright colors.

The field of view will show color in the quadrants only if the dispersion is strong and usually only where the angle of rotation is very large, i.e., near the edge of the field.

The red zones indicate that the rotation is greater for the red light than for the blue light (r > b) and vice versa for the blue zones. For other colors, it can only be deduced that the dispersion is strong.

Even by uncrossing the analyzer, the weakest dispersion can be detected. Generally, a few degrees turn is enough.

If the isogyres are colored blue in the convex part and red in the concave part, it can be deduced that the reflection-rotation is dispersed with r > b, and vice versa for the red in the convex part and the blue in the concave part.

The color that appears on the outermost (concave) side of the isogyre is that of the extinguished light closest to the center and, therefore, the section has greater rotation for this light. The extent of the coloring indicates the degree of dispersion.

In this development, it has been assumed that reflection in oblique incidence has not produced any ellipticity in the reflected vibration. However, in metals such as gold, silver, and copper, there is a marked ellipticity and the angle of incidence becomes significant. Thus, the rotation of the analyser can cause separation of the isogyres, becoming blurred, while the dispersion of the ellipticity (DE) will be shown as a coloration of the isogyres.

– *Dispersion in the polarization figures of anisotropic sections*

The effect of reflection-rotation dispersion is mixed with the dispersion of anisotropic rotation.

In general, the reflection-rotation dispersion is detected by a small rotation of the analyzer in the extinction position, while with crossed polarizers and the section at 45° from the extinction position, the combined effect is observed.

With anisotropic sections in the extinction position, a black cross is formed, and reflection-rotation is the only one that acts. The analyser can be crossed in the usual way in order to detect any dispersion of this rotation.

The analyzer is reset to the crossed position and the microscope stage is turned to the 45° position; this produces the crossing and splitting into two isogyres. Now, the effects observed are due to the combination of two dispersions, the reflection-rotation and the anisotropic rotation.

As the main effect is the coloring of the isogyres, small variations in the angle of rotation for different colored lights must be considered.

Questions

1. The polarization colors are displayed by

   ○ a. the orthorhombic, monoclinic and triclinic opaque minerals and are due to light interference

   ○ b. all opaque minerals and are due to light interference

   ○ c. all opaque minerals and are due to light reflection

   ○ d. opaque orthorhombic, monoclinic and triclinic minerals and are due to light reflection

2. The coloration of an opaque mineral observed under the reflection microscope is due to the

   ○ a. existence of a maximum value of the refractive index for a given wavelength

   ○ b. existence of a maximum value of the bireflectance for a given wavelength

   ○ c. existence of a maximum value of the reflectance for a given wavelength

   ○ d. existence of a maximum value of the absorption coefficient for a given wavelength

3. The polarization figures are used to distinguish between

    ○ a. uniaxial and biaxial anisotropic transparent minerals
    ○ b. isotropic and anisotropic opaque minerals
    ○ c. isotropic and anisotropic transparent minerals
    ○ d. uniaxial and biaxial anisotropic opaque minerals

4. Reflection pleochroism is observed under a polarizing microscope of the

    ○ a. transmission with polarized light
    ○ b. transmission between cross-polarizers
    ○ c. reflection with polarized light
    ○ d. reflection between cross-polarizers

5. What are the names of colors observed in anisotropic minerals between crossed polarizers with a polarizing reflection microscope?

Response: [   ]

6. The coloring of the isogyres of an isotropic section polarization figure reveals reflection-rotation dispersion.

    ○ True
    ○ False

7. Write the name of the device that allows deviation of the light coming from the light source through the objective toward the polished sample located on the stage of the polarizing reflection microscope.

Response: [   ]

8. Can twins be seen on an opaque mineral under a polarizing reflection microscope?

Yes/No Response: [   ]

9. An opaque orthorhombic mineral can be called biaxic because it has two optical axes.

    ○ True
    ○ False

10. Opaque isotropic crystals display polarization figures.

    ○ True
    ○ False

Chapter 16
Electrical, Magnetic, Mechanical, and Elastic Properties

Abstract This chapter is devoted to electrical, magnetic, and mechanical properties of the crystals. Electrical properties of the crystals are important, since the pyroelectric and piezoelectric phenomena are widely used in optic-electronic, radio-electronic, electro-acoustic, energy conversion techniques, etc., mainly in relation to ferroelectric crystals. Pyroelectric and piezoelectric effects are explained. Finally, ferroelectric and anti-ferroelectric crystals and the peculiarities of their electrical properties due to the presence of domains are explained. The basic relations that characterize the behavior of a crystal in a magnetic field are presented, defining magnetic susceptibility and permeability and magnetic moment. Later, and from a crystallographic point of view, using symmetry as a method, magnetic crystals will be classified on the basis of the absence or presence of order of the magnetic moments of the atoms in the crystals. Thus, diamagnetism and paramagnetism and the different types of magnetic structures in crystals, such as ferromagnetism, antiferromagnetism, and ferrimagnetism, are introduced. Finally, the mechanical properties of crystals, such as exfoliation and hardness, are described. Two very important properties in relation to the stability of minerals in the earth's crust and mantle—thermal expansion and compressibility—are also described.

16.1 Electrical Properties

16.1.1 Pyroelectricity

Pyroelectricity consists of the displacement of positive and negative charges in a mineral by a change of temperature, T, that is, the property that possesses a mineral to produce positive and negative charges at the end of its faces when it undergoes a change in temperature (when two of its outermost faces are heated). If the temperature change is done in reverse, the charges on the crystalline faces also change signs.

This property is presented by crystals and minerals whose point group is polar with only one single polar rotational axis or one mirror plane, which allows polarity

Table 16.1 Point groups of crystals exhibiting pyroelectricity

Crystal system	Point group
Triclinic	1
Monoclinic	2, *m*
Orthorhombic	*mm*^{2}
Tetragonal	4, 4*mm*
Trigonal	3, 3*m*
Hexagonal	6, 6*mm*

like point group *m*. Finally, out of 21 non-centrosymmetric point groups, there are only the following 10 pyroelectric point groups (Table 16.1):

An example is tourmaline.

16.1.2   *Piezoelectricity*

Piezoelectricity is the appearance of a dipolar moment in a mineral under stress.

Piezoelectricity is a phenomenon exhibited by certain crystals as low quartz or mineral salts, such as Rochelle[1] or Seignette salt (potassium sodium tartrate tetrahydrate).

When subjected to stress, charges are produced (Fig. 16.1), or mechanical deformation/stress results in electrical potential (e.g., low quartz, with point group 32).

Fig. 16.1 Electric charges or a potential difference is produced when a piezoelectric crystal is subjected to mechanical stress

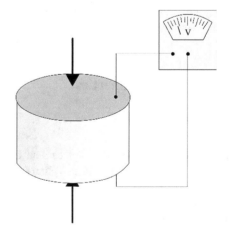

---

[1] "La Rochelle" is the harbour city in France where Pierre Seignette discovered the growth of potassium sodium tartrate.

Fig. 16.2 Orthonormalized physical system X–Y axis and two-fold polar rotation axes of trigonal quartz

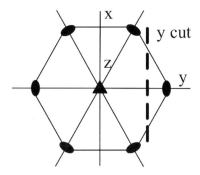

In the second case, the vibration they produce, to a certain and specific frequency, depends on the thickness and orientation of the slice that has been cut from the piece of crystal. In that case, it is immediately identified because it produces an alternating electrical current, and the crystal can replace an oscillating circuit. Since the frequency produced depends on the thickness and orientation of the cut, high-frequency fixed oscillators will be obtained, which are widely used in many electronic devices as clock signals that regulate computers and quartz clocks.

The low quartz appears as trigonal crystals, showing its three axes (Fig. 16.2a).

Figure 16.3a shows a simple scheme of the negative and positive ions in quartz. The polarity of the axes can be seen because the positive charges are shown at one end of the axes and the negative charges at the other end. When the structure is compressed in the direction of the polar axis E1, the structure deforms, displacing the charges in such a way that the positive charges predominate on one side and the negative charges on the opposite side (Fig. 16.3b). When pressure is applied perpendicular to the electric axis (Fig. 16.3c), opposite charges appear again on opposite sides, but in this case, the charges are interchanged with those in Fig. 16.3b.

Only minerals and crystals in 20 point groups that have polar axes present this property (Table 16.2).

Some applications that can be listed are as follows:

- Piezoceramics. Inside they have a piezoelectric crystal that is hit abruptly by the ignition mechanism. This dry blow causes a high electric current capable of creating a voltaic arc or spark that will ignite the lighter.
- Vibration sensor. Each of the pressure variations produced by vibration causes a current pulse proportional to the force exerted. An electrical signal ready to amplify from a mechanical vibration has been obtained, by simply connecting an electrical cable to each side of the crystal and sending this signal to an amplifier, for example, piezo pickups.
- Communication technologies.

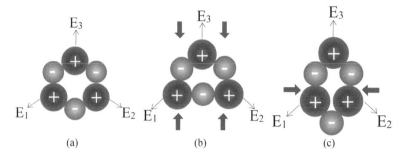

Fig. 16.3 Scheme of the piezoelectric effect of quartz, **a** simple scheme of the negative and positive ions in quartz; **b** structure is compressed in the direction of the polar axis E1; **c** structure is compressed perpendicularly to the direction of the polar axis E1

Table 16.2 Point groups with polar axes

Cubic	Tetragonal	Hexagonal	Trigonal	Orthorhombic	Monoclinic	Triclinic
23, $\overline{4}$ 3m	4, $\overline{4}$ 4mm 422, $\overline{4}m$	6, $\overline{6}$ 6mm 622, $\overline{6}m2$	3 3m 32	mm2 222	2m	1

16.2 Magnetic Properties

16.2.1 Introduction

Certain minerals have the property of behaving like a magnet, due to the interaction of magnetic dipoles on an atomic scale.

Magnetism is the property that certain minerals possess to attract iron and its derivatives.

Only atoms with incomplete $3d$ orbitals, such as the elements of the first transition series, (Cu, Fe, Ti, Ni, Co, Mn, and V, for example) behave as magnetic dipoles.

Natural magnets are permanent because they maintain their ownership of attraction without the need to apply magnetizing forces. The entire area where the magnetic properties of a magnet operate is called a magnetic field, which is characterized by numerous lines of force.

Magnetite is a natural magnet that has been known for a long time. Each electron has two quantum numbers as follows:

– *Orbital moment:* It describes orbit behavior around the nucleus.
– *Spin moment:* It describes the behavior of the electron spin.

Both can generate magnetic dipoles.

Therefore, spin is primarily responsible for the magnetic properties of atoms and molecules.

In 1820, Oersted discovered the relationship between the magnetic properties of a magnet and electricity, when he found that an electric current produced a magnetic field around it. This property and the induction property are used for the construction of various electrical equipment, such as motors, dynamos, measuring devices (voltmeters and ammeters), and electromagnets, among others. The first known magnet was the magnetite, very abundant in the region of Magnesia, from which it gets its name, cited by Platon and Plinio, and in which its natural magnet properties could be observed from very remote times.

16.2.2   Types of Minerals According to Magnetic Properties

A mineral is magnetized when subjected to a magnetic field.

$$I = \chi H, \tag{16.1}$$

where

I   is the magnetic induction,
H   is the magnetic field,
χ is magnetic susceptibility, and
I and H   have the same direction.

In 1948, Taggart established a classification of minerals based on their magnetic permeability relative to iron, which is considered the most magnetic substance, as shown in Table 16.3.

– **Paramagnetic minerals**
Paramagnetic minerals are those that have elements of the first transition series, which are the elements that produce magnetic moments (Fe^{3+} and Mn^{2+}, with five unpaired electrons, between the most magnetic ions). The magnetic dipoles of these minerals align while subjected to a magnetic field.

Examples include Olivine $(Mg, Fe)_2SiO_4$ and augite $(Ca, Na)(Mg, Fe, Al)(Al, Si)_2O_6$.

– **Diamagnetic minerals**
Diamagnetic minerals have no magnetic behavior, as electrons with opposing spins are paired.

Examples include Bismuth (Bi), calcite $(CaCO_3)$, albite $(NaAlSi_3O_8)$, quartz (SiO_2), and apatite $(Ca_5(PO_4)_3(F,Cl,OH))$.

Table 16.3 Relative magnetic permeability of minerals

Magnetism type	Mineral	Relative permeability
Ferromagnetism	Iron	100
	Magnetite	40.18
	Franklinite	35.38
	Ilmenite	24.70
Paramagnetism	Pyrrhotite	6.69
	Siderite	1.82
	Hematite	1.32
	Goethite	0.84
	Pirolusite	0.71
Diamagnetism	Garnets	0.40
	Quartz	0.37
	Cerussite	0.30
	Pyrite	0.23
	Dolomite	0.22
	Arsenopyrite	0.15
	Magnesite	0.15
	Clalcopyrite	0.14
	Gypsum	0.12
	Cinnabar	0.10
	Cuprite	0.08
	Smithsonite	0.07
	Orthoclase	0.05
	Galenite	0.04
	Calcite	0.03

– **Ferromagnetic minerals**

Ferromagnetic minerals present a spontaneous magnetic moment, produced by parallel orientation of the spins of the ions of the structure ↑↑↑↑↑↑↑↑.

Examples include FeO, CoO, NiO, and MnO.

Areas in which dipoles are oriented in a certain way and other areas in which they are oriented differently may appear in substances such as metal iron. These zones, called *domains,* are aligned according to the field when the substance is subjected to an external magnetic field (Fig. 16.4).

– **Ferrimagnetic minerals**

Ferrimagnetic minerals are those minerals in which the moments of ion spin are anti-parallels ↑↓↑↓↑↓↑↓. Examples include magnetite series (Fe_3O_4)—ulvoespinel ($TiFe_2O_4$), hematite series (Fe_2O_3)—ilmenite ($FeTiO_3$), and pyrrhotite series ($Fe_{1-x}S$).

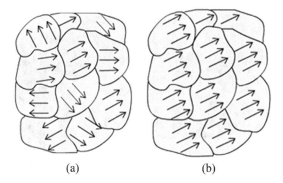

(a)   (b)

Fig. 16.4 Scheme showing random magnetic domains (**a**) and alignment of magnetic domains (**b**)

16.3 Mechanical Properties

16.3.1 Cleavage

Cleavage is the rupture of a crystal or mineral according to certain crystallographic planes, which have the weakest bonds.

Examples

Cubic: {100}
Example: Galenite, halite
pyrite (bottom)

(continued)

(continued)

Octahedral: {111} Examples: Fluorite (top), diamond (center), and pyrite (bottom)	
Rhombohedral Example: Calcite {10$\bar{1}$1}	
Pinacoidal or basal: (001) Example: Mica	

16.3.2 Tenacity

Tenacity is another property, like cleavage, which depends on reticular cohesion and defines the way a mineral deforms under a mechanical action.

The degree of tenacity can be expressed in terms of the following:

- Brittleness—it breaks easily. An example is sulfur (S).
- Malleability—plastic deformation without rupture can be molded into leaves. An example is metals.
- Ductility—it can be plastically deformed (stretched in the form of a yarn). An example is copper (Cu).
- Sectility—it can be cut into chips with a knife. An example is gypsum ($CaSO_4.2H_2O$).
- Flexibility—it can be folded without recovering the original shape. An example is talc.
- Elasticity—it regains the original shape by ceasing the force that deforms it. An example is biotite (K $(Mg, Fe)_3AlSi_3O_{10}(OH)_2$).

16.3.3 Hardness

Hardness is the property that minerals exhibit to resist abrasion and scratching. It depends on molecular cohesion but to a different degree than tenacity.

Hardness is measured in a practical way in relation to the scale designed by the Austrian geologist Friedrich Mohs (1773–1839), with the use of a series of mineral tips; each mineral strikes the one that precedes it and is scratched by the one that follows it.

Mohs scale: 1 Talc, 2 Gypsum, 3 Calcite, 4 Fluorite, 5 Apatite, 6 Orthoclase, 7 Quartz, 8 Topaz, 9 Corundum, and 10 Diamond.

There is also another more empirical but more practical scale, according to which minerals are classified as follows:

- Very soft minerals that can be scratched with the nail (hardness from 1 to 2).
- Soft minerals that scratch with a copper coin (2–3).
- Semi-hard minerals that scratch easily with a pen cutter (5–6.5).
- Very hard minerals that do not scratch with a sheet of steel.

The hardness of a mineral undergoes large changes with direction and will exhibit different values, depending on which direction the measurement is taken when scratching a mineral. Most minerals do not see such variation, except in kyanite and calcite. The kyanite has hardness 5 in its long direction and hardness 7 perpendicular to that direction; the calcite has hardness 3 on all faces except the (0001) which has hardness 2.

16.4   Elastic Properties

16.4.1   Homogeneous Deformation

A crystal undergoes a homogeneous deformation when the symmetry properties of the crystal are preserved at each moment of deformation.

This type of deformation occurs when a crystal is dilated or compressed by the effect of an increase in temperature and hydrostatic pressure.

The two most important properties in relation to mineral stability in the earth's crust and mantle are thermal expansion and compressibility.

16.4.2   Thermal Expansion or Expansion

The dimensions of a crystal change when its temperature changes.

Also, with an increase in temperature, the amplitude of the vibrations of the ions increases, and when the kinetic energy exceeds the attractive force between the ions, either the crystal or mineral forms a different structure (polymorph) or melts.

The linear coefficient of thermal expansion $\alpha_p(T)$ at constant pressure is given by

$$\alpha_P = \frac{\Delta l}{l \Delta T}, \tag{16.1}$$

$$\alpha_P(V) = \frac{\Delta V}{V \Delta T}, \tag{16.2}$$

where l is the length and V is the volume.

16.4.3   Compressibility

The dimensions of a crystal change when subjected to pressure. When a mineral is subject to pressure and is uniform on all sides, it responds by decreasing volume.

The linear compressibility coefficient is given, at a constant temperature, by

$$\beta_T = \frac{\Delta l}{l \Delta P}. \tag{16.3}$$

The volume coefficient of compressibility $\beta_T(V)$ is given by

$$\beta_T(V) = \frac{\Delta V}{V \Delta P}. \tag{16.4}$$

Questions

1. Minerals that have no magnetic response are called

   Response: |

2. Scalar properties depend on the direction in which they are measured.

   ⊙ True
   ⊙ False

3. Thermal expansion is what type of property?

   ⊙ a. Mechanical
   ⊙ b. Electric
   ⊙ c. Magnetic
   ⊙ d. Elastic

4. What elements of the periodic table have dipolar moments?
   ⊙ a. noble gases
   ⊙ b. transitional metals
   ⊙ c. alkaline metals
   ⊙ d. non-metals

5. Will a diamagnetic mineral contain any elements of the first transition series? Yes/No.

   Response: |

6. Cleavage is a rupture according to certain planes of twin.

   ⊙ True
   ⊙ False

7. Pyroelectricity is a property due to a change of pressure and is presented by crystals with a polar point group.

   ⊙ True

   ⊙ False

8. Relate each mineral to the physical property.

Tourmaline (silicate)	Paramagnetism
Hematite (iron oxide)	Diamagnetism
Gypsum (hydrated calcium sulfate)	Pyroelectricity
Pyrite (iron sulfide)	Hardness

9. If you think the mineral in the following photo shows cleavage, then write the name of the type of cleavage.

Response: |

10. The two most important properties in relation to mineral stability in terrestrial crust and mantle are thermal expansion and compressibility.

   ○ True
   ○ False

Chapter 17
Methods and Applications of X-ray Diffraction in Crystallography and Mineralogy

Abstract The fundamentals of X-rays and X-ray diffraction by crystals are discussed. The Laue and Bragg equations that express the conditions for diffraction to occur will be discussed. Ewald construction will be shown, using the concept of the reciprocal lattice, which is very useful for the interpretation of the experimental methods of X-ray diffraction. The different experimental methods will be presented, depending on the type of radiation (monochromatic or polychromatic) and sample (monocrystalline or polycrystalline) and the information they provide. The powder method will be explained in a more detailed way because of its extraordinary usefulness due to the variety of information it provides.

17.1 Nature of X-rays

X-rays are part of the electromagnetic spectrum, and they were discovered by Roentgen in 1895. They occupy the range of frequencies or wavelengths between ultraviolet rays and γ-rays. They can be classified into *hard* and *soft,* based on the greater or lesser capacity of radiation to penetrate matter. The unit used for X-rays is the Ångström (Å). The range of wavelengths used in X-ray diffraction is between 0.5 and 2.5 Å. It includes the most characteristic radiation of the X-ray spectrum, K_α of the Cu with $\lambda = 1.5418$ Å. Early research revealed the similarity between X-rays and light. Both radiations, X-rays and light, propagate in a straight line. They have the property to pass through opaque bodies. The photographic plates are impressed by X-rays. They excite the fluorescence and phosphorescence of certain substances. They do not experience alteration under the action of electric or magnetic fields. They have polarization effects.

© The Author(s), under exclusive license to Springer Nature Switzerland AG 2022
C. Marcos, *Crystallography*, Springer Textbooks in Earth Sciences,
Geography and Environment, https://doi.org/10.1007/978-3-030-96783-3_17

17.2   X-ray Production, X-ray Tube

They originate whenever electrons with sufficient kinetic energy collide with matter.

The X-ray tube is the instrument used to produce X-rays. In X-ray tubes, a high voltage is applied to accelerate a beam of electrons produced by heating a filament by a current. An example is tungsten, the cathode, an electrode that undergoes a reduction reaction, by which a material reduces its oxidation state when receiving electrons. The accelerated electrons collide against the anode, an electrode in which an oxidation reaction takes place, whereby a material, by losing electrons, increases its oxidation state. X-rays are emitted in all directions, but they go outside through one or more windows.

An X-ray tube scheme can be seen in Fig. 17.1.

Fig. 17.1 X-ray tube scheme

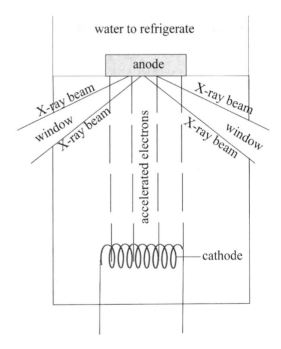

17.3   Spectra Emitted by X-ray Tube

X-ray emits two spectra, continuous, and characteristic spectra.

– *Continuous spectrum*

Continuous spectrum appears below a certain voltage value applied to the X-ray tube. It appears to occur as an effect of electrostatic electron interactions (continuous retardation of the electrons) in the vicinity of the nuclei of the anode atoms. Increasing the voltage increases the intensity of the continuous spectrum and the entire spectrum shifts to higher energies and shorter wavelengths, respectively.

– *Characteristic spectrum*

The characteristic spectrum, superimposed on continuous spectrum, always appears at fixed and determined energy and wavelength values for a given anode material. The characteristic spectrum appears forming spectral series that are designated by the letters K, L, M, N, and so forth, denoting the principle quantum numbers of orbitals. The wavelengths of the lines of each series decrease in the M, L, K, direction. It occurs if the accelerated electron, from the anode, shoots out an electron from an inner orbital (inelastic collision) and the vacancy is filled by an electronic transition (Fig. 17.2) of the outermost orbital electrons to the innermost.

Fig. 17.2 X-ray emission

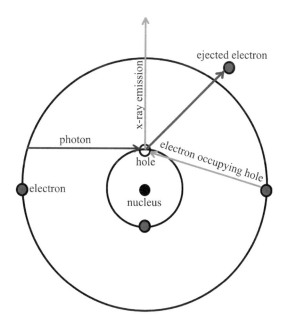

17.4   X-ray Diffraction Theory

Laue and his collaborators demonstrated, in 1912, that X-rays, discovered by Roentgen, were very short wavelength electromagnetic radiation, i.e., high frequency and extremely penetrating. They suggested that if the structure of a crystal is periodic, it could be used to diffract X-rays. This preposition was based on three hypotheses:

1. Crystals are periodic.
2. X-rays are electromagnetic waves.
3. The wavelength of the X-rays is of the same order of magnitude (1 to 3) as the distance repeated by the motifs (ions, atoms, molecules, or assemblies thereof) in the crystals.

X-ray diffraction is a particular case of consistent radiation *scattering*.[1]

When X-rays interact with matter, some part is absorbed, and a decrease in intensity occurs as it passes through more material thickness (Fig. 17.3).

Fig. 17.3 Outline showing the interaction of X-rays with matter

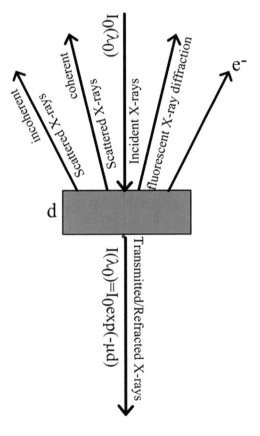

---

[1] Dispersion is defined as spatial separation of characteristic X-rays according to their wavelengths and it is of interest for monochromatisation of higher sophisticated X-ray methods.

This absorption is due to the interaction phenomena that originate and give rise to two general types of radiation:

1. *Fluorescence* radiation has variable wavelengths λ_f, and its emission is always accompanied by the release of electrons.
2. *Scattered* radiation

    – *Coherent scattered radiation.* It consists of that fraction of the primary radiation without shift of waves in time and/or in space, which is needed for interference and diffraction.
    – *Incoherent scattered radiation or Compton scattering.* It consists of that fraction of the primary radiation with some shift in time and/or space of waves relative to each other.
    – X-ray diffraction basically consists of a process of constructive interference of coherent X-ray waves (Fig. 17.4) that occurs in certain directions of space.

These waves must be in phase to be coherent and of the same wavelength. It happens when the path difference between them is zero or an integer multiple of wavelengths, $\Delta x = n\lambda (n = 0, 1, 2...)$.

Destructive *interference* (Fig. 17.5) occurs when interfering waves have path differences $x = \lambda/2, 3\lambda/2, 5\lambda/2, ...$

Fig. 17.4 Wave constructive interference

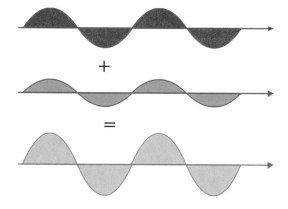

Fig. 17.5 Wave destructive
interference

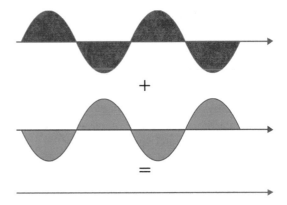

Other types of interference (constructive and destructive) may occur between the
two types of interference.

The intensity of the wave is proportional to the square of its amplitude.

For X-ray diffraction to occur, the following is necessary:

1. The object or crystal on which X-rays are affected is periodic.
2. The distances between crystal atoms are of the same order of magnitude as the
   wavelength of X-rays, similar to what happens with a diffraction grating and
   visible light (Fig. 17.6).

Fig. 17.6 Electromagnetic
waves diffraction by a
diffraction grating

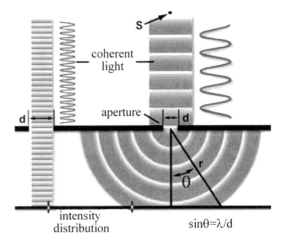

Box: 17.1 Example

Diffraction of visible light

When light passes through a single slit whose width d is on the order of the wavelength of the light, a single slit diffraction pattern can be observed on a screen at a distance $\gg d$ away from the slit. The intensity is a function of angle. According to Huygens' principle, each part of the slit can be considered a wave emitter. All these waves interfere to produce the diffraction pattern (Fig. 17.7). Where the crest meets the crest there is constructive interference, and where the crest meets the trough there is destructive interference. The center beam is diffracted in a center band (zero order) on the detector screen, flanked by several higher order diffraction bands (1st, 2nd, and 3rd) or maximums. Diffraction bands formed by higher order maximums identify the diffraction angles at which wave fronts with the same phase are amplified as bright areas due to constructive interference. Very far from a point source, the wave fronts are essentially plane waves. The diffraction pattern, with maxima and minima, is called Fraunhofer diffraction.

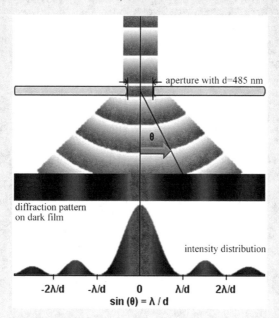

Fig. 17.7 Diffraction pattern on dark film (top) and diffraction maxima (bottom) produced by red monochromatic light

If light is incident on a material with two very small slits separated by a distance d, the waves from each slit interfere behind the material. The waves passing through each slit are diffracted and scattered. For angles at which the diffraction pattern of a single slit produces intensity other than zero, the waves from the two slits can interfere constructively or destructively. The diffraction pattern (pattern of bright and dark bands) is observed on a screen behind the material. The bright bands indicate constructive interference, and the dark bands indicate destructive interference. The bright fringe, zero-order fringe, in the middle of the diagram is caused by constructive interference of the light from the two slits traveling the same distance to the screen. The destructive interference causes the dark fringes on either side of the zero-order fringe. The crest coincides with the crest and the valley with the valley. Destructive interference causes the dark fringes on both sides of the zero-order fringe. When light from one slit travels a distance that is half a wavelength longer than the distance travelled by light from the other slit, the crests coincide with the valleys at these locations. The dark fringes are followed by first-order fringes, one on each side of the zero-order fringe. When the light from one slit travels a distance that is one wavelength longer than the distance travelled by the light from the other slit to reach these positions, the crest again coincides with the crest (Fig. 17.8).

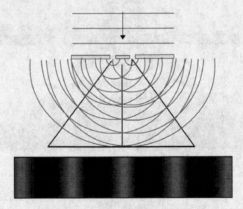

Fig. 17.8 Diffraction pattern produced by interference of the waves from the two slits

17.5   Laue Equations

X-ray diffraction occurs only in certain directions, at certain angles, which are based on the.

- Distance repeated from the periodic structure
- Radiation wavelength

In a crystal, rows of atoms periodically separate according to translations a, b, and c.

First, diffraction of a row of atoms whose translation period is a vector is considered (Fig. 17.9).

The direction of the incident X-ray beam is given by the unit vector S_0.

The direction of the X-ray beam diffracted by the row of atoms is given by the unit vector S.

For the atoms in the reticular row to diffract the X-rays, the following condition must be fulfilled:

The path difference between the incident beam and the diffraction beam must be equal to an integer of wavelengths.

This condition can be expressed as follows:

$$aS - aS_0 = a(S - S_0) = n\lambda$$
$$aS = a \cos \varphi = AD \qquad (17.1)$$
$$aS_0 = a \cos \varphi_0 = BC$$

where

n (0, 1, 2, 3, ...) is the diffraction order
$n = 0 \rightarrow$ order diffraction 0
$n = 1 \rightarrow$ order diffraction 1
$n = 2 \rightarrow$ order diffraction 2

Fig. 17.9 X-rays diffraction by a monoatomic row with interatomic spacing a

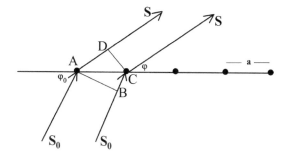

So, it can be written as:

$$a \cos \varphi - a \cos \varphi_0 = n\lambda$$
$$a(\cos \varphi - \cos \varphi_0) = n\lambda \qquad (17.2)$$
$$\cos \varphi = \cos \varphi_0 + {}^{n\lambda}\!/_a$$

In this expression, if the angle φ_0 remains fixed, the diffraction beam can run in any direction of space that forms a diffraction angle compatible with the different values of n.

Since the crystal is three-dimensional, diffraction is also three-dimensional and, therefore, two more equations must be considered—those associated with the reticular rows of b and c periods.

Laue equations
The Laue equations are given by:

$$
\begin{array}{lll}
aS - aS_0 = a(S - S_0) = h\lambda & & a(\cos \varphi - \cos \varphi_0) = h\lambda \\
aS - aS_0 = a(S - S_0) = k\lambda & \text{or} & b(\cos \varphi - \cos \varphi_0) = k\lambda \qquad (17.3)\\
cS - cS_0 = c(S - S_0) = l\lambda & & c(\cos \varphi - \cos \varphi_0) = l\lambda
\end{array}
$$

The geometric idea is that diffraction beams follow directions whose assembly sets up a surface that is a cone.

Each diffracted ray cone corresponds to a solution of the Laue equation that satisfies the values φ and n.

In a single-dimensional network, the cones are arranged as shown in Fig. 17.10.

In a two-dimensional network, the intersection of cones defines the two possible diffraction directions, Oy and Ox (Fig. 17.11).

The diffraction condition in a three-dimensional network requires that the three Laue equations be satisfied simultaneously. In this case, there is only one diffraction direction that is common to the three cones, and it is given by the intersection point of the three cones (Fig. 17.12).

17.6   Bragg's Law and X-ray Reflection

X-ray diffraction can be treated in good approximation with the model[2] of reflection.

---

[2] We can use it in terms of kinematical theory under certain boundary conditions: (1) poor crystal quality (mosaicity); (2) low angular resolution of diffraction (completely sufficient for standard powder diffractometry for phase analysis.

Highly sophisticated X-ray analysis (concentration of phases, defects, strain, etc.) or high quality crystals and/or using X-ray sources with lower divereny than available for the Braggs, or synchrotron radiation, which needs the dynamical theory of diffraction.

Fig. 17.10 Diffraction cones
in a mono-dimensional lattice

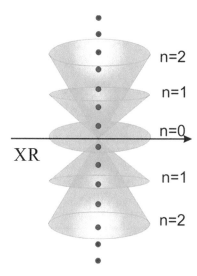

Fig. 17.11 Diffraction cones
in a two-dimensional lattice

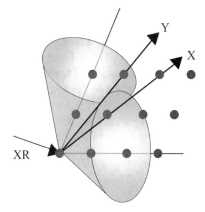

This treatment is simpler and straight forward, and it allows Bragg's law to be
used in powder diffractometry, for example, for phase analysis.

In 1914, the Bragg brothers showed that X-rays diffracted by crystals could be
treated as reflections from atomic planes of the crystal structure, depending on the
diffraction angle for a given wavelength.

Fig. 17.12 Diffraction cones
in a three-dimensional lattice

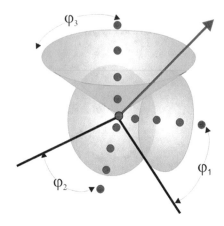

Verification of the reflection laws

1. Angle of incidence is equal to reflection angle.
2. Angle of incidence, reflection angle, incident ray, reflected ray and perpendicular to the separation surface of two media are in the same plane.

They impose the condition for scattered waves at all nodes of the same reticular plane, such as p1 (1st layer at the surface), to be in phase with each other. In general, waves scattered across successive parallel planes will not be in phase with each other, except in the case that their path differences are integer multiples of the wavelength.

To demonstrate, we will consider the p1 and p2 (2nd layer below the surface) planes, belonging to the same family of lattice plane in Fig. 17.13.

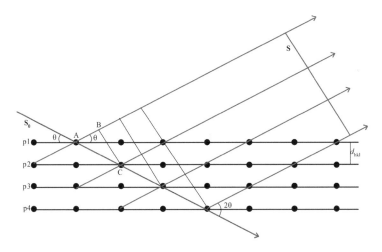

Fig. 17.13 X-rays scattering through the nodes of a family of parallel planes

For the rest of the planes, the same result is reached; that is, the path differences between the waves diffracted by adjacent planes are identical. In this way, the entire plane family (*hkl*) contributes collectively to the production of a common wave front of diffracted rays. This is equivalent to reflecting the incident rays on each plane of the series.

Any solution to Bragg's law constitutes a reflection, whatever the indices of it. Bragg's law is generally expressed as

$$AC - AB = n\lambda$$
$$AC = \frac{d_{hkl}}{sin\theta}$$
$$AB = A cos2\theta = \frac{d_{hkl}}{sin\theta} cos2\theta \qquad (17.4)$$
$$AC - AB = \frac{d_{hkl}}{sin\theta}(1 - cos2\theta) = \frac{d_{hkl}}{sin\theta} 2 sin^2\theta = n\lambda$$
$$2 d_{hkl} sin\theta = n\lambda$$

When *n*, order of diffraction, does not appear in the Bragg law, it means that we assume that dummy planes of indexes *nh, nk, nl* have been interspersed between the true reticular planes (*hkl*), which must be denoted as 'nth order of (*hkl*)', so that the path differences between the waves reflected by each two adjacent planes of the series of planes (*hkl*) is always 1λ.

17.7 Ewald Sphere or Reflection Sphere

Bragg's expression can be put in the form:

$$sin\theta = \frac{1}{d_{hkl}} \Big/ \frac{2}{\lambda} \qquad (17.5)$$

The geometric solution of this expression is that of any right triangle, such as the EOP in Fig. 17.14, which is inscribed on a $2/\lambda$ diameter sphere. This diameter matches the direction of the wave vector $\mathbf{S_0}$ for incident X-rays. The magnitude of $\mathbf{S_0}$ is $1/\lambda$.

This sphere is called the *reflection sphere* or *Ewald sphere* (Fig. 17.14).

Figure 17.14 is interpreted as follows:

– The rotation of a crystal (composed of *hkl* planes) enveloped in a monochromatic X-ray beam causes the points of its reciprocal lattice to pass through the surface of the so-called Ewald sphere.

Fig. 17.14 Ewald sphere

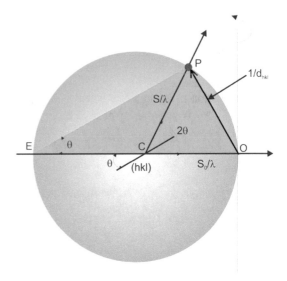

- When a reciprocal point **P**, with Miller indices (*hkl*), collides with the surface of this sphere, a diffracted beam **S** emerges from the center of the sphere and passes through the reciprocal point at that instant.
- The position of the reciprocal point **P** is given by the vector **OP,** which is the reciprocal vector r^*_{hkl} with a modulus of $1/d_{hkl}$.
- The reciprocal vector is perpendicular to the plane (*hkl*) passing through **C** and is parallel to the leg **EP**.
- The reciprocal lattice of the crystal will be built from **O**, at which point the incident X-ray beam leaves the sphere of reflection.
- The incident beam **S₀** and the diffracted S form the angle 2θ, satisfying Bragg's diffraction condition.
- These considerations allow establishing that a maximum diffraction only occurs when the scattering vector **OP** (**S–S₀**) is equal in magnitude and direction to the reciprocal vector. This can also be said as follows: For any plane (*hkl*) to diffract X-rays, its orientation must be such that its representative reciprocal point is located on the surface of the reflection sphere or Ewald sphere. Only in this circumstance does a diffracted ray occur, passing through the reciprocal point.

The only possible solutions to Eq. 17.6 are those that fulfill

$$\sin \theta \leq 1 \tag{17.6}$$

which means that

$$\frac{1}{d_{hkl}} \leq \frac{2}{\lambda} \tag{17.7}$$

This means that the length or modulus of any reciprocal vector $r^*_{hkl} = \frac{1}{d_{hkl}}$ cannot exceed the diameter length of the reflection sphere $2/\lambda$.

Thus, the nodes of the reciprocal lattice contained in a $2/\lambda$ radio sphere with their center at the origin of the reciprocal lattice are those corresponding to the families of lattice planes and their higher orders that can give rise to diffraction. This sphere is called the *limit sphere* (Fig. 17.15). Only planes whose reciprocal nodes remain on the surface of the Ewald sphere will diffract the X-rays.

Fig. 17.15 Limit sphere

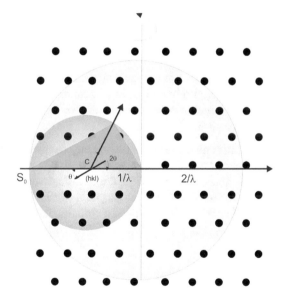

17.8   X-ray Intensity, Atomic Scattering Factor,[3] Structure Factor

- **Intensity of radiation scatters coherently by a free electron**

Thomson's theory (1903) of elastic scattering and the *interaction of polarized X-radiation with a free electron in the plane* are considered.

The electric field exerts a force on the electron and will cause it to oscillate around its equilibrium position, with a frequency equal to that of the wave it receives.

The electron behaves like an oscillator and is, therefore, a radiant energy emitter, because it oscillates in phase with the electrical vector of X-radiation.

It acts as a secondary emitter of a small fraction of the radiation it receives and scatters it consistently.

The intensity of the scattered radiation, calculated at a distance R from the electron, is given by the expression:

$$I_e = I_0 \frac{e^4}{m^2 c^2 R^2} \sin^2 \alpha \qquad (17.8)$$

where

I_0 is the intensity of the incident radiation
c is the speed of radiation in a vacuum
α is the angle between the direction of the scattered X-rays and the direction of the electron acceleration.

In almost all experimental laboratory devices, the interaction of X radiation with a free electron is *non-polarized*.

The vibration direction of the electrical vector is random on a plane perpendicular to the direction of propagation of the wave.

The intensity of the dispersed radiation is given by the expression

$$I_e = I_0 \frac{e^4}{m^2 c^2 R^2} \left(\frac{1 + \cos^2 \alpha}{2} \right) \qquad (17.9)$$

Intensity varies, depending on the direction of scattering.
The intensity is maximum for $2\theta = 0°$ and decreases as it approaches $2\theta = 90°$.

$\left(\frac{1 + \cos^2 \alpha}{2} \right)$ is the *polarization factor* and may vary between 1 and ½.

---

[3] Also known as *form factor*.

– **Atomic scattering factor or form factor**

The atomic scattering factor f is defined by the relationship between the amplitude of the wave scattered by the atom A_a and the amplitude of the wave scattered by a single electron A_e. The radiation scattered by an atom is considered.

$$f = \frac{A_a}{A_e} \tag{17.10}$$

As the intensities are proportional to the square of the amplitudes, it will have

$$f^2 = \frac{A_a}{A_e} \div \frac{I_a}{I_e} \tag{17.11}$$

It is a function of $\sin \theta / \lambda$ because the electrons of the atom do not scatter in phase, except when $\theta = 0°$, in which the electrons scatter completely in phase in the direction of the incident beam. Increasing θ also increases the phase difference and the scattering factor f decreases.

– **Structure factor**

The structure factor specifies the amplitude of the diffraction in *the hkl* reflection due to the contribution of all atoms in the elementary cell. It is symbolized by F_{hkl} and is specified by the following:

– The *modulus,* also called *the amplitude of structure,* is proportional to the amplitude of the beam diffraction by a plane. It can be experimentally calculated from the intensities of diffracted rays.
– The *argument* is the phase of the diffraction ray. It cannot be experimentally calculated and poses problems.

The amplitude of the structure factor $|F_{hkl}|$ and therefore the intensity $I_{(hkl)}$ depend on the class of atoms contained in the cell and the positions of atoms in the cell. These positions or coordinates depend on the relative phase differences between the radiations scattered by the atoms. Here it is necessary to introduce the *electronic density* $\rho_{(xyz)}$, which is the number of electrons per unit volume next to the point of the elementary cell that has *coordinates* x, y, z. Electronic density for the crystal is a periodic function.

If the electronic density $\rho_{(xyz)}$ is known at each point x, y, z, the structure factor F_{hkl} can be calculated.

17.9 Symmetry of Diffraction Effects, Laue Classes

Crystalline materials have an internal order that is usually detected by their X-ray diffraction pattern.

This has symmetry and respects the classical crystallographic constraint, i.e., there are only symmetries of orders 2, 3, 4, and 6.

Box 17.2: Five-fold symmetry and quasi-crystals

However, D. Schechtman[4] discovered in 1984 an aluminum manganese metal alloy whose diffraction pattern has five-fold symmetry. Icosahedrite is the first known naturally occurring quasicrystal phase (Fig. 17.16). It is a mineral ($Al_{63}Cu_{24}Fe_{13}$) approved by the International Mineralogical Association in 2010.[5,6]

Fig. 17.16 Diffraction pattern of icosahedrite (Materialscientist—Own work, CC BY-SA 3.0, https://commons.wikimedia.org/w/index.php?curid=12470027)

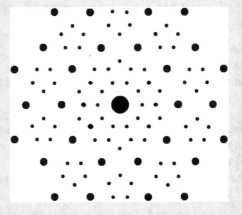

It is a quasi-crystal (contraction of the English terms *quasiperiodic* and *crystal*). Penrose mosaics (Fig. 17.17) have been used as a basis for a mathematical model of the arrangement of quasi-crystal atoms.

---

[4] *Shechtman*, D., Blech, I., Gratias, D. and Cahn, J. W., 1984. Phys. Rev. Lett. 53, 1951.

[5] Bindi, L.; Paul J. Steinhardt; Nan Yao; Peter J. Lu, 2011. Icosahedrite, $Al_{63}Cu_{24}Fe_{13}$, the first natural quasicrystal. Am. Mineral., **96** (5–6): 928–931.

[6] Commission on New Minerals and Mineral Names, Approved as new mineral.

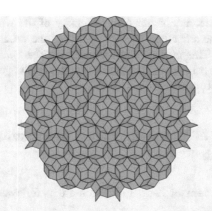

Fig. 17.17 Penrose tiling exhibiting exact five-fold symmetry (Taken from Wikimedia Commons[7])

The diffraction pattern reflects the symmetry of the crystal but, because the X-ray intensity I is proportional, F_{hkl}^2 positive and negative F_{hkl}^2 cannot be distinguished experimentally. As a result, diffraction patterns always have a center of symmetry → *Friedel's law*.

In non-centrosymmetric crystals and in the absence of anomalous scattering, reflection structure factors such as F(*hkl*) and F $\left(\overline{hkl}\right)$ appear as mirror images across the real axis of the so-called Argand diagram, and their corresponding diffracted intensities, I(*hkl*) and I $\left(\overline{hkl}\right)$ are equal. In other words, Friedel's law is fulfilled.

The 11 centrosymmetric point groups are known as *Laue Classes* (Table 17.1):

Table 17.1 Laue classes

Crystal system	Laue class
Cubic	$m\overline{3}m$, $m\overline{3}$
Tetragonal	$4/mm$, $4/m$
Hexagonal	$6/mm$, $6/m$
Rhombohedral	$\overline{3}m$, $\overline{3}$
Orthorhombic	mmm
Monoclinic	$2/m$
Triclinic	$\overline{1}$

[7] Wikimedia Commons, https://commons.wikimedia.org/w/index.php?curid=5839079.

With destructive interference matters, the intensity of the waves diffracted by those planes is null, so the reflections do not appear and the structure factors $F(hkl)$ for those planes is null. This process is called *systematic extinction* or *systematic absence* and is important because information about the spatial group of the crystal is obtained from them (Appendixes IV and V).

17.10 Application of X-ray Diffraction in Crystals and Minerals

The fundamental applications of X-ray diffraction are as follows:

- Qualitative identification of crystalline phases
- Unit cell dimensions
- Determination of the number of atoms or molecules in the unit cell
- Determination of the density referred to the unit cell is given by the expression:

$$\rho = M/V = ZM/V \qquad (17.13)$$

where

M is the mass of all atoms that make up a unit of the chemical formula; that is, molecular weight.
N is the unit formula number contained in the cell.
V is the volume of the unit cell.
Z is the number of formula units contained within unit cell.

- Bravais lattice type
- Crystal system
- Possible space group(s) (often ambiguous)
- Atomic positions from the intensities of the diffracted X-rays and, therefore, the crystal structure.

X-ray diffraction is the most important, non-destructive method for analyzing various materials, including powders, metals, corrosion products, perfect crystals, minerals, alloys, slag and ash, etc.

X-ray diffraction is a very useful technology for material determination, characterization, and quality control.

Whether it is for developing new compounds, materials or processes or optimizing manufacturing processes, non-destructive analysis using X-rays offers several possibilities.

With X-ray diffraction, it is possible to determine a variety of characteristics of macroscopic and microscopic materials, as well as the structure and defects of compounds that make up the materials.

Table 17.2 X-ray diffraction methods

Radiation (λ)	Sample type and characteristics	Method
Polychromatic	Fixed monocrystal	Laue
Monochromatic	Monocrystal with 360° rotation	Rotating crystal
Monochromatic	Monocrystal with partial and oscillating rotation around zone axis ⊥ X-ray	Oscillating crystal
Monochromatic	Monocrystal with oscillation around axis ⊥ or / from X-rays	Weissenberg
Monochromatic	Monocrystal with precession movement	Buerguer Precession
Monochromatic	Crystalline powder	Powder diffractometry
Monochromatic	Monocrystal	Goniometric methods
Synchrotron radiation	Monocrystal Crystalline powder	Synchrotron radiation-based methods

17.11   X-ray Diffraction Methods

The different X-ray diffraction methods, depending on the type of radiation (monochromatic or polychromatic) and the sample (powder or monocrystal) are presented in Table 17.2.

17.11.1   Laue Method

In the Laue method, the crystal has a fixed (not rotated) orientation with respect to a polychromatic X-ray beam. Laue's method is used to determine the cell orientation of a single crystal of known structure.

The Laue diagram (Fig. 17.18) is similar to a stereographic projection of the crystal planes (Fig. 17.19).

17.11.2   Oscillation Method

The crystal and, therefore, the reciprocal net is oscillating a small angle around an axis perpendicular to the plane of the figure and passing through the center. This method allows collecting several reciprocal levels at once over each position of the crystal. By repeating these diagrams, at different starting positions of the crystal, enough data are obtained in a reasonable time. The collection geometry is described in Fig. 17.20. It is used to adjust a single crystal for the Weissenberg method.

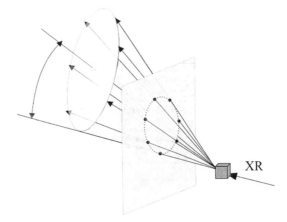

Fig. 17.18 Laue method in transmission mode

Fig. 17.19 Laue diagram of a crystal (www.xtal.iqfr.csic.es/Cristalografia with permission)

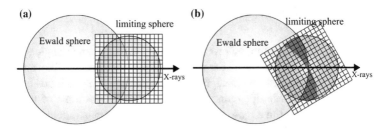

Fig. 17.20 Osscilating method, **a** static cristal—Only those nodes of the reciprocal lattice that coincide with the Ewald sphere and that are within the limiting sphere will diffract; **b** crystal rotated counter clockwise—All the nodes of the reciprocal lattice in the grey area will have diffracted because they passed the Ewald sphere

17.11.3   *Weissenberg Method*

The Weissenberg method is based on the camera of the same name, developed by
Weissenberg (1924). It consists of a metal cylinder that contains X-ray-sensitive
film.

The crystal is mounted on a coaxial axis with said cylinder and is rotated in such
a way that the reciprocal points that intersect the surface of the Ewald sphere are
responsible for the diffraction beams. To separate reflections, the film is moved in
parallel to the crystal rotation.

These beams generate a blackening (spot) on the photographic film that, when
extracted from the metal cylinder, has the appearance shown in Figs. 17.21 and
17.22.

The type of Weissenberg diagrams obtained from the described mode are called
rotation or *oscillation* diagrams, depending on whether the crystal rotation is 360°
or partial (approx. 20°), respectively.

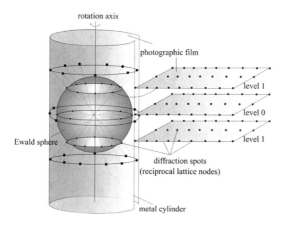

Fig. 17.21 Weissenberg method scheme

Fig. 17.22 Weissenberg
diagram of copper
methaborate (www.xtal.iqfr.
csic.es/Cristalografia with
permission)

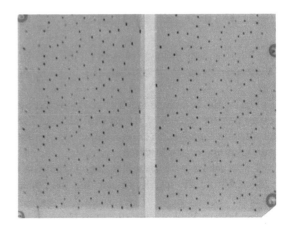

Example

Obtention of cell parameters, crystalline system and possible spatial groups
from the data provided with rotating crystal and Weissenberg techniques of
barite ($BaSO_4$).

Since the rotary diagram has been obtained with the axis of rotation
coincident with the crystallographic axis c, this value is obtained from the
expression

$$c = \frac{h\lambda\sqrt{r^2 + y^2}}{y}$$

where

h refers to the level, in this case 1, 2 and 3

λ is the wavelength used, 1.542 Å

y (mm) is the distance from each level to the center of the plate; the vertical
  y equals to 2θ

r is the radius of the film, and its value is 28.65 mm; 1 mm equals a rotation
  of $2°$

The values of y and c obtained from the three levels of the rotating crystal
are in Table E1.

To obtain the parameters **a** and **b** and the crystalline system, the
Weissenberg diagrams must be interpreted.

Table E1 Values of y and
c obtained from the levels h1,
h2 and h3

y	c
y1 = 6.5 mm	c1 = 6.969 Å
y2 = 14.0 mm	c2 = 7.024 Å
y3 = 25 mm	c3 = 7.036 Å
	c = 7.001 Å

First, the points of the photographs are copied with transparent paper. Each point corresponds to a node in the reciprocal lattice and represents the reflexion of a set of planes in the direct lattice. Any node plane of the reciprocal lattice is perpendicular to the translation of the direct lattice. For example, the plane of the reciprocal lattice nodes defined by a^* and b^* is perpendicular to c; therefore, a^* and b^* are perpendicular to the planes (h00) and (0k0) respectively, which are parallel to c. The location of a node in the reciprocal lattice is specified by the translation vector

$$r = ha^* + kb^* + lc^*$$

In the Weissenberg method, the crystal is rotated around a translation of the direct lattice and the nodes of the reciprocal lattice appear in planes perpendicular to this axis.

Planes (*h00*) and (*0k0*) must be found and the curves that cut to (*h00*) and that must be asymptotes to (*0k0*) and the curves that cut to (0k0) and that must be asymptotes to (*h00*) are plotted for the three levels obtained (Fig. E1).

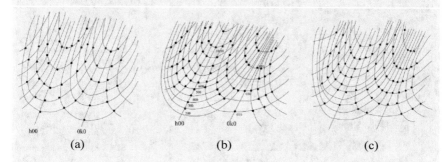

(a)   (b)   (c)

Fig. E1 Weissenberg diagram, **a** Level 1, **b** Level 2, **c** Level 3

As a result of movement of the X-ray film, the Weissenberg pattern is geometrically distorted.

1. The angle of the assymptote

The angle of the assymptote depends on the experimental set-up and the rotation angle, plus translation of crystal and film, respectively.

The angle γ^* is measured between the planes (*h00*) and (*0k0*) which, in this case, is of 45°; therefore, $\gamma = 90°$.

2. Construction of the correct non-distorted reciprocal lattice from the deformed Weissenberg pattern

The notation for the nodes parallel to the $(0k0)$ plane and intersecting the $(h00)$ plane is 100, 200, 300, etc., while the notation for the nodes parallel to $(h00)$ plane and intersecting $0k0$ is 010, 020, 030, etc. The notation of the nodes found at the intersection of the asymptotic curves to the $(h00)$ and $(0k0)$ planes is 110, 120, 130, etc., 210, 220, 230, etc., and 310, 320, 330, etc. In Table E2, all the reflexions obtained are presented.

Table E2 Reflexions obtained from the Weissenberg levels

Reflexions level 0							
$h00$	$0k0$	$hk0$					
200	010	210	220	230	240	250	
400	020	410	620	430	640	450	
600	040	10,1,0	10,2,0	630	840	650	
800	060			930			
10,0,0				10,3,0			

Reflexions level 1							
$h0l$	$0kl$	hkl					
201	011	111	511	531	241	151	
301	031	221	611	631	341	351	
401	051	131	711	731	441	451	
501		231	811	831	541	551	
601		321	911	931	641	651	
10,0,1		421	10,11	10,3,1	841	751	
		521			941		
		10,21					

Reflexions level 2							
$h0l$	$0kl$	hkl					
102	022	112	122	132	142	152	162
302	042	212	422	232	342	352	562
502	062	312	722	332	542	552	
702		512	922	532	742	652	
802		912		732	842	752	
902		10,1,2		932			

3. Calculation of unit cell parameters

Crystalline systems with the calculated angle $\gamma = 90°$ can be

Monoclinic $a{\neq}b{\neq}c$; $\alpha{=}\gamma = 90°$
Orthorhombic $a{\neq}b{\neq}c$; $\alpha{=}\beta{=}\gamma{=}90°$
    Tetragonal $a{=}b \neq c$; $\alpha{=}\beta{=}\gamma = 90°$
    Cubic $a{=}b{=}c$; $\alpha{=}\beta{=}\gamma = 90°$
    The d values for these crystalline systems are calculated using Bragg's law

$$d = \frac{\lambda}{2 \sin \theta}$$

$$d_{100} = \frac{a}{\sqrt{h^2 + k^2 + l^2}}$$

$$a = d_{100} \sqrt{h^2 + k^2 + l^2}$$

2θ for reflection 100 is 11°, so θ is 5.5° and $d_{100} = 8.044$ Å, so a value is 8.044 Å.

This value is different for c, indicating that the crystalline system cannot be cubic.

4. Obtention of symmetry, crystalline system, Bravais lattice type probable or most plausible space group from systematic extinctions

The symmetry shown by the photographs is checked to determine the crystalline system and obtain the possible spatial groups from systematic extinctions (Appendixes IV and V).

The symmetry observed in the plates shows two planes of symmetry, so the monoclinic system is discarded.

To see if it is a tetragonal or orthorhombic crystal, it is calculated b from the d_{010} and, because the value obtained, 5.526 Å, is different from a and c, the crystalline system turns out to be orthorhombic.

By presenting all the reflexions hkl, the lattice is primitive, P (Appendix IV).

The conditions of non-extinction satisfied (Appendix V) are as follows:

For $h00$, $h = 2n$ and presence of 2_1 parallel to [100]

The reflexions $00l$ are not available so it is not possible to obtain the axis 2_1 parallel to [001], although it can be inferred.

For $0kl$, $k + l = 2n$ and presence of n parallel to (100) plane

For $h0l$, $l = 2n$ and presence of a parallel to (010) plane

Since an m plane has been observed on the Weissenberg plates, this plane must be parallel to (001) plane.

With these data, the space group Pmna can be inferred.

17.11.4 Precession Method

The precession method was developed by Martin J. Buerger in the early 1940s. Like the Weissenberg method, it is a method in which crystal moves but the movement of the crystal (and, as a consequence, that of the reciprocal planes) is like that of precession of the planets, hence its name. The film is placed on a flat stand

Fig. 17.23 Precession
method scheme

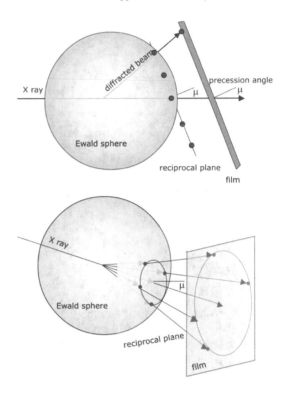

and moves with the crystal. As a result, the non-distorted reciprocal space is imaged.

The crystal must be oriented in such a way that the reciprocal plane to be recorded is perpendicular to the direct beam of the X-rays, i.e., a direct axis matches the direction of the incident X-rays (Fig. 17.23).

These diagrams are much simpler to interpret than those of Weissenberg, as they show the appearance of an undistorted reciprocal plane (Fig. 17.24).

The separation of a given reciprocal plane is achieved by using screens that select certain diffraction beams from that plane.

Similarly, as in the case of Weissenberg, reciprocal distances and diffraction intensities can be measured. However, it is much easier here to observe the symmetry elements of reciprocal space.

The disadvantage of the precession method is the consequence of the film being flat rather than cylindrical, and the solid angle explored is smaller.

Fig. 17.24 Precession
diagram of perosvskite (www.
xtal.iqfr.csic.es/Cristalografia
with permission)

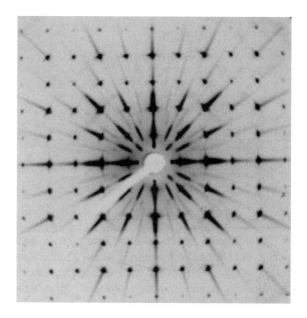

17.11.5   Powder Diffractometry Method

X-ray diffraction in crystalline powder samples or polycrystalline samples was first
revealed in Germany by P. Debye and P. Scherrer in (1916) and, at about the same
time, developed through studies in Hull in the United States.

The historical powder method is based on:

– The use of monochromatic radiation.
– Samples consisting of a powder or polycrystalline aggregate.
– Cylindrical X-ray sensitive film.

The sample should be composed of numerous crystalline fragments of very
small ($\sim$ 1–10 μm) size and ideally *oriented randomly,* relative to each other. As
for orientation, the whole of the powder can be considered isotropic, even if each
small fragment is anisotropic.

Each fragment will present the X-ray beam with a plane, which does not have to
match the plane of another fragment. Thus, we can assume that all the various
lattice planes of the sample in the form of a monocrystal are statistically represented
by the planes that the different fragments expose to X-rays.

The reciprocal vectors associated with these planes undergo multiple
three-dimensional rotations around the point chosen to build the reciprocal lattice.

The geometrical location of these reciprocal vectors is on a sphere of radius
equal to the reciprocal vector modulus $\frac{1}{d_{hkl}}$.

The intersection of the sphere of reciprocal vectors with the Ewald sphere is a
circumference.

Fig. 17.25 Powder
diffractometry method scheme

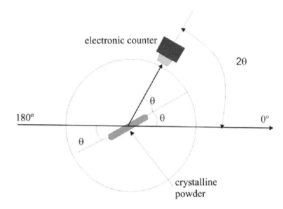

The diffracted rays in the crystalline powder, located in center C of the sphere, pass through the intersection, originating a diffraction cone whose semi-angle is 2θ.

The different reciprocal vectors, each representing a plane of the crystalline powder, originate an equivalent series of concentric spheres, with specific radii that depend on the moduli of the reciprocal vectors.

Thus, the total effect of diffraction consists of the production of a series of diffracted, coaxial ray cones with the direction of the incident X-ray beam, S_0.

Diffractogram registration can be performed using the powder diffractometer.

With the powder diffractometer, when the sample rotates an angle θ, the counter rotates 2θ. This movement θ–2θ causes the diffractometer to be called a two-circle diffractometer (Fig. 17.25). In a commercial diffractometer, the sample is located in the axis center of the precision goniometer, whose angular velocity is synchronized in the previous ratio 2:1 with the detector.

The graphical or diffractogram record consists of peaks distributed according to the angular values, 2θ, and that correspond to those of the reflections they represent (Appendix III). Only their areas constitute very representative magnitudes of the intensities of the corresponding reflections, which can be measured with great accuracy and reproducibility.

The X-ray diffraction diagram consists of a series of peaks corresponding to the reflections of the crystalline planes of the phases present in the sample. Each peak is characterized by its position, related to the θ, Bragg's law angle. The width and height of the peaks in a diffractogram are the result of a combination of instrumental and microstructure-based factors. The width and shape of the peaks in a diffractogram are the result of a combination of instrumental factors and factors based on the microstructure of the sample. When the crystal lattice becomes imperfect, the X-ray diffraction peaks broaden and, therefore, the peak width carries direct information about lattice imperfections. Consequently:

- The position of the diffracted peaks is related to size and dimension of the unit cell.

- Intensity ratios of the diffracted peaks are related to type and location of atoms in the unit cell.
- Full width at half maximum (FWHM) is related to intrinsic properties of the materials of the diffracted peaks (i.e., microstructural analysis).

The discovery of the diffraction phenomenon in this type of sample quickly becomes an experimental technique of widespread use, basically due to the wide field of application we can find for the study of crystalline substances.

Currently, this technique is a common working tool with an extraordinary utility in many different scientific and technological disciplines, because of its multi-faceted nature, in terms of the wide variety of information it provides. The areas of greatest application of this technique are materials research, cement quality control, mineralogy and geology, pharmaceutical, chemistry and catalysis, polymers, archaeology, and new nano- and battery materials and semiconductors.

The information obtained from powder X-ray diffraction includes (a) phase identification, (b) indexing a powder pattern which allows unit cell size and shape determination, (c) quantitative phase analysis, (d) crystallite size and strain, (e) peak intensities, (f) study of solid solutions, (g) texture study, and (h) determination of thermal expansion coefficients. Also included are determination of amorphous content, and crystal structure with high sophisticated computational routines (Rietveld).

(a) Phase identification

The identification of a crystalline phase by this method is based on the fact that each substance in the crystalline state has an X-ray diffractogram that is characteristic to it (Fig. 17.26).

These diagrams are collected in tabs, books, and databases of the *Joint Committee on Powder Diffraction Standards* and grouped into indexes of organic,

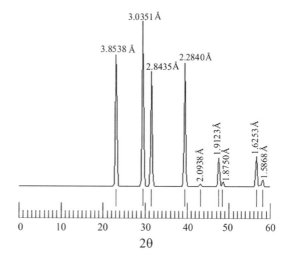

Fig. 17.26 X-ray diffractogram of calcite (CaCO$_3$) showing the interplanar distances corresponding to their diffraction peaks

inorganic, and mineral compounds. It is, therefore, a question of finding the best fit for the problem diagram with one of those collected.

The phase identification using X-ray powder diffraction can be done by:

– Calculating the unit cell and then searching the NIST crystal data database for known compounds with the same or similar unit cells.
– Comparing the measured pattern against the ICDD/JCPDS powder diffraction file data base. This contains powder patterns for a very large number of compounds.

The list of peak positions and intensities are used. Peaks can be located automatically.

The search match can be done:

– Manually, using the *Hanawalt method*, based on the three strongest lines.
– By computer, this should narrow the search down to elements of interest.

With the Hanawalt method, each diagram is characterized by the three most intense diffraction peaks.

It is important to highlight that the angular position 2θ of the diffraction peak is defined in the middle of fullwidth at half-maximum (FWHM). FWHM is sensitive to variation in microstructure and stress–strain accumulation in the material. The full width at half maximum (FWHM) of X-ray diffraction profiles is used to characterize different material properties and surface integrity features.

It contains a system of subgroups resulting from dividing the range of d values into 47 regions, each containing a roughly equal number of diagrams. Each diagram with its three most intense lines is assigned to a group. All the diagrams assigned to each Hanawalt Group are sorted so that, in the first column, the value of d corresponds to the most intense line, in the second column the value of d corresponds to the next line in intensity and, in the third column, the value of d corresponds to the third most intense line. In the five remaining columns, the values of d appear in the same decreasing order, both in value and in intensity. Following the columns corresponding to the values of d, sorted by decreasing values of intensity, the chemical formula, the name, and the number of the corresponding substance sheet appears.

When analyzing unknown diagrams, the steps to follow are listed below:

(1) The values of d are sorted in order of decreasing intensity.

(2) The appropriate Hanawalt group is located in the search manual, with the value of d corresponding to the most intense line.

(3) The d of the second strongest line is verified to fit the d value in the second column of the corresponding Hanawalt group in the search manual.

(4) The d value of the third strongest line in the unknown diagram is taken and checked to see if it is set to the value of d in the third Hanawalt column in the search manual.

(5) Adjusting the d values of the unknown diagram to those of the Hanawalt group in the search manual uses the corresponding tab and checks the setting of all the d values of the unknown diagram with those on the tab.

(6) If the selected lines do not give good fit together, other combinations of lines are chosen in the unknown diagram.

(7) In the event that the unknown diagram corresponds to a mixture of substances, the experimental found diagram is subtracted and the process is repeated again (steps 1 to 6) until all the significant lines of the experimental diagram conform to some diagram of the tabs.

When repeating the procedure, it should be kept in mind that the same diffraction peak can correspond to more than one crystalline phase. One notices, observing the intensities of the peak of the experimental diagram and the tab, if the experimental diagram shows an intensity much higher than that of the tab, it must be suspected that it corresponds to more than one crystalline phase.

Examples

1 The maximum diffraction values of 2θ and the relative intensities for one material composed by one phase, obtained with radiation $K_\alpha Cu$ ($\lambda = 1.540$ Å) are in Table E1.

Question: Identify the phase of the X-ray diffracted material.

Procedure:

The d values in Table E1 (column 3) have been obtained using the Bragg equation.

The values of 2θ and d were sorted (columns 4 and 5) in order of decreasing intensity (column 6).

– The appropriate Hanawalt group is searched in the search manual with the value of d corresponding to the most intense line, in this case 3.34050 Å.

– The d of the second strongest line (4.24857 Å) is verified to fit the d value in the second column of the corresponding Hanawalt group in the search manual.

Table E1 2θ values (column 1), intensity (column 2) and interplanar distances calculated (column 3) of the reflections corresponding to the diffracted powder material. In columns 4, 5 and 6, the values of columns 1, 2 and 3 in decreasing order of I (%). In column 7 the identified crystalline phase

Data provided		Data calculated	Values in decreasing intensity order			
1	2	3	4	5	6	7
2θ(°)	I (%)	d (Å)	2θ(°)	d (Å)	I (%)	Phase
20.8919	17.76	4.24857	26.6641	3.34050	100.00	Quartz
26.6641	100.00	3.34050	20.8919	4.24857	17.76	Quartz
36.5880	7.69	2.45402	50.1810	1.81653	12.57	Quartz
39.5087	6.93	2.27907	59.9988	1.54063	8.44	Quartz
40.3316	3.49	2.23445	36.5880	2.45402	7.69	Quartz
42.4918	5.17	2.12572	39.5087	2.27907	6.93	Quartz
45.8349	3.39	1.97815	68.3584	1.37117	6.05	Quartz
50.1810	12.57	1.81653	68.1798	1.37432	5.99	Quartz
54.9163	3.66	1.67057	42.4918	2.12572	5.17	Quartz
55.3663	1.53	1.65805	67.7808	1.38144	4.83	Quartz
57.2483	0.17	1.60793	54.9163	1.67057	3.66	Quartz
59.9988	8.44	1.54063	40.3316	2.23445	3.49	Quartz
64.0730	1.54	1.45214	45.8349	1.97815	3.39	Quartz
65.8676	0.20	1.41685	75.6961	1.25544	2.33	Quartz
67.7808	4.83	1.38144	79.9187	1.19939	2.27	Quartz
68.1798	5.99	1.37432	81.5293	1.17972	2.03	Quartz
68.3584	6.05	1.37117	81.2058	1.18360	1.82	Quartz
73.5026	1.72	1.28739	73.5026	1.28739	1.72	Quartz
75.6961	2.33	1.25544	64.0730	1.45214	1.54	Quartz
77.7066	1.24	1.22790	55.3663	1.65805	1.53	Quartz
79.9187	2.27	1.19939	83.8729	1.15261	1.28	Quartz
80.1358	1.13	1.19668	77.7066	1.22790	1.24	Quartz
81.2058	1.82	1.18360	80.1358	1.19668	1.13	Quartz
81.5293	2.03	1.17972	65.8676	1.41685	0.20	Quartz
83.8729	1.28	1.15261	57.2483	1.60793	0.17	Quartz

- The d value of the third strongest line (1.81653 Å) in the unknown diagram is taken and checked to see if it is set to the value of d in the third Hanawalt column in the search manual. The selected trio of values of d match those of the card JCPDS 46–1045, corresponding to quartz (SiO_2).

In Fig. E1, the corresponding diffraction pattern of quartz can be observed.

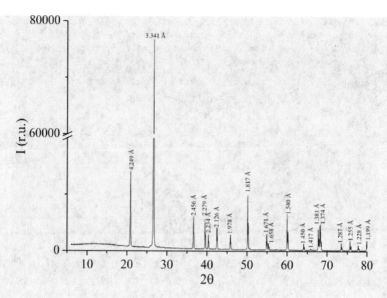

Fig. E1 Diffraction pattern of quartz

2 The maximum diffraction values of 2θ and the relative intensities for one
   material composed by two phases, obtained with radiation $K_\alpha Cu$
   ($\lambda = 1.540$ Å) are in Table E2.

Question: Identify the phases of the X-ray diffracted material.

Procedure

The d values in Table E2 (column 3) have been obtained using the Bragg
equation.

The values of 2θ and d were sorted (column 4 and 5) in order of decreasing
intensity (column 6).

– The appropriate Hanawalt group is located in the search manual, with the
  values of d corresponding to the three most intense lines, 3.34050,
  3.03836, and 2.28647 Å without any coincidence.
– The following Hanawalt groups are located, with the values of d (Å) of the
  second line corresponding to 2.28647, 1.82231, 1.37578, 1.87697,
  1.54484, 1.37642, 4.26546, and 3.34050, 3.03836 and with the values of
  d (Å) of the third line corresponding to 1.82231, 1.37578, 1.87697,
  1.54484, 1.37642, 4.26546, finding a match with the trio of values
  −3.34050, 4.26546, and 1.82231 Å corresponding to quartz (JCPDS 46–
  1045).
– Subsequently, the procedure is repeated with the values of d not assigned
  to the identified phase. The said d (Å) values in decreasing intensity order
  are 3.03836, 2.28647, 1.37578, 2.09618, 1.87697, 1.91291, 1.38506,
  2.49862, 1.67585, 1.98371, 1.25769, 1.52534, 3.85804, 1.28843, and

Table E2 2θ values (column 1), intensity (column 2) and interplanar distances (column 3) of the reflections corresponding to the diffracted powder material and interplanar distances calculated (column 3). In columns 4, 5 and 6, the values of columns 1, 2 and 3 in decreasing order of I (%). In column 7 the identified crystalline phase

Data provided			Data calculated	Values in decreasing intensity order		
1	2	3	4	5	6	7
2θ(°)	I (%)	d (Å)	2θ(°)	d (Å)	I (%)	Phases
20.8257	11.68	4.26546	26.5750	3.35428	100	Quartz
23.0536	3.73	3.85804	29.3972	3.03836	51.04	Calcite
26.5750	100	3.35428	39.4094	2.28647	22.22	Calcite
29.3972	51.04	3.03836	50.0108	1.82231	15.73	Quartz
31.4149	1.21	2.84767	68.0976	1.37578	14.95	Calcite
35.9431	7.28	2.49862	48.502	1.87697	13.62	Calcite
36.4664	9.49	2.46396	59.8729	1.54484	13.54	Quartz
39.4094	22.22	2.28647	68.2543	1.37642	13.32	Quartz
40.2376	4.19	2.24131	43.1578	2.09618	12.08	Calcite
42.3897	6.24	2.13236	20.8257	4.26546	11.68	Quartz
43.1578	12.08	2.09618	47.5340	1.91291	11.5	Calcite
45.7390	4.72	1.98371	36.4664	2.46396	9.49	Quartz
47.5340	11.5	1.91291	67.6431	1.38506	8.82	Calcite
48.5020	13.62	1.87697	35.9431	2.49862	7.28	Calcite
50.0108	15.73	1.82231	57.4004	1.60536	6.97	Quartz
54.7777	5.82	1.67585	42.3897	2.13236	6.24	Quartz
56.6445	2.49	1.62497	54.7777	1.67585	5.82	Calcite
57.4004	6.97	1.60536	45.7390	1.98371	4.72	Calcite
59.8729	13.54	1.54484	64.6812	1.44114	4.61	Quartz
60.7186	4.42	1.52534	75.5371	1.25769	4.47	Calcite
63.1206	1.47	1.47295	60.7186	1.52534	4.42	Calcite
63.9552	3.27	1.45573	40.2376	2.24131	4.19	Quartz
64.6812	4.61	1.44114	23.0536	3.85804	3.73	Calcite
65.6811	3.17	1.42160	73.4335	1.28843	3.61	Calcite
67.6431	8.82	1.38506	63.9552	1.45573	3.27	Quartz
68.0976	14.95	1.37578	65.6811	1.42160	3.17	Quartz
68.2543	13.32	1.37642	77.5400	1.23012	2.97	Quartz
70.3731	1.37	1.33677	56.6445	1.62497	2.49	Calcite
73.4335	3.61	1.28843	63.1206	1.47295	1.47	Quartz
75.5371	4.47	1.25769	70.3731	1.33677	1.37	Quartz
77.5400	2.97	1.23012	31.4149	2.84767	1.21	Quartz

1.62497. The first trio of values 3.03836, 2.28647, 1.37578 Å does not match any card but it is observed that the first duo (3.03836, 2.28647 Å) matches the calcite card (5–586), so the third value is varied until the trio

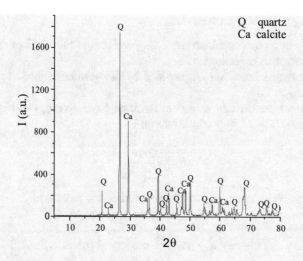

Fig. E2 Diffraction pattern of quartz and calcite

3.03836, 2.28647, 2.09618 Å matches the calcite card. In Fig. E2 the corresponding diffraction pattern of the mixture of quartz (SiO_2) and calcite ($CaCO_3$) can be observed.

Currently, there are software programs that allow this process to be carried out automatically, and they also have JCPDS databases, which greatly facilitate the identification of crystalline phases by the powder method.

Perform background subtraction and peak search. For a PDF card to be considered a good match to the experimental data there should be no strong peaks on the PDF card that are missing from your data. Extra peaks in the data may indicate the presence of additional phases. Carefully check possibilities against the data.

Rietveld refinement is a technique described by Hugo Rietveld[8] for use in the characterization of crystalline materials. Rietveld refinement uses a least-squares approach to refine the lattice parameters, peak width and shape, and preferred orientation to derive a calculated diffraction pattern (from a known or postulated crystal structure) until it matches the measured profile. Once the derived pattern is nearly identical to the unknown sample data, several properties pertaining to that sample can be obtained, including precise quantitative information, crystallite size, and site occupancy factors. This technique uses the complete profile instead of individual reflections and was an important step forward in diffraction analysis of powder samples. Rietveld analysis has the advantage over conventional quantitative methods in that it does not require standards.

---

[8] Hugo M. Rietveld was a Dutch crystallographer who is famous for his publication on the full profile refinement method in powder diffraction, which later became known as the Rietveld refinement method.

(b) **Indexing a powder pattern**

The indexing of a powder pattern consists of assigning Miller indices or interplanar spacings, d, to their reflections.

The only experimental values provided by the powder method are the angular values of $2\theta_{hkl}$.

These values are directly related to the interplanar spacings of d_{hkl}, using the Bragg equation (Eq. 17.4) put in the form:

$$d_{hkl} = \frac{2 \sin \theta_{hkl}}{\lambda} \tag{17.14}$$

where λ is the wavelength of X-radiation used and it is known.

Interplanar spacings are related, using the expression:

$$1/d_{hkl}^2 = 1 \left/ \frac{(1 - \cos^2 \alpha - \cos^2 \beta - \cos^2 \gamma - 2 \cos \alpha \cos \beta \cos \gamma) \cdot}{\left[\begin{array}{c} \left(h^2/a^2\right) \sin^2 \alpha + \left(k^2/b^2\right) \sin^2 \beta + \left(l^2/c^2\right) \sin^2 \gamma \\ + \left(2kl/bc\right)(\cos \beta \cos \gamma - \cos \alpha) \\ + \left(2hl/ac\right)(\cos \gamma \cos \alpha - \cos \beta) \\ + \left(2hk/ab\right)(\cos \alpha \cos \beta - \cos \gamma) \end{array}\right]} \right. \tag{17.15}$$

for each crystal system, with the indices of the corresponding reflections and with the cell or lattice parameters of the corresponding crystalline system.

Therefore, you can obtain the cell parameters, the space group or, if ambiguous, possible space groups, the Bravais lattice type, the crystal system.

Indexing can be done by hand or by computer.

The crystalline powder method uses crystalline powder or polycrystalline samples, which are composed of numerous crystallites. Each crystallite has a reciprocal lattice associated with it, whose origin is in the sphere of Ewald, at the point where the incident ray emerges from it. Since there are many crystallites and they are oriented at random, it is not possible to distinguish the lattice planes of the different monocrystals. On the contrary, vectors whose origin is that of the reciprocal lattice are joined in concentric spheres whose radii are the different possible vectors of the reciprocal lattice. Each sphere of the reciprocal lattice intersects with the reflection sphere, resulting in concentric diffraction cones around the X-ray beam and, in the case of the Debeye-Scherrer method, can be intercepted by a cylindrical chamber in two arcs.

The distance between the arcs can be quantified, so diffraction angles can be measured. In addition, the lengths of the vectors of the reciprocal lattice can be determined, since $\mathbf{r}^*_{hkl} = 1/d_{hkl}$, but not the relative provisions thereof, which explains the difficulties presented by index allocations to the hkl reflections of a powder diagram.

The problem is easier when the system is cubic, tetragonal, or hexagonal. To assign the *hkl* indices to the reflections produced by certain crystal planes in one of the above-mentioned crystal systems, Eqs. 17.4 and 17.5 must be considered.

– Cubic crystal system

If present, 100 reflexion is the first of all, and the problem is easy to solve because, from this hypothesis, progress is made by dividing all the Q_{hkl} by the 1st $(Q_1 = Q_{100})$. For the simplest case; that is, the case of the cubic system it is

$$\frac{1}{d_{hkl}^2} = \frac{h^2 + k^2 + l^2}{a^2} = Q_{hkl} \tag{17.16}$$

$$Q_{hkl}/Q_{100} = \frac{h^2+k^2+l^2}{a^2} \bigg/ \frac{1}{a^2} = h^2 + k^2 + l^2 = N \tag{17.17}$$

N values for the cubic crystal system are in Appendix IV of this chapter.

Because, in the cubic system, the presence of number 7 is not possible, when Q_{1st} is not the Q_{100}, probably the Q_{1st} will be the simplest, immediate to 110 reflexion, i.e., Q_{110}, and subsequently:

$$Q_{hkl}/Q_{111} = \frac{h^2+k^2+l^2}{a^2} \bigg/ \frac{2}{a^2} = \frac{h^2 + k^2 + l^2}{2} = \frac{N}{2} \tag{17.18}$$

where

$$N = 2\left(\frac{Q_{hkl}}{Q_{110}}\right) \tag{17.19}$$

If the 110 reflexion is not present, it will be logical to think that the next simplest immediate that may appear is the 111 reflexion:

$$Q_{hkl}/Q_{111} = \frac{h^2+k^2+l^2}{a^2} \bigg/ \frac{3}{a^2} = \frac{h^2 + k^2 + l^2}{3} = \frac{N}{3} \tag{17.20}$$

where

$$N = 3\left(\frac{Q_{hkl}}{Q_{111}}\right) \tag{17.21}$$

– Tetragonal crystal system

The expressions to be considered are (17.22) and (17.23):

$$\frac{1}{d_{hkl^2}} = \frac{h^2 + k^2 + l^2}{a^2} + \frac{1}{c^2} = Q_{hkl} \tag{17.22}$$

Each Q_{hkl}/Q_{1st} will be composed of an entire part and a decimal part. If we consider that the Q_{1st} is the Q_{100}, it follows that the entire part of Q_{hkl}/Q_{100} is

$$\frac{Q_{hkl}}{Q_{100}} = h^2 + k^2 = N \tag{17.23}$$

The decimal part n.X (n = 0, 1, 2, …) contributes to the value of l.

$$\text{for } l = 1 \quad \text{n.X} \cdot 1^2$$
$$\text{for } l = 2 \quad \text{n.X} \cdot 2^2$$
$$\text{for } l = 3 \quad \text{n.X} \cdot 3^2$$

Accordingly, the value of l is obtained as follows: It may be useful to have a table with all values that have decimals. For example, if we suppose three values with decimals—n.X, n.Y, n.Z—there must be three different values of l. We start working with the first decimal and with n = 0, and we check that with 0.X the rest of the decimal places are satisfied. If not, continue with the next degree of complication 1.X, etc.

– Hexagonal crystal system

The expressions to be considered are (17.24) and (17.25):

$$\frac{1}{d_{hkl}^2} = \frac{4(h^2 + k^2 + l^2)}{3a^2} + \frac{1}{c^2} = Q_{hkl} \tag{17.24}$$

$$\frac{Q_{hkl}}{Q_{100}} = h^2 + k^2 + hk = N \tag{17.25}$$

The procedure for assigning indices to different reflections is the same as just describing the case of the tetragonal system.

– Monoclinic and triclinic crystal systems

The resolution is much more complicated because more parameters appear, which makes it necessary to process the data using iterative calculation processes. Because of this complication, it is not appropriate to address such problems in this context.

Examples

1 Maximum diffraction values of 2θ obtained with radiation $K_\alpha Cu$ λ = 1.540 Å for two crystals, 1 and 2, are in Tables E1 and E2 (column 1), respectively:

1. Deduce the crystal system.
2. Assign indices to reflections.
3. Calculate cell parameters and axial angles.
4. Extract the possible information about the space group.

Solution crystal 1.

1. In the interpretation of the powder diagram, the crystalline system must be decided in advance. To do this, the crystal planes that correspond to the reflections found must be known.
   From the values of 2θ, using Bragg's law, the values of d (second column of Tables E1 and E2) are obtained. From the expression $1/d^2_{hkl}$ the values

Table E1 Maximum diffraction values of 2θ (column 1), interplanar distances (column 2), Q/Q_{1st} (column 3), and hkl indexes of reflections (column 4) for crystal 1

1	2	3	4
2θ(°)	d (Å)	Q/Q_{1st}	hkl
23.03	3.862	1	100
32.81	2.730	2	110
40.38	2.234	3	111
47.07	1.931	4	200
53.00	1.728	5	210
58.54	1.577	6	211
68.78	1.365	8	220
73.61	1.287	9	221 or 300
78.31	1.221	10	310
82.96	1.164	11	311
87.49	1.115	12	222
92.09	1.071	13	320

Table E2 Maximum diffraction values of 2θ (column 1), interplanar distances (column 2), Q/Q_{1st} (column 3), and hkl indexes of reflections (column 4) for crystal 2

1	2	3	4
2θ(°)	d (Å)	Q/Q_{1st}	hkl
38.523	2.3342	1	110
55.618	1.6505	2	200
69.694	1.3476	3	211
82.562	1.1671	4	220
95.056	1.0439	5	222
107.813	0.9529	6	310
121.575	0.8822	7	321
137.849	0.8252	8	400
163.451	0.7781	9	411

Q_{hkl} are obtained but the reflection to which the Q_{100} corresponds is not known, so it is based on an initial hypothesis that is to divide all the Q_{hkl} by the Q_{1^a} (third column of the Tables E1 and E2). This is assumed to be the Q_{100} since, from this relationship; you get a series of integers, as many as reflections. In the case of crystal 2, Q_{1st} does not correspond to reflection 100, since the n° 7 that cannot exist in the cubic system appears, which makes it assume, as the 2nd hypothesis that it corresponds to the simplest immediate reflexion, i.e., the 110. Therefore, the Q_{hkl}/Q_{110} must be multiplied by 2, obtaining the values that appear in Table E2.

2. In the cubic system, the values of h, k and l (fourth column of the Tables E1 and E2) are obtained from the expression $h^2 + k^2 + l^2 = N$, with N being an integer.

3. For the cubic system

$$d = \frac{a}{\sqrt{h^2 + k^2 + l^2}}$$

where

$a = d_{100} = 3.862$ Å for crystal 1. In the cubic system $a = b = c = 3.862$ Å
$a = d_{100} = 3.301$ Å for crystal 2. In the cubic system $a = b = c = 3.301$ Å
$\alpha = \beta = \gamma = 90°$

4. In crystal 1, the lattice is primitive as there is no non-extinction condition. In crystal 2, the lattice is cubic body centered since, for all reflections, it is fulfilled that $k + k + l = 2n$.

(c) Quantitative phase analysis

The quantitative phase analysis is based on the fact that the integral intensities of the reflections of a crystalline phase contained in a sample depend on the relative concentration of that phase in the sample. The relationship between intensity and concentration is not linear, due to absorption effects. The relative amount of the phases in a mixture can be determined by comparing the intensities of peaks in the sample with those from reference materials or by using an internal standard. In this method, the integrated intensity of a peak of the analysed phase is compared with the intensity of a peak of a phase added in known proportions (Eq. 17.26).

$$\frac{I_{ij}}{I_{is}} = k \frac{c_j}{c_s} \qquad (17.26)$$

where

i   is the reflexion
j   is the phase
k   is the slope of the straight line of I_{ij}/I_{is} versus the added quantity c_j is the weight
    fraction of initial j
c_s is the weight fraction of the standard

The material used as standard must fulfil a number of requirements. It must be chemically stable, have no overlapping peaks with the analysed phase, and have no preferential orientation, among other requirements.

Equation 17.27 is the basis for the Reference Intensity Ratio (RIR) method.

$$c_i = \frac{I_i I_j^{rel} c_j}{I_j I_i^{rel} RIR_{i,j}} \qquad (17.27)$$

It is necessary to obtain a calibration curve from standards, which is rather tedious since each component sought needs a calibration curve, each calibration curve needs at least three standards, and each standard must contain exactly the same percentage of pure reference material chosen.

In the method developed by Chung,[9] no calibration curve is needed. The method is based on previous knowledge, or measurement, of relative (reference) intensities of (the strongest) diffraction lines for each pair of phases that are present in the system, or rather for each phase and a reference phase (corundum, α-Al_2O_3). All information related to the quantitative composition of the system can be decoded directly from its diffraction pattern.

With the Rietveld method, the total diffractogram is considered the sum of the individual patterns of each phase, and the information is extracted without separating into components. It is necessary to know the crystalline structure of the component phases, and the difference between the experimental and calculated diffractogram is minimized (Eq. 17.28)

$$R = \sum w_i |Y_i(o) - Y_i(c)|^2 \qquad (17.28)$$

where

$Y_i(o)$ and $Y_i(c)$ are the observed and calculated intensity, respectively, at the i-th point of the data set.

---

[9] Chung, H. (1973). Quantitative Interpretation of X-ray Diffraction Patterns of Mixtures. I. Matrix-Flushing Method for Quantitative Multicomponent Analysis. Journal of Applied Crystallography, 7, 519–529.

Chung, F.H. (1975). Quantitative interpretation of X-ray diffraction patterns of mixtures. III. Simultaneous determination of a set of reference intensities. Journal of Applied Crystallography, 8, 17–19.

Quantitative information for each phase is obtained from the values of the scaling factors.

(d) Crystallite size

X-ray diffraction reflections broaden when the crystal lattice becomes imperfect and, therefore, the reflection broadening carries direct information about the lattice imperfections. In general, these lattice imperfections are classified into two types: nanometer-sized crystals and lattice defects. Thus, the crystal sizes and lattice microstrains of materials are evaluated using the line-broadening theory of X-ray powder diffractometry. The crystallite size can be calculated in a range of about 50 to 300 nm from the full width half maximum (FWHM) measurements using the Scherrer Eq. 17.29[10]

$$crystallite\ size = \frac{K\lambda}{FWHM\cos\theta} \tag{17.29}$$

where K is the shape factor, which is usually close to the default value 0.9, λ is the wavelength of X-ray, and θ is the diffraction angle or Bragg's angle. The values of FWHM and θ are obtained from the diffraction patterns.

The lattice strain arising from crystal imperfection and distortion was calculated using the empirical Eq. 17.30[11]

$$lattice\ strain = \frac{FWHM}{\tan\theta} \tag{17.30}$$

(e) Peak intensities

Phase identification is done primarily by comparing peak positions, although if a match between a database pattern and an experiment is to be considered good, the relative intensities of the peaks should be similar.

Peak positions are determined by the lattice.

Peak intensities are determined by the positions of the atoms in the unit cell.

We can use intensities to figure out where the atoms are (structure analysis).

(f) Study of solid solutions

Variation in the chemical composition of a known substance involves the substitution of atoms, usually of slightly different size, in specific positions in the structure.

As a result of this substitution, the cell dimensions (*a, b,* and *c*) change slightly and, therefore, so do the lattice spacings.

---

[10] Scherrer, P., 1918. Bestimmung der Gröss und der Inneren Struktur von Kolloidteilchen Mittels Röntgenstrahlen. Nachrichten von der Gesellschaft der Wissenschaften, Göttingen. Mathematisch-Physikalische Klasse, 2, 98–100.

[11] Stokes, A.R. and Wilson, A.J.C., 1944. The diffraction of X rays by distorted crystal aggregates. Proceedings of the Physical Society, 56 (3), 174–181.

The positions of the reflections (peaks in a graphical register of the powder diffractometer or lines in the film with Debye–Scherrer camera), i.e., the values of 2θ to which the reflections occur, corresponding to these spacings change as well.

By measuring these small changes in the position of the peaks in a diagram or lines in a film, i.e., changes in the values of 2θ to which reflections occur in a powder pattern of well-known structure substances, changes in chemical composition can be detected.

If the lattice parameters of the pure species in a mixed crystal system are known, the lattice parameters for the solid solution often follows Vegard's law, which means a linear relation between lattice parameter and composition. With a precise measurement of the lattice parameters, the composition of the mixed crystal can be calculated.

(g) **Texture study**

A polycrystalline aggregate like a crystalline powder is supposed to have all the crystallites ideally oriented randomly, relative to each other, and behaves like cone a substance of isotropic characteristics. However, there are a large number of polycrystalline substances that contain a certain portion of crystallites oriented in a specific direction. This has the consequence of highlighting anisotropic characteristics and tends to behave like a monocrystal. It is then said that the substance has preferential orientation.

(h) **Determination of thermal expansion coefficients**

The linear expansion coefficient of a crystal is different in the different crystallographic directions. When the spacing of the planes undergoes a variation, the effect will be revealed in an offset of the 2θ values of the corresponding reflections of the diffraction diagram, according to the Bragg equation.

The expansion coefficient is given by Eq. 17.28

$$\alpha = \frac{\Delta d}{d}\frac{1}{\Delta T} = \frac{d_2 - d_1}{d_1(T_2 - T_1)} \tag{17.28}$$

where

d is displacement
T is the temperature

This expression related to Bragg's equation is put into the form

$$d = \frac{\lambda}{2\sin\theta} \tag{17.29}$$

So, if two powder diagrams of a substance such as silver are obtained at 18 °C and 500 °C, the expansion causes the peaks in both diagrams to be displaced, so that d_{hkl} can be obtained for the two temperatures, in general d_1 for reflections at 18 °C and d_2 for reflections at 500 °C. Thus, we will have two Bragg equations such as the one shown above, in which the values of d_1 and d_2 obtained from the coefficient of thermal expansion are replaced.

17.11.6 Goniometric Methods: 4-circle Diffractometer

Goniometric methods displaced the earlier X-ray diffraction methods after the introduction of digital computers in the late 1970s. At that time, automatic four-circle diffractometers were designed. They have a goniometric system with very precise mechanics and, by means of three rotations, allow the crystal to be placed in any orientation in space, thus fulfilling the requirements of Ewald's construction for diffraction to occur. With a fourth axis of rotation by the electronic detector, the diffracted beam is recorded. All these movements can be programmed to be performed automatically.

Goniometric methods are used to obtain very detailed information about the X-ray diffraction patterns of single crystals.

With gonometric methods, two detectors are used, (1) point detectors (2) area detectors.

(1) Point detectors

Point detectors were first used and were later replaced by area detectors. In point detectors, the detection of each diffracted beam (reflection) is carried out individually, requiring the four angular values of the goniometer to be changed automatically and programmed for each diffracted beam. Measurement times in this equipment are usually in the order of one minute per reflection.

(2) Area detectors

Area detectors allow the detection of many diffraction beams simultaneously, thus saving time in the experiment. This technology is particularly useful in the case of proteins and, in general, of any material that may deteriorate during exposure to X-rays, since the detection of each of the images collected (with several hundred or thousands of reflections) is done in a minimum time, in the order of seconds.

One of the most commonly used area detectors (Fig. 17.27) is based on the so-called CCDs (Charge Coupled Device) which are normally installed on goniometric equipment with Kappa (κ) geometry. Because of their speed; their use is widespread in the field of protein crystallography, associated with rotating anode generators or in large synchrotron installations. The most modern technology involves the use of area detectors based on the so-called CMOS (Complementary Metal-Oxide Semiconductor) technology, which allows a very short readout time, leading to a considerable increase in diffraction imaging rate during the data collection process.

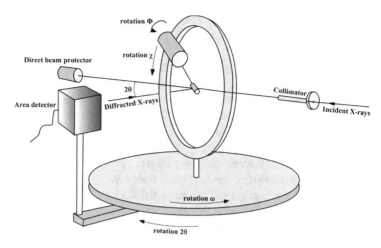

Fig. 17.27 Area detector

Two geometries can be used with the goniometric methods:

– Eulerian geometry, where the crystal is oriented by means of the three Eulerian angles, as follows:

Φ represents the rotation about the axis of the goniometer head,
χ allows swinging about the closed circle, and
ω allows full rotation of the goniometer.

The fourth circle represents the detector rotation, 2θ.

– Kappa geometry, which does not have a closed circle equivalent to the χ. Instead, the function is performed by the so-called κ (kappa) and ω_κ axes, so that by combining them, χ is obtained within the range from -90 to $+90°$. The main advantage of this geometry is wide accessibility to the crystal. The angles Φ and 2θ are identical to those presented in Eulerian geometry.

17.11.7 Synchrotron Radiation-Based Methods

The principles of the synchrotron were described in 1945 almost simultaneously by McMillan (1907–1991) at the University of California and Veksler (1907–2006) in the former Soviet Union. The first synchrotron was built in California by McMillan.

The synchrotron is an accelerator of charged particles traveling in a toroidal shaped tube (magnets are used along the tube so that the Lorentz magnetic force[12] maintains the curved trajectory of the particles). There are different types of accelerators, including electron synchrotron, proton synchrotron or tevatron, storage rings, and particle colliders.

In particle accelerator synchrotrons, particles are kept in a closed orbit. In them, two different particle beams are accelerated in opposite directions to study the products of their collision. In other synchrotrons, known as storage rings, a beam of particles of a single type is kept circulating indefinitely at a fixed energy. It is used as synchrotron light sources to study materials at atomic radius resolution in medicine and in manufacturing processes and materials characterization. A third use of synchrotrons is as a pre-accelerator of particles prior to their injection into a storage ring. These synchrotrons are known as boosters.

In the synchrotron, the electric and magnetic fields vary. The maximum speed at which particles can be accelerated is given by the point at which the synchrotron radiation emitted by the spinning particles is equal to the energy supplied.

In large synchrotron installations, the generation of X-rays is different. A synchrotron installation contains a very large ring (on the order of kilometers) through which electrons are circulated at very high speed inside rectilinear channels that occasionally break to adapt to the curvature of the ring (Fig. 17.28). These electrons are made to change direction to move from one channel to other using high-energy magnetic fields. It is at that moment, in the change of direction, that the electrons emit a very energetic radiation called synchrotron radiation. This radiation is composed of a continuum of wavelengths ranging from microwaves to so-called hard X-rays.

Fig. 17.28 Scheme of the trajectories followed by the electrons inside the synchrotron

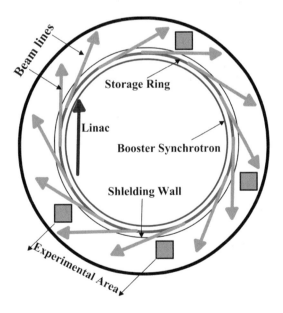

---

[12] The Lorentz force is the force exerted by the electromagnetic field that receives a charged particle or an electric current.

The X-rays obtained in synchrotron installations have two great advantages for crystallography:

(1) The wavelength can be adjusted at will, and
(2) their brightness (magnitude related to intensity) is at least one trillion times (10^{21}) higher than that of conventional X-ray sources.

Examples of synchrotron applications include the following:

– In chemistry, to follow a chemical reaction with nanosecond intervals. This technique has been used, for example, to study the behavior of catalytic materials under real operating conditions.
– In the study of matter under special conditions, for example, diffraction experiments at very high pressures using a diamond cell, consisting of two diamond crystal materials of interest, which is confined in a special capsule. The high hardness of diamond allows pressures millions of times higher than atmospheric pressure. These pressures are comparable to those attained at the center of the earth and, therefore, make it possible to study the state of matter at 5500 °C and 3.6 million bars, in order to try to answer the question of what is the real structure of Fe–Ni alloys in the center of the earth.
– In structural biology, to investigate the position in space of each of the molecules that form small viruses or proteins, this is the first step towards the realization of specific drugs. For example, recently, D. Stuart and his group (Oxford, UK) at the European synchrotron in Grenoble have managed to measure 50.000 reflections in an X-ray diffraction pattern of a crystal of the bluetongue virus, which causes many deaths in cattle. The precision in the position of the molecules is 0.35 nm. This is the largest molecular structure ever solved.
– In medical uses, such as mammography and angiography.
– In industry. Some multinational companies that invest in their own research install lines where they can study faults in some of the electronic components they manufacture, to make masks by means of lithographic treatments for both microelectronic and mechanical applications, or to calibrate instruments.

These devices do not produce any type of residue or environmental radioactivity, since the electrons cannot leave the vacuum tube without interacting with the air and recombining with its molecules. Therefore, a failure, for example, of electricity or in the vacuum system or in acceleration only causes loss of the beam.

Exercises

1 The maximum diffraction values of 2θ for two crystals, 1 and 2, obtained with radiation $K_\alpha Cu$ ($\lambda = 1.540$ Å) are in Tables E1 and E2.

Table E1 Maximum diffraction values of 2θ (column 1), interplanar distances (column 2), Q/Q_{1st} (column 3), and hkl indexes of reflections (column 4) for crystal 1

1	2	3	4
$2\theta(°)$	d (Å)	Q/Q_{1st}	hkl
23.03	3.862	1	100
32.81	2.730	2	110
40.38	2.234	3	111
47.07	1.931	4	200
53.00	1.728	5	210
58.54	1.577	6	211
68.78	1.365	8	220
73.61	1.287	9	221 or 300
78.31	1.221	10	310
82.96	1.164	11	311
87.49	1.115	12	222
92.09	1.071	13	320

Table E2 Maximum diffraction values of 2θ (column 1), interplanar distances (column 2), Q/Q_{1st} (column 3), and hkl indexes of reflections (column 4) for crystal 2

1	2	3	4
$2\theta(°)$	d (Å)	Q/Q_{1st}	hkl
38.523	2.3342	1	110
55.618	1.6505	2	200
69.694	1.3476	3	211
82.562	1.1671	4	220
95.056	1.0439	5	222
107.813	0.9529	6	310
121.575	0.8822	7	321
137.849	0.8252	8	400
163.451	0.7781	9	411

1. Deduce the crystal system.
2. Assign indices to reflections.
3. Calculate cell parameters and axial angles.
4. Extract the possible information about the space group.

2 The maximum diffraction values of 2θ for one crystal, obtained with radiation $K_\alpha Cu$ ($\lambda = 1.540$ Å) are in Table E2.

1. Identify the phase of the X-ray diffracted material.

2θ(°)	I (%)	2θ(°)	I (%)
20.9026	12.83	50.1321	20.2
24.0353	0.37	51.0180	2.13
26.6522	100	54.8530	5.36
29.4308	3.66	55.3376	2.95
30.8973	7.78	57.3400	0.63
33.2707	0.21	59.9500	13.92
35.0835	0.42	6353760	0.66
36.5581	8.22	64.0231	2.50
39.4586	9.76	65.7206	1.06
40.3046	4.27	67.7073	9.49
41.0689	2.40	68.1200	15.23
42.4686	7.32	68.3164	13.25
43.1755	0.92	70.3056	0.43
44.8809	1.33	73.4268	3.43
45.7866	5.08	75.5979	4.32
47.5290	1.00	77.6061	2.39
48.5332	0.89		

Questions

1. What does an X-ray diffraction diagram represent?
   - O a. direct lattice
   - O b. crystalline planes
   - O c. crystalline structure
   - O d. reciprocal lattice
2. Laue suggested that the periodic structure of a crystal could be used to diffract X-rays because

   O a. Crystals are periodic, X-rays are waves, and the length of X-rays is the same order of magnitude as the distance that the motifs in the crystals repeat.

   O b. Crystals are homogeneous, X-rays are waves, and the length of X-rays is the same order of magnitude as the distance repeated by the motifs in the crystals.

   O c. Crystals are periodic, X-rays are waves, and the length of X-rays is the same order of magnitude as the distance that the planes repeat in the crystals.

   O d. Crystals are symmetrical, X-rays are waves, and the length of X-rays is the same order of magnitude as the distance that the planes repeat in the crystals.

3. The range of wavelengths used in X-ray diffraction is between

   ○ a. 0.5 and 2.5 nm
   ○ b. 0.5 and 2.5 Ångstroms (Å)
   ○ c. 0.5 and 2.5 cm
   ○ d. 0.5 and 2.5 m

4. For a plane to diffract the X-rays

   ○ a. Its orientation should be such that its reciprocal point is on the diameter of the Ewald sphere.
   ○ b. Its orientation should be such that its reciprocal point is outside the sphere of Ewald.
   ○ c. Your guidance should be such that your reciprocal point is located anywhere.
   ○ d. Its orientation must be such that its reciprocal point is on the sphere of Ewald.

5. For any plane (*hkl*) to diffract X-rays, its orientation must be such that its representative reciprocal point is located on the surface of the ___

   Response: [   ]

6. In almost all experimental devices, the interaction of X-radiation with a free electron is

   (a) polarized
   (b) non-polarized

   Choose a or b

   Response: [   ]

7. What is the name of the factor referring to the scattering of radiation by the electrons of an atom?
   In the answer, do not repeat terms and, if more than one is required, separate them by a blank

   Response: [   ]

8. How many point groups correspond to Laue classes?

   Response: [   ]

9. What collides with matter for X-rays to occur?

   Response: [   ]

10. The scattering vector OP $(S–S_0)$ must have equal magnitude and direction as which vector?

   Response: [   ]

11. Can diffracted X-rays follow in any direction? (Yes or No)

   Response: [   ]

12. Continuous spectrum occurs as an effect of electrostatic electron interactions in the vicinity of the nuclei of the anode atoms.

   ○ True
   ○ False

13. The point group obtained from a diffraction diagram has a ternary axis of its own rotation
    Select one:

   ○ True
   ○ False

14. Destructive interferences in X-ray diffraction occurs when interfering waves are out of phase and the trajectory difference is half the wavelength.

   ○ True
   ○ False

15. Laue groups are spatial groups with a center of symmetry.

   ○ True
   ○ False

16. The rate of spread of X-rays relative to that of light is different

   ○ True
   ○ False

17. Write an x if the point group coincides with a Laue class.

222	
mmm	
422	
$m\,\bar{3}m$	
mm2	
$\bar{1}$	

18. Pair the sample type with the X-ray technique

Laue	crystal powder method
oscillating crystal	monocrystal
method of powder	crystal powder method

19. Describe the sample used in the crystalline powder method in X-ray diffraction.

    (a) many randomly oriented crystallites
    (b) monocrystal
    (c) many crystallites oriented in a certain direction

Respond with a, b or c.

Response: [   ]

20. The sphere of reflection containings nodes of the reciprocal network that correspond to planes that diffract the X-rays.

   O True
   O False

Appendix I Correspond to Chapter 4

See Tables A.1, A.2, A.3, A.4, A.5, A.6 and A.7.

Table A.1 Triclinic crystalline system

Crystal system + Crystalline class				
Crystalline class	Punctual group	Special forms	General forms	Stereographic projection
Triclinic + Pedial				
Pedial (Hemiedry)	1	No special form	Pedion (*hkl*)	
Triclinic + Pinacoidal				
Pinacoidal (Holohedry)	$\bar{1}$	No special form	Pinacoid $(hkl)(\overline{hkl})$	

© The Editor(s) (if applicable) and The Author(s), under exclusive license to
Springer Nature Switzerland AG 2022
C. Marcos, *Crystallography*, Springer Textbooks in Earth Sciences,
Geography and Environment, https://doi.org/10.1007/978-3-030-96783-3

Table A.2 Monoclinic crystalline system

Crystal system + Crystalline class				
Crystalline class	Punctual group	Special forms	General forms	Stereographic projection
Monoclinic + Sphenoidal				
Sphenoidal (Enantiomorphic hemiedry)	2	Pedion (010) Pinacoid {h0l}	Sphenoid $(hkl)(\bar{h}k\bar{l})$	
Monoclinic + Domatic				
Domatic (Hemimorphic hemiedry)	m	Pedion (h0l) Pinacoid {010}	Dome $(hkl)(h\bar{k}l)$	
Monoclinic + Prismatic				
Prismatic (Holohedry)	2/m	Pinacoids {010}, {h0l}	Rhombic prism $(hkl)(\bar{h}k\bar{l})\overline{(hkl)}(h\bar{k}l)$	

Table A.3 Orthorhombic crystalline system

Crystal system + Crystalline class				
Crystalline class	Punctual group	Special forms	General forms	Stereographic projection
Orthorhombic + Disphenoidal				
Disphenoidal (Enantiomorphic hemiedry)	222	Pinacoids {100}, {010}, {001} Rhombic prisms {hk0}, {0kl}, {h0l}	Rhombic diesphenoid $(hkl)(\bar{h}\bar{k}l)(\bar{h}k\bar{l})(h\bar{k}\bar{l})$	
Orthorhombic + Pyramidal				
Pyramidal (Hemimorphic hemiedry)	mm2	Pedion (001) Pinacoid {001} Orthorhombic prism {hk0} Domes {h0l}, {0kl}	Rhombic pyramid $(hkl)(\bar{h}kl)(\bar{h}\bar{k}l)(h\bar{k}l)$	
Orthorhombic + Dipyramidal				
Dipyramidal (Holohedry)	mmm or $2/m\,2/m\,2/m$	Pinacoids {100}, {010}, {001} Orthorhombic prisms {hk0}, {h0l}, {0kl}	Rhombic dipyramid $(hkl)(\bar{h}kl)(\bar{h}\bar{k}l)(h\bar{k}l)$ $(\bar{h}k\bar{l})(h\bar{k}\bar{l})(\bar{h}\bar{k}\bar{l})(hk\bar{l})$	

Table A.4 Tetragonal crystalline system

Crystal system + Crystalline class				
Crystalline class	Punctual group	Special forms	General forms	Stereographic projection
Tetragonal + Pyramidal				
Pyramidal (Tetartohedry)	4	Pedion (001) Tetragonal prism$\{hk0\}$	Tetragonal pyramid $(hkl)(\overline{hkl})(\overline{k}hl)(k\overline{h}l)$	
Tetragonal + Disphenoidal				
Disphenoidal (Tetartohedry with inversion)	$\overline{4}$	Pinacoid $\{001\}$ Tetragonal prism $\{hk0\}$	Tetragonal diesphenoid or tetragonal tetrahedron $(hkl)(\overline{hkl})(\overline{k}hl)(k\overline{h}l)$	
Tetragonal + Dipyramidal				
Dipyramidal (Enantiomorphic hemiedry)	$4/m$	Pinacoid $\{001\}$ Tetragonal prism $\{hk0\}$	Tetragonal dipyramid $(hkl)(\overline{hkl})(\overline{k}hl)(k\overline{h}l)$ $(\overline{hk}l)(h\overline{k}l)(k\overline{hl})(\overline{k}h\overline{l})$	

(continued)

Table A.4 (continued)

Crystal system + Crystalline class

Crystalline class	Punctual group	Special forms	General forms	Stereographic projection
Tetragonal + Trapezohedral				
Trapezohedral (Enantiomorphic hemiedry)	422	Pinacoid {001} Tetragonal prisms {100}, {110} Ditetragonal prism {hk0} Tetragonal dipyramids {hhl}, {h0l}	Tetragonal trapezohedron $(hkl)\,(\overline{h}\overline{k}l)\,(\overline{k}hl)\,(k\overline{h}l)$ $(\overline{h}kl)\,(h\overline{k}l)\,(\overline{k}\overline{h}l)\,(\overline{k}hl)$	
Tetragonal + Ditetragonal-pyramidal				
Ditetragonal-pyramidal (Hemimorphic hemiedry)	4mm	Pedion (001) Tetragonal prisms {100}, {110} Ditetragonal prism {hk0} Tetragonal pyramids {hhl}, {h0l}	Ditetragonal pyramid $(hkl)\,(\overline{h}\overline{k}l)\,(\overline{k}hl)\,(k\overline{h}l)$ $(\overline{h}kl)\,(h\overline{k}l)\,(\overline{k}\overline{h}l)\,(khl)$	
Tetragonal + Scalenohedral				
Scalenohedral (Hemidry with inversion)	$\overline{4}2m$	Pinacoid {001} Tetragonal prisms {100}, {110} Ditetragonal prism{hk0} Tetragonal dipyramid {h0l} Tetragonal diesphenoids or tetragonal tetrahedron {hhl}	Tetragonal scalenohedron $(hkl)\,(\overline{h}\overline{k}l)\,(\overline{k}hl)\,(k\overline{h}l)$ $(\overline{h}kl)\,(h\overline{k}l)\,(\overline{k}\overline{h}l)\,(khl)$	

(continued)

Table A.4 (continued)

Crystal system + Crystalline class

Crystalline class	Punctual group	Special forms	General forms	Stereographic projection
Tetragonal + Scalenohedral				
Scalenohedral (Hemidry with inversion)	$\bar{4}\,m2$	Pinacoid {001} Tetragonal prisms {100}, {110} Ditetragonal prism {hk0} Tetragonal dipyramid {hhl} Tetragonal diesphenoid or tetragonal tetrahedron {h0l}	Tetragonal scalenohedron $(hkl)\,(\bar{h}kl)\,(k\bar{h}l)\,(\bar{k}\bar{h}l)$ $(\bar{h}kl)\,(h\bar{k}l)\,(\bar{k}hl)\,(khl)$	
Tetragonal + Ditetragonal-dipyramidal				
Ditetragonal-dipyramidal (Holohedry)	$4/m\,mm$ or $4/m\,2/m\,2/m$	Pinacoid {001} Tetragonal prisms {100}, {110} Ditetragonal prism {hk0} Tetragonal dipyramids {h0l}, {hhl}	Ditetragonal dipyramid $(hkl)\,(\bar{h}kl)\,(kh l)\,(\bar{k}hl)$ $(\bar{h}kl)\,(h\bar{k}l)\,(k\bar{h}l)\,(\bar{k}\bar{h}l)$ $(h\bar{k}l)\,(\bar{h}\bar{k}l)\,(\bar{k}\bar{h}l)\,(k\bar{h}l)$ $(\bar{h}\bar{k}l)\,(h\bar{k}l)\,(\bar{k}hl)\,(khl)$	

Table A.5 Trigonal or rhombohedral crystalline system (With hexagonal axes, Bravais-Miller indices $(hk\,\bar{\imath}\,l)$ are used, with $h + k + i = 0$)

Crystal system + Crystalline class				
Crystalline class	Punctual group	Special forms	General forms	Stereographic projection
Trigonal + Pyramidal tetartohedral				
Pyramidal tetartohedral (Tetartohedry)	3	Pedion (0001) Trigonal prism $\{hki0\}$	Trigonal pyramid $(hkil)(ihkl)(kihl)$	
Trigonal + Rhombohedral				
Rhombohedral (Enantiomorphic hemiedry)	$\bar{3}$	Pinacoid $\{0001\}$ Hexagonal Prism $\{hki0\}$	Rhombohedron $(hkil)(ihkl)(kihl)$ $(\bar{h}\bar{k}\bar{\imath}\bar{l})(\bar{\imath}\bar{h}\bar{k}\bar{l})(\bar{k}\bar{\imath}\bar{h}\bar{l})$	
Trigonal + Trapezohedral				
Trapezohedral (Enantiomorphic hemiedry)	321	Pinacoid $\{0001\}$ Hexagonal prism $\{10\bar{1}0\}$ Trigonal prism $\{11\bar{2}0\}$ or $\{\bar{1}\bar{1}20\}$ Ditrigonal prism $\{hki0\}$ Rhombohedron $\{h0\bar{h}l\}$ Trigonal dipyramid $\{hh\bar{2}hl\}$	Trigonal trapezohedron $(hkil)(ihkl)(kihl)$ $(kh\bar{\imath}\bar{l})(ih\bar{k}\bar{l})(ikh\bar{l})$	

(continued)

Table A.5 (continued)

Crystal system + Crystalline class

Crystalline class	Punctual group	Special forms	General forms	Stereographic projection
Trigonal + Trapezohedral				
Trapezohedral (Enantiomorphic hemiedry)	312	Pinacoid $\{0001\}$ Hexagonal Prism $\{10\bar{1}0\}$ Trigonal prism $\{11\bar{2}0\}$ or $\{\bar{1}1\bar{2}0\}$ Ditrigonal Prism $\{hki0\}$ Rhombohedron $\{h0\bar{h}l\}$ Trigonal dipyramid $\{hh\overline{2h}l\}$	Trigonal trapezohedron $(hkil)$ $(ihkl)$ $(kihl)$ $(\overline{khil})$ $(\overline{hikl})$ $(\overline{ikhl})$	
Trigonal + Ditrigonal pyramidal				
Ditrigonal pyramidal (Hemimorphic hemiedry)	3*m*1	Pedion (0001) Trigonal prism $\{10\bar{1}0\}$ o $\{\bar{1}010\}$ Hexagonal prism $\{11\bar{2}0\}$ Ditrigonal prism $\{hki0\}$ Trigonal pyramid $\{h0\bar{h}l\}$ Hexagonal pyramid $\{hh\overline{2h}l\}$	Ditrigonal pyramid $(hkil)$ $(ihkl)$ $(kihl)$ $(\overline{khil})$ $(\overline{hikl})$ $(\overline{ikhl})$	

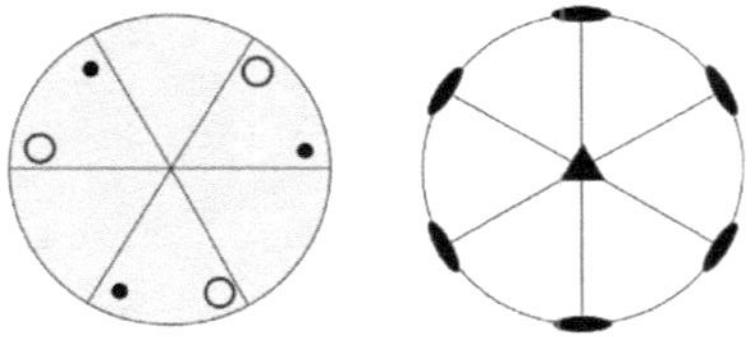

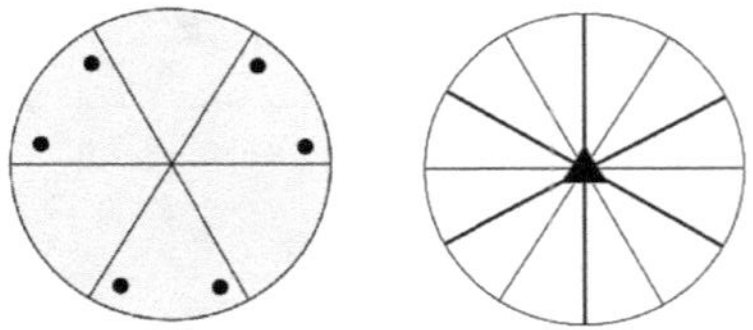

(continued)

Table A.5 (continued)

Crystal system + Crystalline class

Crystalline class	Punctual group	Special forms	General forms	Stereographic projection
Trigonal + Ditrigonal pyramidal				
Ditrigonal pyramidal (Hemimorphic hemiedry)	$31m$	Pedion (0001) or $(000\overline{1})$ Trigonal prism $\{11\overline{2}0\}$ or $\{11\overline{2}0\}$ Hexagonal prism $\{10\overline{1}0\}\{01\overline{1}0\}$ Ditrigonal prism $\{hki0\}$ Trigonal pyramid $\{hh\overline{2}hl\}$ Hexagonal pyramid $\{h0\overline{h}l\}$	Ditrigonal pyramid $(hkil)\,(ihkl)\,(kihl)$ $(khil)\,(hikl)\,(ikhl)$	
Trigonal + Ditrigonal scalenohedral				
Ditrigonal scalenohedral (Holohedry)	$\overline{3}m1$ or $\overline{3}\,^2/_m1$	Pinacoid $\{0001\}$ Hexagonal prisms $\{10\overline{1}0\}$, $\{11\overline{2}0\}$ Dihexagonal prism $\{hki0\}$ Hexagonal dipyramid $\{hh\overline{2}hl\}$ Rhombohedron $\{h0\overline{h}l\}$	Ditrigonal scalenohedron $(hkil)\,(ihkl)\,(kihl)$ $(khil)\,(hikl)\,(ikhl)$ $(\overline{hkil})\,(\overline{ihkl})\,(\overline{kihl})$ $(\overline{khil})\,(\overline{hikl})\,(\overline{ikhl})$	

(continued)

Table A.5 (continued)

Crystal system + Crystalline class				
Crystalline class	Punctual group	Special forms	General forms	Stereographic projection
Trigonal + Ditrigonal scalenohedral				
Ditrigonal scalenohedral (Holohedry)	$\bar{3}1m$ or $\bar{3}1\frac{2}{m}$	Pinacoid $\{0001\}$ Hexagonal prisms $\{10\bar{1}0\}$, $\{11\bar{2}0\}$ Dihexagonal Prism $\{hki0\}$ Hexagonal dipyramid $\{hh\bar{2}hl\}$ Rhombohedron $\{h0\bar{h}l\}$	$(hki l)\,(ihkl)\,(kihl)$ $(khil)\,(hik l)\,(ikhl)$ $(\bar{h}k\bar{i}l)\,(\bar{i}hkl)\,(\bar{k}i\bar{h}l)$ $(\bar{k}h\bar{i}l)\,(\bar{h}i\bar{k}l)\,(\bar{i}k\bar{h}l)$	

Table A.6 Hexagonal crystalline system

Crystal system + Crystalline class				
Crystalline class	Punctual group	Special forms	General forms	Stereographic projection
Hexagonal + Pyramidal				
Pyramidal (Tetartohedry)	6	Pedion {0001} Hexagonal prism {$hki0$}	Hexagonal pyramid $(hkil)(ihkl)(kihl)$ $(\overline{h}ki\overline{l})(i\overline{h}k\overline{l})(\overline{k}i\overline{h}l)$	
Hexagonal + Trigonal dipyramidal				
Trigonal dipyramidal (Tetartohedry with inversion)	$\overline{6}$	Pinacoid {0001} Trigonal prism {$hki0$}	Trigonal dipyramid $(hkil)(ihkl)(kihl)$ $(hki\overline{l})(ihk\overline{l})(kih\overline{l})$	
Hexagonal + Dipyramidal				
Dipyramidal (Paramorphic hemiedry)	$6/m$	Pinacoid {0001} Hexagonal prism {$hki0$}	Hexagonal dipyramid $(hkil)(ihkl)(kihl)$ $(hki\overline{l})(ihk\overline{l})(kih\overline{l})$ $(\overline{h}ki\overline{l})(i\overline{h}k\overline{l})(\overline{k}i\overline{h}l)$ $(\overline{h}ki\overline{l})(\overline{i}hk\overline{l})(\overline{k}i\overline{h}l)$	

(continued)

Table A.6 (continued)

Crystal system + Crystalline class				
Crystalline class	Punctual group	Special forms	General forms	Stereographic projection
Hexagonal + Trapezohedral				
Trapezohedral (Enantiomorphic hemiedry)	622	Pinacoid {0001} Hexagonal prism {10$\bar{1}$0} Dihexagonal prism {hki0} Hexagonal Prisms {hk$\bar{i}$0}, {11$\bar{2}$0} Hexagonal dipyramids {h0$\bar{h}$l}, {hh$\overline{2h}$l}	Hexagonal trapezohedron ($hki\bar{l}$)($ihkl$)($kihl$) ($khil$)($hikl$)($ikhl$) ($\bar{k}\bar{i}h\bar{l}$)($\bar{h}\bar{i}k\bar{l}$)($\bar{i}\bar{k}h\bar{l}$) ($\bar{h}\bar{k}i\bar{l}$)($\bar{i}h\bar{k}l$)($\bar{k}\bar{i}h\bar{l}$)	
Hexagonal + Dihexagonal pyramidal				
Dihexagonal pyramidal (Hemimorphic hemiedry)	6mm	Pedion {0001} Hexagonal prisms {10$\bar{1}$0}, {11$\bar{2}$0} Hexagonal Dihexagonal prism {hki0} Hexagonal pyramids {h0$\bar{h}$l}, {hh$\overline{2h}$l}	Dihexagonal pyramid ($hki\bar{l}$)($ihkl$)($kihl$) ($khil$)($hikl$)($ikhl$) ($\bar{k}\bar{i}hl$)($\bar{h}\bar{i}kl$)($\bar{i}\bar{k}hl$) ($\bar{h}\bar{k}il$)($\bar{i}h\bar{k}l$)($\bar{k}\bar{i}hl$)	
Hexagonal + Ditrigonal dipyramidal				
Ditrigonal dipyramidal (Hemiedry with inversion)	$\bar{6}$m2	Pinacoid {0001} Trigonal prism {10$\bar{1}$0} or {$\bar{1}$010} Hexagonal prism {11$\bar{2}$0} Ditrigonal prism {hki0} Trigonal dipyramid {h0$\bar{h}$l} Hexagonal dipyramid {hh$\overline{2h}$l}	Ditrigonal dipyramid ($hki\bar{l}$)($ihkl$)($kihl$) ($hk\bar{i}l$)($ih\bar{k}l$)($ik\bar{h}l$) ($\bar{k}hil$)($\bar{h}\bar{i}kl$)($\bar{i}kh\bar{l}$)	

(continued)

Table A.6 (continued)

Crystal system + Crystalline class				
Crystalline class	Punctual group	Special forms	General forms	Stereographic projection
Hexagonal + Ditrigonal dipyramidal				
Ditrigonal dipyramidal (Hemiedry with inversion)	$\bar{6}2m$	Pinacoid $\{0001\}$ Trigonal prism $\{10\bar{1}0\}$ or $\{\bar{1}010\}$ Hexagonal prism $\{11\bar{2}0\}$ Ditrigonal prism $\{hki0\}$ Trigonal dipyramid $\{hh\bar{2}hl\}$ Hexagonal dipyramid $\{h0\bar{h}l\}$	Ditrigonal dipyramid $(hkil)\,(ihkl)\,(kihl)$ $(hki\bar{l})\,(ihk\bar{l})\,(kih\bar{l})$ $(\bar{k}hi\bar{l})\,(\bar{h}ik\bar{l})\,(\bar{i}kh\bar{l})$ $(\bar{k}hi\bar{l})\,(\bar{h}ik\bar{l})\,(\bar{i}kh\bar{l})$	
Hexagonal + Dihexagonal dipyramidal				
Dihexagonal dipyramidal (Holohedry)	$6/mmm$ or $6/m2/m2/m$	Pinacoid $\{0001\}$ Hexagonal prisms $\{10\bar{1}0\}$, $\{11\bar{2}0\}$ Dihexagonal prism $\{hki0\}$ Hexagonal dipyramids $\{h0\bar{h}l\}$, $\{hh\bar{2}hl\}$	Dihexagonal dipyramids $(hkil)\,(ihkl)\,(kihl)$ $(hki\bar{l})\,(ihk\bar{l})\,(kih\bar{l})$ $(\bar{h}i\bar{k}l)\,(\bar{i}h\bar{k}l)\,(\bar{k}ih\bar{l})$ $(\bar{h}i\bar{k}l)\,(\bar{i}h\bar{k}l)\,(\bar{k}ih\bar{l})$ $(\bar{k}hi\bar{l})\,(\bar{h}ik\bar{l})\,(\bar{i}kh\bar{l})$ $(\bar{k}hi\bar{l})\,(\bar{h}ik\bar{l})\,(\bar{i}kh\bar{l})$ $(khil)\,(hikl)\,(ikhl)$	

Table A.7 Cubic crystalline system

Crystal system + Crystalline class

Crystalline class	Punctual group	Special forms	General forms	Stereographic projection
Cubic + Tetrahedral-Pentagondodecahedral				
Tetrahedral-Pentagondodecahedral (tetartohedy)	23	Cube or hexahedron {100} Rombododecahedron {011} Pentagon-dodecahedron or pyritohedron {0kl} Tetrahedron {111} Tetragon-tritetrahedron or deltohedron {hhl}h > l Trigon-tritetrahedron or tristetrahedron {hhl}h < l	Pentagon- tritetrahedron or tetartoid $(hkl)\,(\bar{h}\bar{k}l)\,(h\bar{k}\bar{l})\,(\bar{h}k\bar{l})\,(lhk)\,(l\bar{h}\bar{k})$ $(\bar{l}h\bar{k})\,(\bar{l}\bar{h}k)\,(klh)\,(\bar{k}l\bar{h})\,(\bar{k}\bar{l}h)\,(k\bar{l}\bar{h})$	
Cubic + Disdodecahedral				
Disdodecahedral (Parametric hemiedry)	$m\bar{3}$	Cube or hexahedron {100} Rombododecahedron {110} Pentagon-dodecahedron or pyritohedron {0kl} Octahedron {111} Tetragon-trioctahedron or trapezohedron or deltoid-icositetrahedron {hhl}h < l Trigon-trioctahedron or trisoctahedron {hhl}h > l	Didodecahedron or diploid $(hkl)\,(\bar{h}\bar{k}l)\,(h\bar{k}\bar{l})\,(\bar{h}k\bar{l})\,(lhk)\,(l\bar{h}\bar{k})$ $(\bar{l}h\bar{k})\,(\bar{l}\bar{h}k)\,(klh)\,(\bar{k}l\bar{h})\,(\bar{k}\bar{l}h)\,(k\bar{l}\bar{h})$ $(\bar{h}\bar{k}\bar{l})\,(hk\bar{l})\,(\bar{h}kl)\,(h\bar{k}l)\,(\bar{l}\bar{h}\bar{k})\,(\bar{l}hk)$ $(lh\bar{k})\,(l\bar{h}k)\,(\bar{k}\bar{l}\bar{h})\,(k\bar{l}h)\,(kl\bar{h})\,(\bar{k}lh)$	

(continued)

Table A.7 (continued)

Crystal system + Crystalline class				
Crystalline class	Punctual group	Special forms	General forms	Stereographic projection
Cubic + Pentagon-icositetrahedral				
Pentagon-icositetrahedral (Enantiomorphic hemiedry)	432	Cube or hexahedron {100} Rombododecahedron {110} Tetrahexahedron or tetraquishexahedron {0kl} Octahedron {111} Tetragon-trioctahedron or trapezohedron or deltoid-icositetrahedron {hhl}h < l Trigon-trioctahedron or trisoctahedron {hhl}h > l	Pentagon-trioctahedron or Gyroid or pentagon-icositetrahedron $(hkl)\,(h\bar{k}l)\,(\bar{h}kl)\,(hk\bar{l})\,(lhk)\,(\bar{l}hk)$ $(\bar{l}hk)\,(l\bar{h}k)\,(klh)\,(\bar{k}l\bar{h})\,(\bar{k}lh)\,(kl\bar{h})$ $(kh\bar{l})\,(\bar{k}h\bar{l})\,(\bar{k}hl)\,(khl)\,(lkh)\,(\bar{l}kh)$ $(lk\bar{h})\,(l\bar{k}h)\,(hlk)\,(\bar{h}lk)\,(\bar{h}l\bar{k})\,(hl\bar{k})$	
Cubic + Hexakistetrahedral				
Hexakistetrahedral (Hemimorphic hemiedry)	$\bar{4}3m$	Cube or hexahedron {100} Rombododecahedron {110} Tetrahexahedron or tetraquishexahedron {0kl} Tetrahedron {111} or {$\bar{1}11$} Trigon-tritetrahedron or tristetrahedron {hhl}h < l Tetragon-tritetratrahedron or deltohedron or deltoid-dodecahedron {hhl}h > l	Hexatetrahedron or hexakystetrahedron $(hkl)\,(h\bar{k}l)\,(\bar{h}kl)\,(hk\bar{l})\,(lhk)\,(\bar{l}hk)$ $(\bar{l}hk)\,(l\bar{h}k)\,(klh)\,(\bar{k}l\bar{h})\,(\bar{k}lh)\,(kl\bar{h})$ $(kh\bar{l})\,(\bar{k}h\bar{l})\,(\bar{k}hl)\,(khl)\,(lkh)\,(\bar{l}kh)$ $(lk\bar{h})\,(l\bar{k}h)\,(hlk)\,(\bar{h}lk)\,(\bar{h}l\bar{k})\,(hl\bar{k})$	

(continued)

Table A.7 (continued)

Crystal system + Crystalline class

Crystalline class	Punctual group	Special forms	General forms	Stereographic projection
Cubic + Hexakisoctahedral				
Hexakisoctahedral (Holohedry)	$m\bar{3}m$ or $4/m\,\bar{3}\,2/m$	Cube or hexahedron $\{100\}$ Rombododecahedron $\{011\}$ Tetrahexahedron or tetraquishexahedron $\{0kl\}$ Octahedron $\{111\}$ Tetragon-trioctahedron or trapezohedron or deltoid-icositetrahedron $\{hhl\}h<l$ Trigon-trioctahedron or trisoctahedron $\{hhl\}h>l$	Hexaoctahedron $(hkl)\,(\bar{h}\bar{k}l)\,(h\bar{k}\bar{l})\,(\bar{h}k\bar{l})\,(lhk)\,(\bar{l}\bar{h}k)$ $(\bar{l}hk)\,(l\bar{h}k)\,(klh)\,(\bar{k}\bar{l}h)\,(\bar{k}lh)\,(k\bar{l}h)$ $(\bar{h}kl)\,(h\bar{k}l)\,(\bar{h}\bar{k}\bar{l})\,(hkl)\,(\bar{l}\bar{h}\bar{k})\,(lh\bar{k})$ $(lh\bar{k})\,(\bar{l}h\bar{k})\,(\bar{k}lh)\,(k\bar{l}\bar{h})\,(kl\bar{h})\,(\bar{k}\bar{l}\bar{h})$ $(kh\bar{l})\,(\bar{k}\bar{h}l)\,(\bar{k}hl)\,(k\bar{h}l)\,(\bar{h}lk)\,(h\bar{l}k)$ $(\bar{k}\bar{h}\bar{l})\,(khl)\,(\bar{l}hk)\,(\bar{l}\bar{h}\bar{k})\,(lhk)\,(\bar{l}hk)$ $(lk\bar{h})\,(\bar{l}\bar{k}h)\,(\bar{l}kh)\,(l\bar{k}h)\,(\bar{h}kl)\,(hk\bar{l})$ $(\bar{l}k\bar{h})\,(l\bar{k}\bar{h})\,(lk\bar{h})\,(\bar{l}\bar{k}\bar{h})\,(hlk)\,(\bar{h}\bar{l}k)$	

Appendix II Correspond to Chapter 4

Stereographic Projections of Minerals

(Drawn with Shape software[1])

Mineral	Figures and point groups
Andalusite	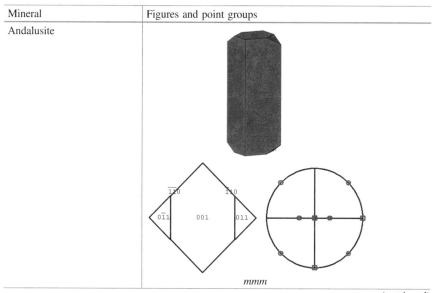
	mmm

(continued)

---

[1]Shape software, V7.1.2, 2004.

(continued)

Mineral	Figures and point groups
Apatite	
Biotite	
Calcite	

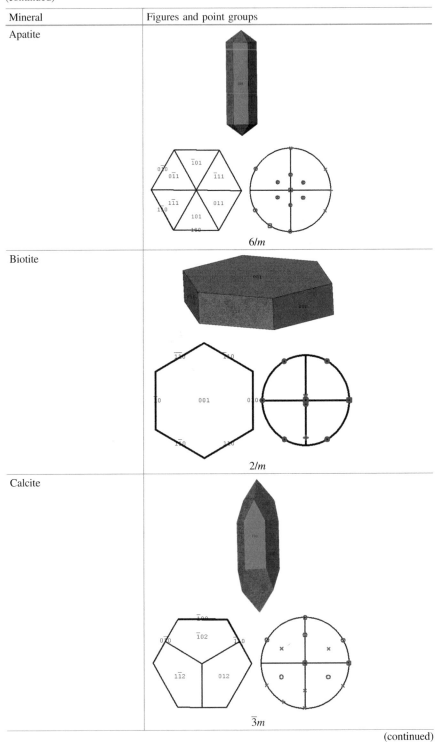

(continued)

Mineral	Figures and point groups
Cordierite	*mmm*
Epidote	*2/m*
Fluorite	$m\,\overline{3}\,m$

(continued)

Mineral	Figures and point groups
Hornblende	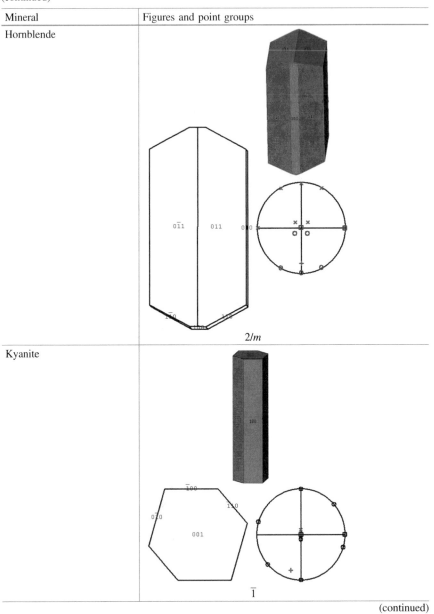
Kyanite	

(continued)

(continued)

Mineral	Figures and point groups
Pyrope garnet	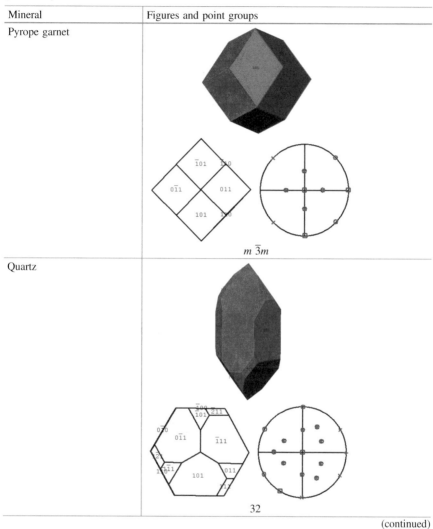
Quartz	

(continued)

Mineral	Figures and point groups
Tourmaline	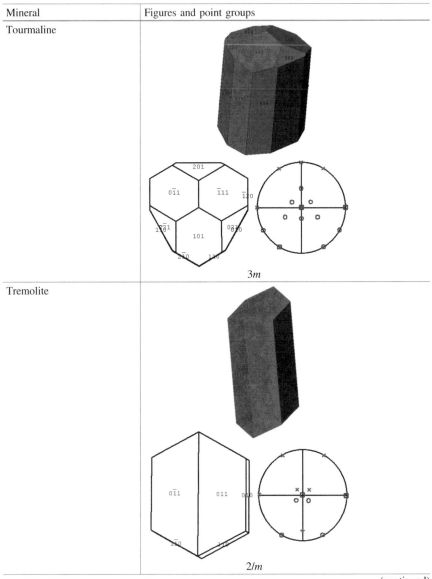

(continued)

Mineral	Figures and point groups
Sillimanite	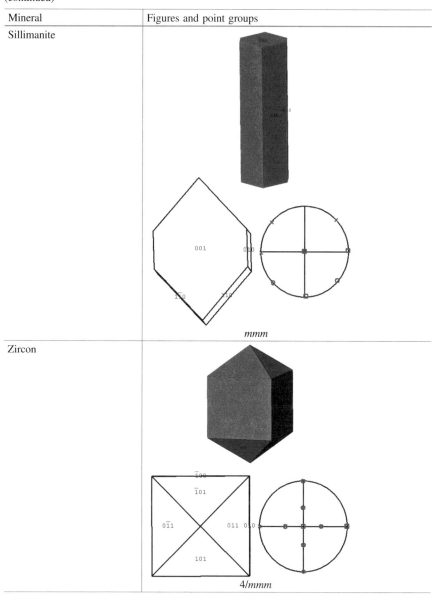
Zircon	

Appendix III Correspond to Chapter 9

Structural distortion and order of Al and Si in the structure of alkali feldspars

In an alkali feldspar ($KAlSi_3O_8$), the ideal structure is monoclinic, C2/m. The framework is constituted by rings of four tetrahedra, 2 T1 and 2 T2, alternating and occupied by Si^{4+} and Al^{3+}, with a disordered distribution. Two tetrahedra have their vertices pointing upwards and two have their vertices pointing downwards (Fig. A.1).

- These rings are joined to form a layer. In this layer, the rings of four tetrahedra are joined to another ring of four other tetrahedra, so that both rings are related by a plane of symmetry.
- The tetrahedra are joined along axis a, giving rise to a chain in the form of a crankshaft.
- Between the rings of tetrahedra, there are large gaps in which the K^+ ions are located, just above the planes of symmetry. They also form a chain. These ions have coordination number 10.

At low temperature, there is an increase in order and, consequently, a loss of symmetry of the structure. The ordered distribution of Al^{3+} in microcline destroys the symmetry plane and binary axis present in sanidine and orthoclase, leaving only

Fig. A.1 Tetrahedra with vertices pointing upwards and downwards alternating and occupied by Si^{4+} and Al^{3+}

C. Marcos, *Crystallography*, Springer Textbooks in Earth Sciences, Geography and Environment, https://doi.org/10.1007/978-3-030-96783-3

one center of symmetry. The low-temperature structure is the most ordered and least symmetrical, C $\bar{1}$.

– High-temperature structures are more expanded than low-T structures.
– As T decreases, the structure tends to contract around the cations that occupy the large interstitials.
– Large cations such as K^+ allow the structure to remain expanded.
– The T at which the structure contracts decreases from Na to K feldspars.
– K-rich feldspars maintain a stable C2/m structure at room T. At high T, Al^{3+} and Si^{4+} are stable.
– At high T, the Al^{3+} and Si^{4+} are disordered, while at low T, there is a tendency toward order.
– Order is a slow process involving the breaking of Si-O and Al-O bonds.
– To distribute the Al^{3+} and Si^{4+} in an orderly way, it is not possible with two different tetrahedral positions. There must be a reduction in symmetry so that there are four T positions where the Al^{3+} can be placed in one position and the Si^{4+} in the other three T positions. This process of order implies distortion in the tetrahedra, since the Al^{3+} tetrahedron is somewhat larger than the Si^{4+} tetrahedron.

Sanidine appears in rapidly cooled rocks.
Orthoclase occurs in rocks cooled to a moderate T.
Microcline will be found in plutonic igneous rocks (slowly cooled deep in the earth) and in metamorphic rocks.

The main characteristics observed in the phase diagram (Fig. A.2), in relation to alkali feldspars, are the following:

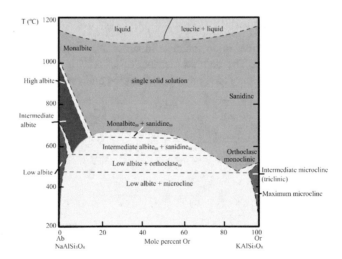

Fig. A.2 Phase diagram of feldspars

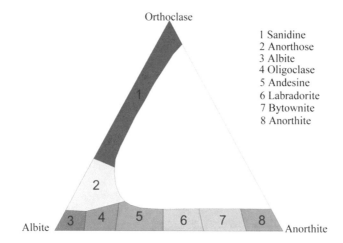

Fig. A.3 Ternary diagram of feldspars (according to Deer et al., 2012)[2]

- In K-rich feldspars, the ordering of Al and Si results in the transformation from monoclinic to triclinic feldspars.
- A solvus produces a miscibility gap between Na-rich feldspars and K-rich feldspars.
- The intermediate compositions between the two feldspars, when cooled slowly, unmix (exfoliate) and give rise to two intergrowths.

Chemical composition of alkaline feldspars

The chemical formula of alkali feldspars can be expressed as MT_4O_8 where $M = K^+$, Na^+ (alkaline metals), and T is the tetrahedral position where Si^{4+} and Al^{3+} are located. Thus, Orthoclase (Or) = $KAlSi_3O_8$ and Albite (Al) = $Na\ AlSi_3O_8$ can be differentiated. The ternary diagram (Fig. A.3) shows the alkali feldspars, which are so named because the alkali metals that form Na^+ and K^+ are substituted for each other following the continuous spectrum of compositions in the series.[2]

In the case of alkali feldspars, it can be observed that there is a small solid solution between Or-Ab. This is because the charge is identical between the substituted K^+ and Na^+ atoms; however, the size is sufficiently different so that the solid solution is limited at low temperatures, being the size of the cations $K^+ = 1.3$ Å and $Na^+ = 1.0$ Å (Griffen, 1992).[3]

[2]Deer, W.A., Howie, R.A., y Zussman, J. 2012. An introduction to the rock-forming minerals, (3rd. edition). Longmans, Londres, 248–309.
[3]Griffen, D.T. 1992. Silicate Crystal Chemistry. Oxford University Press, New York, 3–82.

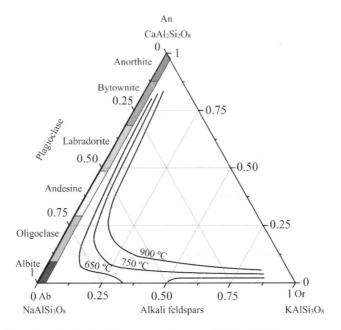

Fig. A.4 Solvus lines in the Or-Ab-An ternary diagram (Griffen, 1992)

The result is a complete solid solution (Fig. A.4) located above the 750C°
isotherm at P H$_2$O = 1 kbar, and below this temperature, the exsolution of the alkali
feldspars occurs where they cool slowly and crystallize. This zone is known as the
miscibility gap (Griffen, 1992).

In this ternary diagram, the regions corresponding to the solid solutions are
restricted by isotherms, so that outside these isotherms, i.e., toward the sides representing Or-Ab and Ab-An, the solid solution occurs for the corresponding temperatures while, inside the isotherms themselves, the exsolution processes occur.

Appendix IV

Conditions of systematic non-extinction due to Bravais lattices.

Reflection	Condition	Interpretation	Lattice	Crystalline system
hkl	$h + k + l = 2n$	Interior or body-centered lattice	I	2, 3, 4, 7
	$h + k = 2n$	C-face-centered lattice	C	2,3
	$h + l = 2n$	B-face-centered lattice	B	3
	$k + l = 2n$	A-face-centered lattice	A	2, 3
	h,k,l, all even or all odd	Face-centered lattice	F	3, 7
	$-h + k + l = 3n$	Rhombohedral lattice indexed according to hexagonal axes, orientation +	R	5
	$h–k + l = 3n$	Rhombohedral lattice indexed according to hexagonal axes, orientation	R	5
	$h + k + l = 3n$	Hexagonal lattice indexed according to rhombohedral axes	H	6
	any value	Primitive lattice	P	All

Crystal systems: 1-triclinic, 2-monoclinic, 3-orthorhombic, 4-tetragonal, 5-rhombohedral, 6-hexagonal, and 7-cubic

C. Marcos, *Crystallography*, Springer Textbooks in Earth Sciences, Geography and Environment, https://doi.org/10.1007/978-3-030-96783-3

Appendix V

Conditions of selected systematic non-extinction (= reflection condition) due to elements of symmetry with translation.

Reflection	Condition	Interpretation	Vector	Symbol	Crystal system
$0kl$	$k = 2n$ $l = 2n$ $k + l = 2n$ $k + l = 4n$ and $k,l = 2n$	glide plane parallel to (100)	$b/2$ $c/2$ $(b + c)/2$ $(b + c)/4$	b c n d	3, 4, 7 3, 4, 7 3, 4, 7 3, 4, 7
$h0l$	$h = 2n$ $l = 2n$ $h + l = 2n$ $h + l = 4n$ and $h,l = 2n$	(010)	$a/2$ $c/2$ $(a + c)/2$ $(a + c)/4$	a c n d	2, 3, 4, 7 2, 3, 4, 7 2, 3, 4, 7 3, 4, 7
$hk0$	$h = 2n$ $k = 2n$ $h + k = 2n$ $h + k = 4n$ and $h,k = 2n$	(001)	$a/2$ $b/2$ $(a + b)/2$ $(a + b)/4$	a b n d	3, 4, 7 3, 4, 7 3, 4, 7 3, 4, 7
$h\bar{h}0l$	$l = 2n$	$\{11\bar{2}0\}$	$c/2$	c	5, 6,
$hh\bar{2}l1$	$l = 2n$	$\{1\bar{1}00\}$	$c/2$	c	5, 6
$hhl, h\bar{h}1$	$l = 2n$ $2h + l = 4n$	$(110), (1\bar{1}0)$	$c/2$ $(a + b + c)/4$	c, n d	4, 7 4, 7
$hkk, hk\bar{k}$	$h = 2n$ $2k + h = 4n$	$(011), (01\bar{1})$	$a/2$ $(a + b + c)/4$	a, n d	7 7
$hkh, \bar{h}kh$	$k = 2n$ $2h + k = 4n$	$(101), (10\ \bar{1})$	$c/2$ $(a + b + c)/4$	b, n d	7 7
		Screw axis parallel to			

(continued)

C. Marcos, *Crystallography*, Springer Textbooks in Earth Sciences, Geography and Environment, https://doi.org/10.1007/978-3-030-96783-3

(continued)

Reflection	Condition	Interpretation	Vector	Symbol	Crystal system
$h00$	$h = 2n$	[100]	$a/2$	2_1	3, 4, 7
	$h = 4n$		$a/4$	4_2	7
				$4_1, 4_3$	7
$0k0$	$k = 2n$	[010]	$b/2$	2_1	2, 3, 4, 7
	$k = 4n$		$b/4$	4_2	7
				$4_1, 4_3$	7
$00\,l$	$l = 2n$	[001]	$c/2$	2_1	3, 7
	$l = 4n$		$c/4$	4_2	4, 7
				$4_1, 4_3$	4, 7
$000\,l$	$l = 2n$	[0001]	$c/2$	6_3	6
	$l = 3n$		$c/3$	$3_1, 3_2, 6_2,$	5, 6,
	$l = 6n$		$c/6$	6_4	6
				$6_1, 6_5$	

Crystal systems: 1-triclinic, 2-monoclinic, 3-orthorhombic, 4-tetragonal, 5-trigonal, 6-hexagonal, and 7-cubic

n = odd integer and $2n$ = even integer

Appendix VI

Each mineral corresponds to a crystal structure and an X-ray powder diffraction pattern.

The structures were made with Atoms software.[4] The table of 2, intensity, d-spacing, and *hkl* values for each mineral is XPOW Copyright 1993 Bob Downs, Ranjini Swaminathan, and Kurt Bartelmehs.[5] X-ray diffraction data were obtained from the database of the RRUFF Project,[6] Department of Geosciences, University of Arizona.

Note: In the X-ray powder diffraction patterns, *hkl* numbers of the reflections have been indicated.

Andalusite (Al_2SiO_5), orthorhombic (P*nnm*).

Structure

Winter, J.K., Ghose, S. (1979). Thermal expansion and high-temperature crystal chemistry of the Al_2SiO_5 polymorphs. American Mineralogist, 64, 573–586.

Cell parameters: 7.798000 Å, 7.903100 Å, 5.556600 Å, 90°, 90°, 90°.

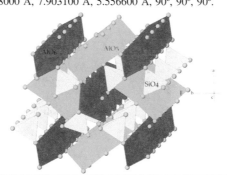

(continued)

---

[4]Atoms software V6.1.2 (2004).
[5]Downs, B. Swaminathan, R. and Bartelmehs, K. (1993) American Mineralogist 78, 1104–1107.
[6]RRUFF Project (Department of Geosciences, University of Arizona), with permission.

© The Editor(s) (if applicable) and The Author(s), under exclusive license to   469
Springer Nature Switzerland AG 2022
C. Marcos, *Crystallography*, Springer Textbooks in Earth Sciences,
Geography and Environment, https://doi.org/10.1007/978-3-030-96783-3

(continued)

X-ray diffraction
Diffraction data computed using the structure data from the above paper listed, along with the
cell parameters refined from the powder pattern of R050258 (RRUFF database).
X-ray wavelength: 1.541838 Å.

2θ	INTENSITY	D-SPACING	H	K	L
16.01	99.56	5.5467	1	1	0
19.57	69.89	4.5433	0	1	1
19.65	100.00	4.5234	1	0	1
22.54	3.26	3.9481	0	2	0
22.68	54.03	3.9250	1	1	1
22.85	4.41	3.8966	2	0	0
25.31	31.11	3.5219	1	2	0
25.51	25.21	3.4943	2	1	0
32.25	22.78	2.7774	0	0	2
32.29	71.82	2.7733	2	2	0
36.03	5.52	2.4937	1	3	0
36.18	33.15	2.4835	1	1	2
36.21	4.37	2.4813	2	2	1
36.42	24.18	2.4676	3	1	0
37.83	11.39	2.3786	0	3	1
38.26	10.56	2.3531	3	0	1
39.63	18.28	2.2750	1	3	1
39.69	37.12	2.2716	0	2	2
39.99	27.35	2.2551	3	1	1
41.41	26.13	2.1809	1	2	2
41.54	25.98	2.1743	2	1	2
41.62	32.00	2.1701	3	2	0
45.98	6.24	1.9741	0	4	0
46.62	3.68	1.9483	4	0	0
47.52	1.17	1.9136	1	4	0
48.10	5.00	1.8916	4	1	0
49.10	8.49	1.8555	1	3	2
50.44	6.87	1.8093	1	4	1
50.63	7.71	1.8027	0	1	3
50.67	3.70	1.8015	1	0	3
51.00	9.10	1.7906	4	1	1
52.07	1.99	1.7563	1	1	3
52.13	9.81	1.7543	3	3	1
53.41	3.11	1.7154	2	3	2
54.67	2.59	1.6786	2	4	1
57.25	4.38	1.6090	0	4	2
57.80	17.51	1.5950	4	0	2
59.73	3.45	1.5478	1	5	0
60.07	1.12	1.5399	2	2	3
60.11	20.71	1.5391	3	3	2
60.54	6.38	1.5291	5	1	0
60.98	1.32	1.5191	0	5	1
61.19	2.49	1.5144	0	3	3
61.28	3.26	1.5124	3	4	1
61.49	2.68	1.5078	3	0	3

(continued)

61.51	3.76	1.5072	4	3	1
61.81	1.61	1.5007	5	0	1
62.26	11.88	1.4910	1	5	1
62.43	51.35	1.4872	2	4	2
62.46	6.49	1.4866	1	3	3
62.72	5.44	1.4810	3	1	3
62.82	13.73	1.4789	4	2	2
63.04	11.38	1.4743	5	1	1
63.55	4.47	1.4636	2	5	0
67.42	22.60	1.3887	0	0	4
67.53	12.40	1.3867	4	4	0
69.51	1.28	1.3520	1	5	2
69.79	2.01	1.3471	1	1	4
70.25	2.90	1.3395	5	1	2
70.78	2.90	1.3307	1	4	3
71.25	1.19	1.3232	4	1	3
72.18	3.90	1.3083	3	3	3
72.48	1.13	1.3037	5	3	1
72.79	1.92	1.2989	6	0	0
72.87	1.30	1.2977	1	6	0
73.05	8.56	1.2948	2	5	2
73.24	1.90	1.2919	1	2	4
73.34	1.83	1.2905	2	1	4
73.69	10.56	1.2852	5	2	2
73.93	1.21	1.2816	6	1	0
76.35	3.80	1.2469	2	6	0
76.72	9.36	1.2417	2	2	4
76.80	11.55	1.2406	4	4	2
77.31	4.67	1.2338	6	2	0
78.83	4.89	1.2138	3	5	2
78.87	1.69	1.2133	1	3	4
79.10	4.12	1.2102	3	1	4
80.26	1.67	1.1957	4	3	3
80.34	2.09	1.1947	5	4	1
80.52	1.19	1.1924	5	0	3
80.78	2.92	1.1893	0	6	2
80.92	6.66	1.1875	1	5	3
81.63	4.38	1.1790	5	1	3
81.84	4.98	1.1766	6	0	2
82.05	1.37	1.1740	3	6	0
82.42	2.87	1.1697	3	2	4
87.30	1.10	1.1164	1	7	0
88.95	1.21	1.0999	1	0	5
89.50	2.06	1.0945	1	7	1
89.50	2.06	1.0945	1	7	1

(continued)

(continued)

X-ray diffraction pattern

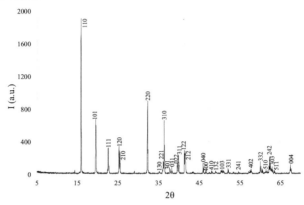

Apatite ($Ca_5(PO_4)_3F$), hexagonal ($P6_3/m$).

Structure
Hughes, J.M., Cameron, M., Crowley, K.D. (1989). Structural variations in natural F, OH, and Cl apatites. American Mineralogist, 74, 870–876.
Cell parameters: 9.3925 Å, 9.3925 Å, 6.8839 Å, 90°, 90°, 120°.

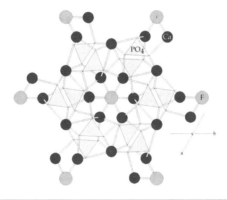

X-ray diffraction
XPOW Copyright 1993 Bob Downs, Ranjini Swaminathan, and Kurt Bartelmehs
For reference, see Downs et al. (1993) American Mineralogist 78, 1104–1107
Diffraction data computed using the structure data from the above paper listed, along with the cell parameters refined from the powder pattern of R040098 (RRUFF database).
X-ray wavelength: 1.541838 Å

(continued)

(continued)

2θ	INTENSITY	D-SPACING	H	K	L
10.88	9.87	8.1341	1	0	0
16.87	3.69	5.2547	1	0	1
18.90	1.50	4.6962	1	1	0
21.85	5.36	4.0671	2	0	0
22.92	6.54	3.8795	1	1	1
25.44	2.02	3.5016	2	0	1
25.89	37.51	3.4420	0	0	2
28.15	12.03	3.1698	1	0	2
29.04	14.05	3.0744	1	2	0
31.88	69.57	2.8072	1	2	1
31.88	30.43	2.8072	2	1	1
32.25	40.25	2.7762	1	1	2
33.04	57.71	2.7114	3	0	0
34.13	25.69	2.6274	2	0	2
35.59	3.82	2.5227	3	0	1
39.29	4.77	2.2929	1	2	2
39.96	4.47	2.2560	1	3	0
39.96	16.66	2.2560	3	1	0
40.59	2.86	2.2224	2	2	1
42.15	1.94	2.1438	1	3	1
42.15	3.88	2.1438	3	1	1
43.92	4.41	2.0617	1	1	3
44.56	1.26	2.0335	4	0	0
45.38	3.60	1.9985	2	0	3
46.84	26.86	1.9397	2	2	2
48.23	5.66	1.8868	1	3	2
48.23	7.88	1.8868	3	1	2
48.80	3.43	1.8661	3	2	0
49.57	10.06	1.8389	2	1	3
49.57	23.70	1.8389	1	2	3
50.68	8.30	1.8011	2	3	1
50.68	7.77	1.8011	3	2	1
51.48	6.18	1.7750	1	4	0
51.48	7.19	1.7750	4	1	0
52.22	1.17	1.7516	3	0	3
52.25	13.10	1.7508	4	0	2
53.23	15.89	1.7210	0	0	4
56.06	7.54	1.6405	2	3	2
57.27	1.93	1.6087	1	3	3
57.27	1.54	1.6087	3	1	3
58.28	1.65	1.5832	5	0	1
60.20	4.45	1.5372	4	2	0
60.67	4.10	1.5264	3	3	1
61.78	2.13	1.5017	2	1	4
61.78	1.05	1.5017	1	2	4
61.84	2.06	1.5003	4	2	1
63.22	10.80	1.4708	5	0	2
63.70	1.15	1.4609	1	5	0

(continued)

(continued)

64.09	8.28	1.4530	3	0	4
64.35	5.44	1.4478	2	3	3
65.29	6.00	1.4291	1	5	1
65.50	1.37	1.4250	3	3	2
69.96	1.58	1.3448	5	1	2
71.93	4.14	1.3127	3	4	1
72.58	2.63	1.3025	2	5	0
74.08	1.54	1.2798	2	5	1
74.26	3.34	1.2771	4	2	3
75.69	1.60	1.2565	2	1	5
75.69	3.57	1.2565	1	2	5
76.41	2.40	1.2465	4	3	2
76.85	1.14	1.2404	1	6	0
77.21	2.66	1.2356	1	4	4
77.21	4.43	1.2356	4	1	4
77.45	3.58	1.2324	1	5	3
78.52	5.03	1.2182	5	2	2
81.40	1.04	1.1822	5	0	4
83.71	3.89	1.1554	3	4	3
84.43	1.79	1.1473	0	0	6
84.51	1.30	1.1465	2	4	4
84.51	2.02	1.1465	4	2	4
85.78	1.19	1.1327	2	5	3
87.53	3.29	1.1145	1	1	6
87.61	1.82	1.1137	1	5	4
88.19	1.55	1.1079	3	2	5
88.89	1.17	1.1010	5	3	2
88.89	1.03	1.1010	7	0	2

X-ray diffraction pattern
$Ca_5(P_{0.98}O_{0.01})_3F$

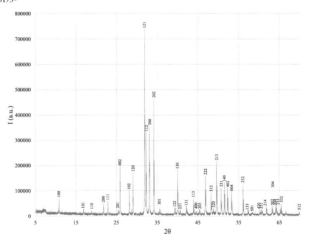

Calcite ($CaCO_3$), trigonal ($R\bar{3}c$).

Structure
Markgraf, S.A., Reeder, R.J. (1985). High-temperature structure refinements of calcite and magnesite. American Mineralogist, 70, 590–600.
Cell parameters: 4.9869 Å, 4.9869 Å, 17.0496 Å, 90°, 90°, 120°.

ATOM	X	Y	Z	OCCUPANCY	ISO(B)
Ca	0.00000	0.00000	0.00000	1.000	0.936
C	0.00000	0.00000	0.25000	1.000	0.907
O	0.25670	0.00000	0.25000	1.000	1.512

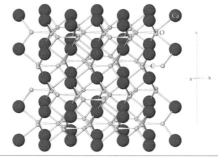

X-ray diffraction
Diffraction data computed using the structure data from the above paper listed, along with the cell parameters refined from the powder pattern of R040098 (RRUFF database).
X-ray wavelength: 1.541838 Å.

2θ	INTENSITY	D-SPACING	H	K	L
23.09	8.36	3.8526	0	1	2
29.44	100.00	3.0337	1	0	4
31.48	2.07	2.8416	0	0	6
36.02	14.36	2.4934	1	1	0
39.46	18.81	2.2834	1	1	3
43.22	14.64	2.0933	2	0	2
47.18	6.50	1.9263	0	2	4
47.58	19.73	1.9112	0	1	8
48.58	20.46	1.8742	1	1	6
56.65	3.42	1.6249	2	1	1
57.48	9.78	1.6032	1	2	2
60.76	5.68	1.5244	2	1	4
61.09	2.63	1.5169	2	0	8
61.47	2.79	1.5084	1	1	9
63.15	2.06	1.4723	1	2	5
64.76	6.70	1.4396	3	0	0
65.72	3.76	1.4208	0	0	12
69.30	1.33	1.3560	2	1	7
70.36	2.30	1.3381	0	2	10
73.01	2.98	1.2959	1	2	8
77.29	2.23	1.2345	1	1	12
81.66	2.54	1.1791	2	1	10
83.91	1.58	1.1531	1	3	4

(continued)

(continued)

X-ray diffraction pattern

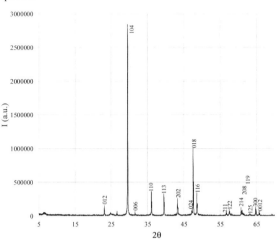

Clinochlore ($Mg_5Al(AlSi_3O_{10})(OH)_8$), monoclinic (C2/c).

Structure
McMurchy R C. (1934). The crystal structure of the chlorite minerals. Zeitschrift fur Kristallographie, 88, 420–432
Locality: Miles City, Montana, USA.
Cell parameters: 5.304843 Å, 9.179822 Å, 14.33853 Å, 90°, 97.57009°, 90°.

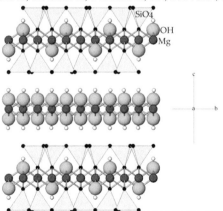

(continued)

(continued)

X-ray diffraction
XPOW Copyright 1993 Bob Downs, Ranjini Swaminathan, and Kurt Bartelmehs
For reference, see Downs et al. (1993) American Mineralogist 78, 1104–1107
Diffraction data computed using the structure data from the above paper listed, along with the
cell parameters refined from the powder pattern of R060725 (RRUFF database).
X-ray wavelength: 1.541838 Å.

2θ	INTENSITY	D-SPACING	H	K	L
6.23	91.50	14.2136	0	0	1
12.46	100.00	7.1068	0	0	2
18.73	96.60	4.7379	0	0	3
19.34	19.29	4.5899	0	2	0
19.46	3.26	4.5630	1	1	0
19.75	67.36	4.4968	$\bar{1}$	1	1
20.33	1.50	4.3678	0	2	1
21.12	54.00	4.2069	1	1	1
21.91	3.69	4.0563	$\bar{1}$	1	2
23.07	38.13	3.8557	0	2	2
24.35	18.07	3.6544	1	1	2
25.06	92.19	3.5534	0	0	4
25.50	18.01	3.4921	$\bar{1}$	1	3
27.04	13.17	3.2966	0	2	3
28.67	6.67	3.1136	1	1	3
30.05	19.59	2.9732	$\bar{1}$	1	4
31.46	21.96	2.8427	0	0	5
33.68	8.54	2.6600	1	1	4
33.83	1.88	2.6486	$\bar{2}$	0	1
33.88	5.24	2.6448	1	3	0
34.77	18.39	2.5790	$\bar{2}$	0	2
34.90	35.28	2.5697	1	3	1
35.41	77.37	2.5342	$\bar{1}$	3	2
35.52	32.55	2.5266	2	0	1
36.83	29.87	2.4394	$\bar{2}$	0	3
37.03	49.17	2.4267	1	3	2
37.19	1.36	2.4167	0	2	5
37.83	26.39	2.3773	$\bar{1}$	3	3
38.01	14.66	2.3666	2	0	2
39.26	1.34	2.2941	$\bar{2}$	2	1
39.86	10.06	2.2608	$\bar{2}$	0	4
40.12	17.95	2.2468	1	3	3
40.09	2.29	2.2484	$\bar{2}$	2	2

(continued)

(continued)

40.75	2.20	2.2134	2	2	1
40.85	4.72	2.2082	$\bar{1}$	1	6
41.16	1.11	2.1923	$\bar{1}$	3	4
41.32	1.01	2.1839	0	4	2
42.98	1.76	2.1034	2	2	2
43.69	2.02	2.0710	$\bar{2}$	0	5
43.99	6.29	2.0573	1	3	4
44.61	6.73	2.0305	0	0	7
44.66	1.25	2.0281	$\bar{2}$	2	4
45.07	1.77	2.0106	1	1	6
45.24	48.97	2.0037	$\bar{1}$	3	5
45.52	23.31	1.9919	2	0	4
47.12	1.06	1.9278	0	4	4
48.18	9.28	1.8880	$\bar{2}$	0	6
48.52	16.30	1.8754	1	3	5
49.03	1.73	1.8569	0	2	7
49.88	2.53	1.8272	2	2	4
49.93	16.57	1.8257	$\bar{1}$	3	6
50.25	8.39	1.8147	2	0	5
51.13	1.49	1.7857	0	4	5
52.37	3.97	1.7460	$\bar{2}$	2	6
52.69	3.79	1.7363	$\bar{3}$	1	1
52.93	1.06	1.7290	2	4	0
52.91	4.10	1.7297	$\bar{1}$	5	1
53.11	1.47	1.7236	$\bar{1}$	1	8
53.21	2.27	1.7205	$\bar{2}$	0	7
53.42	2.20	1.7144	$\bar{2}$	4	2
53.59	3.01	1.7092	1	3	6
53.87	3.26	1.7011	$\bar{1}$	5	2
54.48	2.84	1.6835	3	1	1
54.90	3.13	1.6715	$\bar{2}$	4	3
55.14	7.84	1.6647	$\bar{1}$	3	7
55.50	4.44	1.6549	2	0	6
55.77	3.41	1.6475	2	4	2
55.74	2.93	1.6483	0	4	6
56.58	1.91	1.6259	3	1	2
57.77	1.08	1.5951	1	1	8
58.35	1.25	1.5807	2	4	3
58.40	1.36	1.5793	0	0	9
58.73	13.76	1.5713	$\bar{2}$	0	8
59.14	24.46	1.5614	1	3	7
58.98	1.25	1.5652	$\bar{3}$	1	5
59.33	1.66	1.5568	2	2	6

(continued)

(continued)

60.43	46.39	1.5310	$\bar{3}$	3	1
60.48	24.56	1.5300	0	6	0
60.82	2.18	1.5222	$\bar{1}$	3	8
61.20	1.26	1.5135	2	0	7
61.87	7.64	1.4989	$\bar{3}$	3	3
62.07	7.42	1.4944	3	3	1
62.01	5.77	1.4957	0	6	2
62.92	2.43	1.4763	3	1	4
63.71	1.84	1.4600	$\bar{3}$	3	4
63.90	3.13	1.4559	0	6	3
64.01	2.77	1.4537	3	3	2
65.12	1.11	1.4316	1	3	8
65.27	1.66	1.4288	$\bar{1}$	5	6
65.65	4.30	1.4214	0	0	10
66.26	2.81	1.4098	$\bar{3}$	3	5
66.50	3.33	1.4052	0	6	4
66.66	3.36	1.4023	3	3	3
66.52	1.19	1.4049	0	4	8
66.93	25.87	1.3973	$\bar{1}$	3	9
67.34	13.02	1.3896	2	0	8
68.07	1.45	1.3766	$\bar{2}$	4	7
69.76	1.21	1.3472	0	6	5
69.97	1.11	1.3438	3	3	4
71.75	5.08	1.3147	4	0	0
71.68	7.61	1.3158	$\bar{2}$	6	2
71.92	1.95	1.3121	$\bar{4}$	0	3
72.13	3.93	1.3087	2	6	1
72.94	4.90	1.2961	$\bar{2}$	6	3
73.11	2.71	1.2935	4	0	1
73.38	3.62	1.2895	$\bar{4}$	0	4
73.69	5.98	1.2848	2	6	2
73.47	2.21	1.2881	$\bar{1}$	3	10
73.91	1.13	1.2815	3	3	5
74.46	1.45	1.2735	$\bar{2}$	2	10
75.13	1.20	1.2638	1	7	1
75.94	1.21	1.2523	3	5	1
76.41	2.26	1.2456	1	7	2
76.91	1.49	1.2388	$\bar{1}$	7	3
77.71	1.37	1.2281	3	5	2
77.92	2.09	1.2253	$\bar{2}$	0	11

(continued)

(continued)

78.42	4.83	1.2188	1	3	10
78.35	1.27	1.2197	$\bar{4}$	0	6
78.83	2.67	1.2133	2	6	4
79.22	1.35	1.2084	0	4	10
80.80	3.61	1.1887	$\bar{2}$	6	6
81.25	1.92	1.1833	4	0	4
81.15	1.83	1.1845	0	0	12
81.81	1.39	1.1765	$\bar{4}$	0	7
82.39	3.13	1.1697	2	6	5
84.73	1.52	1.1433	$\bar{2}$	6	7
85.26	1.79	1.1375	$\bar{2}$	0	12
85.80	3.44	1.1318	1	3	11
85.93	1.43	1.1304	$\bar{4}$	0	8
86.35	1.07	1.1259	0	4	11
86.59	2.44	1.1234	2	6	6
89.31	4.46	1.0962	$\bar{2}$	6	8
89.93	2.16	1.0902	4	0	6
89.93	2.16	1.0902	4	0	6

X-ray diffraction pattern

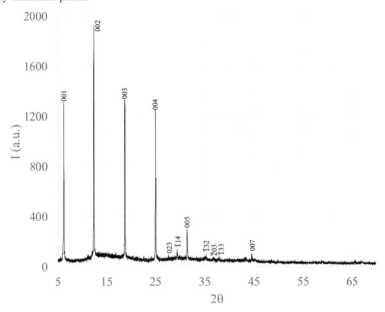

Cordierite $(Mg_2Al_4Si_5O_{18})$, orthorhombic (C*ccm*).

Structure
Wallace, J.H., Wenk, H.R. (1980). Structure variation in low cordierites. American Mineralogist, 65, 96–111.
Cell parameters: 17.0777 Å, 9.7300 Å, 9.3470 Å, 90°, 90°, 90°.

ATOM	X	Y	Z	OCCUPANCY	ISO(B)
Mg	0.33742	0.00000	0.25000	0.900	0.537
Fe	0.33742	0.00000	0.25000	0.068	0.537
Al	0.25000	0.25000	0.25006	0.962	0.413
Si	0.00000	0.50000	0.25000	0.876	0.343
Al	0.00000	0.50000	0.25000	0.097	0.343
Si	0.19252	0.07803	0.00000	0.963	0.326
Si	0.13516	-0.23731	0.00000	0.970	0.326
Al	0.05083	0.30791	0.00000	0.911	0.344
Si	0.05083	0.30791	0.00000	0.048	0.344
O	0.24722	-0.10301	0.35881	1.000	0.704
O	0.06220	-0.41620	0.34910	1.000	0.717
O	-0.17326	-0.31006	0.35854	1.000	0.725
O	0.04321	-0.24815	0.00000	1.000	1.008
O	0.12231	0.18464	0.00000	1.000	1.010
O	0.16468	-0.07957	0.00000	1.000	1.028
OH	0.00000	0.00000	0.25000	0.740	24.395
OH	0.00000	0.00000	0.00000	0.120	59.672

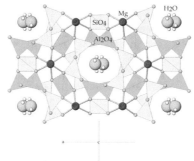

X-ray diffraction
Diffraction data computed using the structure data from the above paper listed, along with the cell parameters refined from the powder pattern of R040081 (RRUFF database).
X-ray wavelength: 1.541838 Å.

2θ	INTENSITY	D-SPACING	H	K	L
10.36	61.08	8.5388	2	0	0
10.46	100.00	8.4541	1	1	0
18.05	17.06	4.9134	3	1	0
18.24	7.62	4.8650	0	2	0
18.99	11.17	4.6735	0	0	2
20.81	1.84	4.2694	4	0	0
21.68	7.66	4.0996	2	0	2
21.73	50.48	4.0901	1	1	2
26.32	51.69	3.3863	3	1	2
26.45	32.80	3.3703	0	2	2
27.80	1.44	3.2090	4	2	0

(continued)

(continued)

28.31	23.19	3.1521	4	0	2
28.47	64.51	3.1350	2	2	2
29.31	39.17	3.0467	5	1	1
29.43	32.30	3.0351	4	2	1
29.62	38.49	3.0159	1	3	1
31.43	1.18	2.8463	6	0	0
33.78	12.31	2.6531	5	1	2
33.89	11.92	2.6454	4	2	2
34.05	5.63	2.6327	1	3	2
36.58	4.35	2.4567	6	2	0
36.95	3.58	2.4325	0	4	0
36.98	1.09	2.4309	6	0	2
37.26	1.15	2.4133	3	3	2
38.53	13.97	2.3368	0	0	4
39.27	1.64	2.2940	7	1	1
39.51	1.58	2.2808	5	3	1
40.26	1.86	2.2400	5	1	3
40.35	1.43	2.2354	4	2	3
40.49	1.34	2.2277	1	3	3
41.53	6.50	2.1746	6	2	2
41.87	1.71	2.1577	0	4	2
42.83	2.54	2.1112	7	1	2
42.85	3.23	2.1103	3	1	4
42.94	1.59	2.1064	0	2	4
43.06	3.91	2.1009	5	3	2
43.25	6.06	2.0920	2	4	2
44.29	1.67	2.0451	2	2	4
46.45	1.69	1.9548	8	2	0
46.58	3.38	1.9497	7	3	0
46.78	2.82	1.9417	8	0	2
47.00	1.60	1.9335	1	5	0
47.20	2.81	1.9258	4	4	2
47.52	1.39	1.9134	8	2	1
48.30	4.03	1.8845	7	1	3
48.30	1.85	1.8843	1	3	4
48.50	6.38	1.8771	5	3	3
48.67	6.05	1.8708	2	4	3
48.90	2.56	1.8624	9	1	0
49.28	2.39	1.8492	6	4	0
49.50	2.66	1.8414	3	5	0
50.30	2.96	1.8140	6	4	1
50.52	2.26	1.8067	3	5	1
50.54	3.47	1.8061	6	0	4
50.75	6.89	1.7988	3	3	4
53.27	1.13	1.7195	6	4	2
53.67	5.38	1.7078	10	0	0
54.17	19.49	1.6932	6	2	4
54.25	6.25	1.6908	5	5	0

(continued)

(continued)

54.45	11.52	1.6852	0	4	4
55.25	2.21	1.6627	7	1	4
55.49	1.88	1.6559	8	2	3
55.97	1.12	1.6429	1	5	3
56.16	1.47	1.6378	9	3	0
56.77	1.47	1.6217	0	6	0
57.16	2.51	1.6114	10	2	0
57.43	1.36	1.6045	8	4	0
57.66	3.47	1.5986	9	1	3
57.88	1.55	1.5932	2	6	0
58.00	6.32	1.5902	6	4	3
58.20	3.60	1.5852	3	5	3
58.92	1.03	1.5675	4	4	4
59.32	3.13	1.5578	0	0	6
59.84	1.92	1.5456	9	3	2
60.42	1.78	1.5321	0	6	2
60.80	1.49	1.5234	10	2	2
61.88	3.09	1.4993	8	2	4
61.99	4.17	1.4970	7	3	4
62.02	1.01	1.4964	4	6	1
62.33	2.81	1.4897	1	5	4
62.55	3.18	1.4850	3	1	6
62.61	1.52	1.4836	0	2	6
63.66	2.87	1.4617	2	2	6
63.90	1.81	1.4567	11	1	2
64.41	5.09	1.4466	7	5	2
64.63	2.02	1.4420	4	6	2
66.34	2.21	1.4090	6	6	0
66.69	2.09	1.4026	5	1	6
66.75	2.49	1.4014	4	2	6
66.85	1.49	1.3995	1	3	6
68.50	1.05	1.3698	5	5	4
68.72	1.20	1.3659	12	2	0
68.98	6.42	1.3614	12	0	2
69.15	1.06	1.3586	9	5	0
69.56	1.90	1.3515	12	2	1
69.63	1.44	1.3503	3	7	0
69.70	9.52	1.3490	6	6	2
69.98	2.51	1.3444	9	5	1
70.46	2.44	1.3365	3	7	1
71.06	4.91	1.3266	10	2	4
71.30	2.86	1.3227	8	4	4
71.51	1.09	1.3194	9	1	5
71.70	2.63	1.3163	2	6	4
71.74	3.28	1.3156	6	2	6
71.80	1.59	1.3147	6	4	5
71.98	1.15	1.3119	0	4	6
72.03	1.14	1.3111	12	2	2

(continued)

(continued)

72.67	1.22	1.3012	7	1	6
72.82	1.18	1.2988	5	3	6
72.92	1.14	1.2973	3	7	2
72.96	1.24	1.2967	2	4	6
74.25	3.38	1.2773	11	3	3
74.39	1.07	1.2753	10	4	3
75.03	2.71	1.2659	1	7	3
75.56	1.37	1.2584	8	0	6
75.87	1.55	1.2540	4	4	6
76.49	1.14	1.2453	9	5	3
77.36	2.63	1.2336	5	1	7
77.41	2.07	1.2328	4	2	7
77.51	2.49	1.2315	1	3	7
79.99	3.48	1.1995	10	4	4
80.62	1.50	1.1917	1	7	4
80.92	3.65	1.1880	12	4	2
81.65	2.18	1.1792	12	2	4
81.83	1.63	1.1770	0	8	2
82.05	1.22	1.1745	9	5	4
82.51	2.00	1.1692	3	7	4
82.57	4.66	1.1684	0	0	8
84.46	1.14	1.1470	14	2	2
86.15	1.04	1.1288	9	3	6
86.66	1.21	1.1234	0	6	6

X-ray diffraction pattern

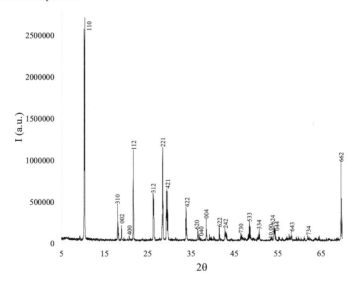

Diopside, pyroxene ($CaMgSi_2O_6$), monoclinic (C2/c).

Structure
Cameron, M., Sueno, S., Prewitt, C.T., Papike, J.J. (1973). High-temperature crystal chemistry of acmite, diopside, hedenbergite, jadeite, spodumene, and ureyite. American Mineralogist, 58, 594–618.
Cell parameters: 9.7481 Å, 8.9230 Å, 5.2508 Å, 90°, 105.886°, 90°.

ATOM	X	Y	Z	OCCUPANCY	ISO(B)
Si	0.28620	0.09330	0.22930	1.000	0.230
Mg	0.00000	0.90820	0.25000	1.000	0.262
Ca	0.00000	0.30150	0.25000	1.000	0.518
O	0.11560	0.08730	0.14220	1.000	0.336
O	0.36110	0.25000	0.31800	1.000	0.464
O	0.35050	0.01760	0.99530	1.000	0.385

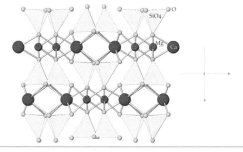

X-ray diffraction
Diffraction data computed using the structure data from the above paper listed, along with the cell parameters refined from the powder pattern of R040009 (RRUFF database).
X-ray wavelength: 1.541838 Å.

2θ	INTENSITY	D-SPACING	H	K	L
19.90	3.12	4.4615	0	2	0
24.33	2.72	3.6587	1	1	1
26.66	10.45	3.3436	0	2	1
27.60	25.02	3.2318	2	2	0
29.88	100.00	2.9904	$\bar{2}$	2	1
30.30	26.84	2.9496	3	1	0
30.91	36.29	2.8932	$\bar{3}$	1	1
31.56	1.22	2.8351	1	3	0
34.99	24.27	2.5644	$\bar{1}$	3	1
35.47	2.17	2.5311	$\bar{2}$	0	2
35.55	41.21	2.5251	0	0	2
35.70	41.36	2.5152	2	2	1
37.65	1.30	2.3892	1	3	1
39.15	17.38	2.3012	3	1	1
40.74	9.51	2.2146	1	1	2
41.00	1.28	2.2015	$\bar{2}$	2	2

(continued)

(continued)

41.07	9.46	2.1976	0	2	2
41.93	10.05	2.1546	3	3	0
42.39	18.55	2.1323	$\bar{3}$	3	1
42.92	8.70	2.1072	$\bar{4}$	2	1
43.62	1.06	2.0750	4	2	0
44.39	19.27	2.0406	0	4	1
44.99	11.95	2.0148	$\bar{4}$	0	2
45.21	7.36	2.0057	2	0	2
46.12	7.84	1.9683	$\bar{1}$	3	2
49.00	2.19	1.8592	3	3	1
49.68	6.69	1.8351	5	1	0
49.85	3.06	1.8293	2	2	2
50.34	3.61	1.8126	1	3	2
52.17	13.78	1.7531	1	5	0
53.27	1.91	1.7196	$\bar{5}$	1	2
54.52	1.43	1.6831	$\bar{1}$	5	1
54.92	4.82	1.6718	0	4	2
55.42	6.50	1.6580	$\bar{3}$	1	3
56.65	14.39	1.6247	$\bar{5}$	3	1
56.69	17.47	1.6238	$\bar{2}$	2	3
56.99	5.21	1.6159	4	4	0
58.16	2.09	1.5862	5	3	0
59.12	3.79	1.5626	6	0	0
59.66	4.77	1.5497	3	5	0
60.55	5.23	1.5292	$\bar{6}$	0	2
60.70	2.49	1.5257	$\bar{6}$	2	1
60.84	6.59	1.5226	4	0	2
61.41	1.42	1.5099	$\bar{5}$	3	2
61.71	13.51	1.5031	$\bar{1}$	3	3
62.25	1.87	1.4915	2	4	2
62.45	4.67	1.4872	0	6	0
63.69	1.63	1.4611	4	4	1
64.41	2.66	1.4466	$\bar{6}$	2	2
64.69	1.45	1.4410	4	2	2
65.42	2.41	1.4266	0	6	1
65.65	18.68	1.4221	5	3	1
66.35	9.65	1.4088	$\bar{3}$	5	2
66.46	4.36	1.4067	1	5	2
67.40	4.57	1.3894	2	2	3
68.15	1.28	1.3759	$\bar{7}$	1	1
68.28	1.20	1.3737	$\bar{2}$	4	3

(continued)

(continued)

			h	k	l
70.02	1.64	1.3437	0	4	3
70.82	6.93	1.3304	$\bar{7}$	1	2
71.07	4.28	1.3265	6	2	1
71.14	1.13	1.3252	5	1	2
71.19	1.18	1.3246	7	1	0
71.68	3.40	1.3166	$\bar{5}$	3	3
73.58	3.89	1.2872	$\bar{3}$	1	4
73.69	2.20	1.2856	$\bar{1}$	1	4
73.92	3.72	1.2822	$\bar{2}$	6	2
73.97	6.31	1.2814	0	6	2
75.06	1.83	1.2656	$\bar{4}$	0	4
75.25	1.05	1.2628	$\bar{4}$	6	1
75.27	3.56	1.2626	0	0	4
75.36	1.57	1.2612	$\bar{7}$	3	1
75.61	1.05	1.2576	4	4	2
76.34	7.80	1.2475	3	5	2
77.17	1.37	1.2361	$\bar{1}$	7	1
78.78	2.25	1.2149	1	7	1
78.95	1.24	1.2126	$\bar{7}$	1	3
82.26	1.91	1.1720	8	0	0
84.13	1.05	1.1507	$\bar{6}$	4	3
84.14	1.10	1.1506	6	2	2
84.41	3.51	1.1476	$\bar{6}$	0	4
88.25	1.84	1.1073	2	2	4

X-ray diffraction pattern
$Ca_{0.97}Na_{0.01}Mg_{0.97}Fe_{0.02}Al_{0.01}Si_2O_6$

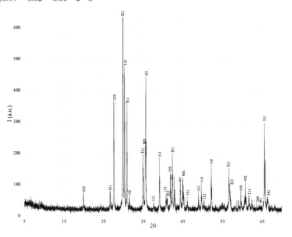

Fluorite (CaF_2), cubic $(Fm\,\overline{3}\,m)$.

Structure
Speziale, S., Duffy, T.S. (2002). Single-crystal elastic constants of fluorite (CaF_2) to 9.3 GPa.
Physics and Chemistry of Minerals, 29, 465–472
Cell parameters: 5.4639 Å, 5.4639 Å, 5.4639 Å, 90°, 90°, 90°.

ATOM	X	Y	Z	OCCUPANCY	ISO(B)
Ca	0.00000	0.00000	0.00000	1.000	1.300
F	0.25000	0.25000	0.25000	1.000	1.000

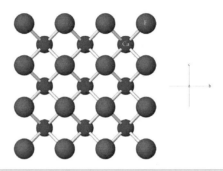

X-ray diffraction
Diffraction data computed using the structure data from the above paper listed, along with the
cell parameters refined from the powder pattern of R040099 (RRUFF database).
X-ray wavelength: 1.541838 Å.

2θ	INTENSITY	D-SPACING	H	K	L
28.29	91.58	3.1546	1	1	1
47.04	100.00	1.9318	2	2	0
55.80	28.89	1.6474	3	1	1
68.72	10.85	1.3660	4	0	0
75.90	8.76	1.2535	3	3	1
87.45	17.34	1.1153	4	2	2

X-ray diffraction pattern

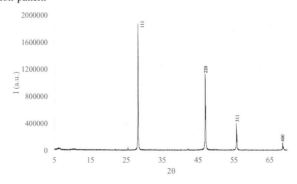

Forsterite (Mg$_2$SiO$_4$), orthorhombic (P*bnm*).

Structure
Birle, J.D., Gibbs, G.V., Moore, P.B., Smith, J.V. (1968). Crystal structures of natural olivines.
American Mineralogist, 53, 807–824
Cell parameters: 4.7617 Å, 10.2255 Å, 5.9927 Å, 90°, 90°, 90°.

ATOM	X	Y	Z	OCCUPANCY	ISO(B)
Mg	0.00000	0.00000	0.00000	0.900	0.330
Fe	0.00000	0.00000	0.00000	0.100	0.330
Mg	0.98975	0.27743	0.25000	0.900	0.360
Fe	0.98975	0.27743	0.25000	0.100	0.360
Si	0.42693	0.09434	0.25000	1.000	0.200
O	0.76580	0.09186	0.25000	1.000	0.350
O	0.22012	0.44779	0.25000	1.000	0.420
O	0.27810	0.16346	0.03431	1.000	0.410

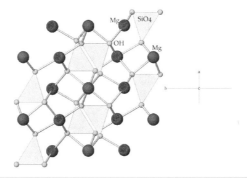

X-ray diffraction
Diffraction data computed using the structure data from the above paper listed, along with the
cell parameters refined from the powder pattern of R040018 (RRUFF database).
X-ray wavelength: 1.541838 Å.

2θ	INTENSITY	D-SPACING	H	K	L
17.34	19.25	5.1128	0	2	0
22.86	53.85	3.8895	0	2	1
23.87	18.46	3.7281	1	0	1
25.43	19.99	3.5026	1	1	1
25.56	12.82	3.4845	1	2	0
29.66	5.73	3.0123	1	2	1
29.82	15.56	2.9964	0	0	2
32.30	66.77	2.7716	1	3	0
34.70	2.62	2.5851	0	2	2
35.69	77.71	2.5156	1	3	1
36.50	100.00	2.4615	1	1	2
37.79	1.42	2.3809	2	0	0

(continued)

(continued)

38.28	12.52	2.3514	0	4	1
38.84	9.44	2.3188	2	1	0
39.67	37.15	2.2719	1	2	2
40.03	26.66	2.2523	1	4	0
41.77	17.27	2.1626	2	1	1
44.53	4.98	2.0346	1	3	2
46.53	2.65	1.9518	2	3	0
46.71	2.62	1.9448	0	4	2
48.44	6.08	1.8791	1	5	0
48.86	1.38	1.8640	2	0	2
50.33	4.40	1.8129	1	1	3
50.93	4.18	1.7930	1	5	1
52.23	68.17	1.7513	2	2	2
52.52	20.26	1.7423	2	4	0
52.83	4.30	1.7330	1	2	3
54.88	13.64	1.6730	2	4	1
56.11	13.40	1.6393	0	6	1
56.25	2.64	1.6355	2	3	2
56.81	18.93	1.6205	1	3	3
57.93	4.93	1.5920	1	5	2
58.65	9.39	1.5740	0	4	3
58.88	3.49	1.5685	3	1	0
60.32	1.37	1.5343	3	0	1
61.07	3.52	1.5173	3	1	1
61.14	2.40	1.5159	3	2	0
61.25	3.43	1.5134	2	1	3
61.57	4.76	1.5062	2	4	2
61.77	5.04	1.5018	2	5	1
61.94	27.83	1.4982	0	0	4
62.72	39.06	1.4814	0	6	2
63.28	1.28	1.4696	3	2	1
63.45	1.46	1.4660	2	2	3
64.79	3.47	1.4389	3	3	0
64.85	1.18	1.4377	0	2	4
66.87	7.76	1.3991	3	3	1
67.01	11.41	1.3965	1	7	0
67.39	8.26	1.3896	3	1	2
69.49	18.15	1.3526	3	2	2
69.64	1.16	1.3502	2	6	1
69.74	6.38	1.3485	3	4	0
71.60	9.86	1.3180	1	3	4

(continued)

(continued)

71.75	4.44	1.3156	3	4	1
71.91	1.76	1.3130	2	4	3
72.97	4.73	1.2965	0	6	3
74.89	2.31	1.2680	2	0	4
75.56	1.56	1.2584	2	1	4
75.60	1.81	1.2578	2	6	2
76.09	1.15	1.2510	1	6	3
76.34	2.08	1.2474	1	4	4
76.68	2.36	1.2427	3	0	3
77.65	1.25	1.2297	3	4	2
77.98	2.96	1.2252	2	5	3
80.72	5.05	1.1904	4	0	0
80.88	1.05	1.1884	2	3	4
82.31	2.64	1.1714	1	5	4
82.65	3.90	1.1675	3	3	3
82.70	2.16	1.1669	0	2	5
83.10	1.29	1.1623	1	0	5
83.59	5.05	1.1567	3	5	2
84.21	2.62	1.1498	2	7	2
84.97	2.00	1.1414	1	8	2
85.08	3.09	1.1403	3	6	1
85.48	5.75	1.1360	2	4	4
86.40	3.13	1.1262	2	8	0
87.22	4.09	1.1176	3	4	3
88.98	5.04	1.1001	1	3	5
89.00	2.67	1.0999	4	1	2

X-ray diffraction pattern
$Mg_{1.81}Fe_{0.18}Ni_{0.01}SiO4$

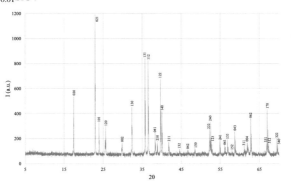

Kyanite (Al$_2$SiO$_5$), triclinic (P $\overline{1}$).

Structure
Guggenheim, S., Chang, Y.H., Koster van Groos, A.F. (1987). Muscovite dehydroxylation:
High-temperature studies. American Mineralogist, 72, 537–550
From Diamond mine, Keystone, South Dakota.
Cell parameters: 5.2010 Å, 9.0220 Å, 20.0430 Å, 90°, 95.780°, 90°.

ATOM	X	Y	Z	OCCUPANCY	ISO(B)
K	0.00000	0.09820	0.25000	0.930	1.682
Na	0.00000	0.09820	0.25000	0.070	1.682
Al	0.25110	0.08360	0.00010	0.915	0.628
Fe	0.25110	0.08360	0.00010	0.080	0.628
Al	0.46480	0.92940	0.13565	0.225	0.621
Si	0.46480	0.92940	0.13565	0.775	0.621
Al	0.45170	0.25840	0.13545	0.225	0.632
Si	0.45170	0.25840	0.13545	0.775	0.632
O	0.41700	0.09260	0.16830	1.000	1.198
O	0.25100	0.81120	0.15820	1.000	1.424
O	0.25100	0.37050	0.16900	1.000	1.239
O	0.45790	0.94380	0.05360	1.000	0.843
O	0.38580	0.25280	0.05370	1.000	1.052
OH	0.45650	0.56360	0.05030	0.915	0.944
F	0.45650	0.56360	0.05030	0.085	0.944

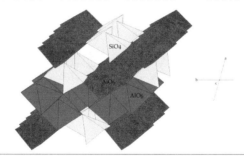

X-ray diffraction
Diffraction data computed using the structure data from the above paper listed, along with the
cell parameters refined from the powder pattern of R040104 (RRUFF database).
X-ray wavelength: 1.541838 Å.

2θ	INTENSITY	D-SPACING	H	K	L
8.87	96.93	9.9705	0	0	2
17.79	26.00	4.9853	0	0	4
19.68	6.64	4.5110	0	2	0
19.78	37.48	4.4887	1	1	0
19.89	98.21	4.4634	$\overline{1}$	1	1
20.18	12.06	4.3998	0	2	1
20.66	25.48	4.2994	1	1	1
21.62	23.06	4.1099	0	2	2
22.42	10.75	3.9654	1	1	2

(continued)

(continued)

22.92	64.02	3.8805	$\bar{1}$	1	3
23.84	60.74	3.7326	0	2	3
24.89	2.89	3.5778	1	1	3
25.52	84.25	3.4909	$\bar{1}$	1	4
26.65	61.18	3.3449	0	2	4
26.82	73.93	3.3235	0	0	6
27.88	78.40	3.1997	1	1	4
28.61	6.75	3.1198	$\bar{1}$	1	5
29.90	82.58	2.9879	0	2	5
31.28	56.17	2.8599	1	1	5
32.08	43.17	2.7902	$\bar{1}$	1	6
34.49	4.95	2.6001	1	3	0
34.56	37.50	2.5952	$\bar{1}$	3	1
34.67	18.70	2.5873	2	0	0
34.94	58.90	2.5680	$\bar{2}$	0	2
34.97	13.59	2.5660	1	1	6
35.03	100.00	2.5617	1	3	1
35.83	8.59	2.5063	$\bar{1}$	1	7
36.03	9.26	2.4926	0	0	8
36.46	24.80	2.4641	$\bar{1}$	3	3
36.76	12.89	2.4452	2	0	2
37.33	4.05	2.4086	0	2	7
37.52	16.65	2.3973	$\bar{2}$	0	4
37.78	40.08	2.3809	1	3	3
39.97	1.47	2.2555	0	4	0
40.03	10.24	2.2522	$\bar{2}$	2	1
40.18	3.69	2.2443	2	2	0
40.24	4.66	2.2412	0	4	1
40.44	9.34	2.2304	$\bar{1}$	3	5
40.85	10.42	2.2089	2	2	1
40.88	3.67	2.2073	2	0	4
41.03	6.33	2.1999	0	4	2
41.32	10.15	2.1850	$\bar{2}$	2	3
42.03	3.62	2.1497	2	2	2
42.04	22.47	2.1492	$\bar{2}$	0	6
42.32	4.33	2.1359	0	4	3
42.45	46.94	2.1294	1	3	5
43.68	6.50	2.0722	2	2	3
44.07	11.17	2.0550	0	4	4
44.56	4.09	2.0333	$\bar{2}$	2	5
45.49	34.46	1.9941	0	0	10
45.76	1.17	1.9827	2	2	4
46.06	25.31	1.9707	$\bar{1}$	3	7
46.61	12.59	1.9487	2	0	6
46.82	3.31	1.9402	$\bar{2}$	2	6

(continued)

(continued)

48.07	2.44	1.8929	$\bar{2}$	0	8
48.59	2.95	1.8738	1	3	7
48.80	3.66	1.8663	0	4	6
49.35	1.66	1.8466	$\bar{1}$	3	8
50.01	1.28	1.8239	0	2	10
51.69	1.65	1.7683	0	4	7
52.42	4.92	1.7454	$\bar{2}$	2	8
52.78	1.56	1.7343	$\bar{1}$	1	11
52.92	14.42	1.7302	$\bar{1}$	3	9
53.56	5.95	1.7111	2	0	8
53.81	1.22	1.7038	1	5	0
53.85	1.21	1.7024	$\bar{1}$	5	1
53.93	4.50	1.7002	2	4	0
54.00	1.31	1.6982	$\bar{3}$	1	2
54.18	1.33	1.6928	1	5	1
54.19	2.07	1.6925	2	2	7
54.56	4.96	1.6820	$\bar{3}$	1	3
54.84	2.60	1.6739	$\bar{2}$	4	3
55.26	19.46	1.6625	$\bar{2}$	0	10
55.42	6.61	1.6579	2	4	2
55.86	40.16	1.6458	1	3	9
55.94	5.97	1.6436	3	1	2
55.98	5.21	1.6427	$\bar{2}$	4	4
56.17	2.89	1.6374	1	5	3
56.50	11.83	1.6288	$\bar{1}$	5	4
56.77	2.26	1.6216	2	4	3
56.91	10.45	1.6180	$\bar{3}$	1	5
57.45	5.15	1.6041	3	1	3
57.50	8.57	1.6027	$\bar{2}$	4	5
57.62	3.94	1.5998	2	2	8
57.76	4.95	1.5961	1	5	4
58.67	3.34	1.5737	$\bar{3}$	1	6
59.24	3.34	1.5599	$\bar{2}$	2	10
59.26	1.22	1.5593	0	2	12
59.32	8.91	1.5578	3	1	4
59.40	5.79	1.5559	$\bar{2}$	4	6
59.72	2.99	1.5484	1	5	5
60.21	4.87	1.5370	$\bar{1}$	5	6
60.60	2.65	1.5279	2	4	5
60.79	22.33	1.5237	$\bar{1}$	3	11
61.14	2.06	1.5159	1	1	12
61.50	9.57	1.5077	2	0	10
61.69	23.80	1.5037	0	6	0

(continued)

(continued)

61.77	50.38	1.5019	$\bar{3}$	3	1
61.88	2.17	1.4994	0	6	1
62.03	3.40	1.4962	1	5	6
62.13	1.41	1.4940	0	4	10
62.42	1.83	1.4878	$\bar{3}$	3	3
62.46	2.64	1.4869	0	6	2
62.67	3.31	1.4824	3	3	1
63.04	1.25	1.4746	$\bar{2}$	2	11
63.04	4.35	1.4746	2	4	6
63.41	2.44	1.4670	$\bar{2}$	0	12
64.09	5.63	1.4530	1	3	11
64.12	3.57	1.4523	0	2	13
64.67	1.63	1.4413	1	5	7
65.25	5.57	1.4300	2	2	10
65.54	2.24	1.4244	0	0	14
66.02	1.68	1.4151	$\bar{3}$	1	9
66.07	1.79	1.4142	1	1	13
66.13	1.92	1.4130	0	4	11
67.14	1.26	1.3941	$\bar{2}$	4	9
67.19	2.11	1.3932	$\bar{1}$	1	14
68.20	1.99	1.3750	$\bar{3}$	3	7
68.49	1.58	1.3700	0	6	6
69.43	5.51	1.3536	2	2	11
69.56	27.88	1.3514	$\bar{1}$	3	13
70.35	12.98	1.3383	2	0	12
70.37	9.36	1.3379	0	4	12
71.18	1.56	1.3247	1	1	14
71.39	5.97	1.3213	$\bar{2}$	2	13
72.35	1.53	1.3061	$\bar{1}$	1	15
72.64	2.60	1.3016	$\bar{2}$	6	1
72.74	5.79	1.3001	2	6	0
72.76	3.98	1.2997	$\bar{4}$	0	2
72.90	13.67	1.2976	$\bar{2}$	6	2
73.16	6.81	1.2936	4	0	0
73.17	7.73	1.2935	$\bar{3}$	3	9
73.21	1.83	1.2928	1	3	13
73.56	6.04	1.2875	0	6	8
74.11	8.31	1.2794	3	3	7
74.49	7.45	1.2738	$\bar{2}$	6	4
74.86	3.12	1.2684	0	4	13
74.98	4.35	1.2667·	4	0	2
75.93	2.15	1.2532	$\bar{2}$	2	14
76.15	5.32	1.2501	$\bar{1}$	7	1

(continued)

(continued)

76.24	1.41	1.2489	$\bar{4}$	2	2
76.24	3.06	1.2488	$\bar{4}$	0	6
76.27	1.32	1.2484	$\bar{3}$	5	2
76.42	3.97	1.2463	0	0	16
76.58	1.63	1.2442	0	6	9
76.63	1.29	1.2435	4	2	0
76.68	5.50	1.2427	2	6	4
76.74	3.66	1.2419	$\bar{3}$	5	3
77.28	2.08	1.2346	$\bar{1}$	7	3
77.91	3.38	1.2263	3	5	2
78.10	3.49	1.2237	1	7	3
78.29	5.27	1.2212	$\bar{4}$	2	5
78.55	3.73	1.2178	2	2	13
78.73	1.53	1.2155	$\bar{3}$	5	5
79.40	3.09	1.2069	$\bar{3}$	3	11
79.46	2.46	1.2061	1	7	4
79.67	1.16	1.2035	$\bar{4}$	2	6
79.83	1.18	1.2015	4	2	3
79.90	1.85	1.2006	0	6	10
80.24	2.09	1.1964	$\bar{3}$	5	6
80.54	2.21	1.1926	3	3	9
80.73	1.80	1.1903	$\bar{2}$	2	15
80.81	4.66	1.1894	3	5	4
81.16	2.31	1.1851	1	7	5
81.38	2.17	1.1824	$\bar{4}$	2	7
81.51	2.87	1.1809	3	1	11
81.58	1.88	1.1800	4	2	4
81.72	1.07	1.1783	$\bar{2}$	4	13
83.29	1.09	1.1601	1	3	15
83.31	1.12	1.1600	$\bar{1}$	1	17
83.67	1.29	1.1558	4	2	5
83.69	3.04	1.1555	$\bar{1}$	7	7
85.88	1.10	1.1317	3	1	12
86.08	1.26	1.1295	2	6	8
86.11	3.75	1.1292	$\bar{2}$	4	14
86.78	1.30	1.1222	4	4	0
87.40	1.22	1.1159	$\bar{4}$	4	4
87.47	10.24	1.1152	$\bar{2}$	6	10
87.87	1.81	1.1111	$\bar{1}$	5	14
88.61	5.50	1.1037	4	0	8

(continued)

(continued)

X-ray diffraction pattern
($K_{0.91}Na_{0.09}Al_{1.70}$ $Mg_{0.08}Ti_{0.02}$ $Fe^{3+}_{0.14}Fe^{2+}_{0.06}$ $(Si_{3.12}Al_{0.88})O_{10}(OH)_2$)

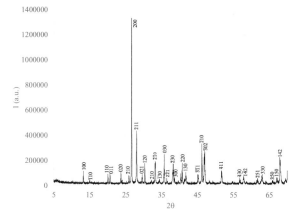

Muscovite ($KAl_2(Si_3Al)O_{10}(OH,F)_2$), monoclinic ($C2/m$).

Structure
Guggenheim, S., Chang, Y.H., Koster van Groos, A.F. (1987). Muscovite dehydroxylation:
High-temperature studies. American Mineralogist, 72, 537–550
From Diamond mine, Keystone, South Dakota.
Cell parameters: 5.2010 Å, 9.0220 Å, 20.0430 Å, 90°, 95.780°, 90°.

ATOM	X	Y	Z	OCCUPANCY	ISO(B)
K	0.00000	0.09820	0.25000	0.930	1.682
Na	0.00000	0.09820	0.25000	0.070	1.682
Al	0.25110	0.08360	0.00010	0.915	0.628
Fe	0.25110	0.08360	0.00010	0.080	0.628
Al	0.46480	0.92940	0.13565	0.225	0.621
Si	0.46480	0.92940	0.13565	0.775	0.621
Al	0.45170	0.25840	0.13545	0.225	0.632
Si	0.45170	0.25840	0.13545	0.775	0.632
O	0.41700	0.09260	0.16830	1.000	1.198
O	0.25100	0.81120	0.15820	1.000	1.424
O	0.25100	0.37050	0.16900	1.000	1.239
O	0.45790	0.94380	0.05360	1.000	0.843
O	0.38580	0.25280	0.05370	1.000	1.052
OH	0.45650	0.56360	0.05030	0.915	0.944
F	0.45650	0.56360	0.05030	0.085	0.944

(continued)

(continued)

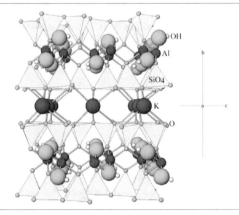

X-ray diffraction
Diffraction data computed using the structure data from the above paper listed, along with the
cell parameters refined from the powder pattern of R040104 (RRUFF database).
X-ray wavelength: 1.541838 Å.

2θ	INTENSITY	D-SPACING	H	K	L
8.87	96.93	9.9705	0	0	2
17.79	26.00	4.9853	0	0	4
19.68	6.64	4.5110	0	2	0
19.78	37.48	4.4887	1	1	0
19.89	98.21	4.4634	$\bar{1}$	1	1
20.18	12.06	4.3998	0	2	1
20.66	25.48	4.2994	1	1	1
21.62	23.06	4.1099	0	2	2
22.42	10.75	3.9654	1	1	2
22.92	64.02	3.8805	$\bar{1}$	1	3
23.84	60.74	3.7326	0	2	3
24.89	2.89	3.5778	1	1	3
25.52	84.25	3.4909	$\bar{1}$	1	4
26.65	61.18	3.3449	0	2	4
26.82	73.93	3.3235	0	0	6
27.88	78.40	3.1997	1	1	4
28.61	6.75	3.1198	$\bar{1}$	1	5
29.90	82.58	2.9879	0	2	5
31.28	56.17	2.8599	1	1	5
32.08	43.17	2.7902	$\bar{1}$	1	6
34.49	4.95	2.6001	1	3	0
34.56	37.50	2.5952	$\bar{1}$	3	1
34.67	18.70	2.5873	2	0	0
34.94	58.90	2.5680	$\bar{2}$	0	2
34.97	13.59	2.5660	1	1	6
35.03	100.00	2.5617	1	3	1

(continued)

(continued)

35.83	8.59	2.5063	$\bar{1}$	1	7
36.03	9.26	2.4926	0	0	8
36.46	24.80	2.4641	$\bar{1}$	3	3
36.76	12.89	2.4452	2	0	2
37.33	4.05	2.4086	0	2	7
37.52	16.65	2.3973	$\bar{2}$	0	4
37.78	40.08	2.3809	1	3	3
39.97	1.47	2.2555	0	4	0
40.03	10.24	2.2522	$\bar{2}$	2	1
40.18	3.69	2.2443	2	2	0
40.24	4.66	2.2412	0	4	1
40.44	9.34	2.2304	$\bar{1}$	3	5
40.85	10.42	2.2089	2	2	1
40.88	3.67	2.2073	2	0	4
41.03	6.33	2.1999	0	4	2
41.32	10.15	2.1850	$\bar{2}$	2	3
42.03	3.62	2.1497	2	2	2
42.04	22.47	2.1492	$\bar{2}$	0	6
42.32	4.33	2.1359	0	4	3
42.45	46.94	2.1294	1	3	5
43.68	6.50	2.0722	2	2	3
44.07	11.17	2.0550	0	4	4
44.56	4.09	2.0333	$\bar{2}$	2	5
45.49	34.46	1.9941	0	0	10
45.76	1.17	1.9827	2	2	4
46.06	25.31	1.9707	$\bar{1}$	3	7
46.61	12.59	1.9487	2	0	6
46.82	3.31	1.9402	$\bar{2}$	2	6
48.07	2.44	1.8929	$\bar{2}$	0	8
48.59	2.95	1.8738	1	3	7
48.80	3.66	1.8663	0	4	6
49.35	1.66	1.8466	$\bar{1}$	3	8
50.01	1.28	1.8239	0	2	10
51.69	1.65	1.7683	0	4	7
52.42	4.92	1.7454	$\bar{2}$	2	8
52.78	1.56	1.7343	$\bar{1}$	1	11
52.92	14.42	1.7302	$\bar{1}$	3	9
53.56	5.95	1.7111	2	0	8
53.81	1.22	1.7038	1	5	0
53.85	1.21	1.7024	$\bar{1}$	5	1
53.93	4.50	1.7002	2	4	0
54.00	1.31	1.6982	$\bar{3}$	1	2
54.18	1.33	1.6928	1	5	1
54.19	2.07	1.6925	2	2	7
54.56	4.96	1.6820	$\bar{3}$	1	3

(continued)

(continued)

54.84	2.60	1.6739	$\bar{2}$	4	3
55.26	19.46	1.6625	$\bar{2}$	0	10
55.42	6.61	1.6579	2	4	2
55.86	40.16	1.6458	1	3	9
55.94	5.97	1.6436	3	1	2
55.98	5.21	1.6427	$\bar{2}$	4	4
56.17	2.89	1.6374	1	5	3
56.50	11.83	1.6288	$\bar{1}$	5	4
56.77	2.26	1.6216	2	4	3
56.91	10.45	1.6180	$\bar{3}$	1	5
57.45	5.15	1.6041	3	1	3
57.50	8.57	1.6027	$\bar{2}$	4	5
57.62	3.94	1.5998	2	2	8
57.76	4.95	1.5961	1	5	4
58.67	3.34	1.5737	$\bar{3}$	1	6
59.24	3.34	1.5599	$\bar{2}$	2	10
59.26	1.22	1.5593	0	2	12
59.32	8.91	1.5578	3	1	4
59.40	5.79	1.5559	$\bar{2}$	4	6
59.72	2.99	1.5484	1	5	5
60.21	4.87	1.5370	$\bar{1}$	5	6
60.60	2.65	1.5279	2	4	5
60.79	22.33	1.5237	$\bar{1}$	3	11
61.14	2.06	1.5159	1	1	12
61.50	9.57	1.5077	2	0	10
61.69	23.80	1.5037	0	6	0
61.77	50.38	1.5019	$\bar{3}$	3	1
61.88	2.17	1.4994	0	6	1
62.03	3.40	1.4962	1	5	6
62.13	1.41	1.4940	0	4	10
62.42	1.83	1.4878	$\bar{3}$	3	3
62.46	2.64	1.4869	0	6	2
62.67	3.31	1.4824	3	3	1
63.04	1.25	1.4746	$\bar{2}$	2	11
63.04	4.35	1.4746	2	4	6
63.41	2.44	1.4670	$\bar{2}$	0	12
64.09	5.63	1.4530	1	3	11
64.12	3.57	1.4523	0	2	13
64.67	1.63	1.4413	1	5	7
65.25	5.57	1.4300	2	2	10
65.54	2.24	1.4244	0	0	14
66.02	1.68	1.4151	$\bar{3}$	1	9
66.07	1.79	1.4142	1	1	13
66.13	1.92	1.4130	0	4	11
67.14	1.26	1.3941	$\bar{2}$	4	9

(continued)

			h	k	l
67.19	2.11	1.3932	$\bar{1}$	1	14
68.20	1.99	1.3750	$\bar{3}$	3	7
68.49	1.58	1.3700	0	6	6
69.43	5.51	1.3536	2	2	11
69.56	27.88	1.3514	$\bar{1}$	3	13
70.35	12.98	1.3383	2	0	12
70.37	9.36	1.3379	0	4	12
71.18	1.56	1.3247	1	1	14
71.39	5.97	1.3213	$\bar{2}$	2	13
72.35	1.53	1.3061	$\bar{1}$	1	15
72.64	2.60	1.3016	$\bar{2}$	6	1
72.74	5.79	1.3001	2	6	0
72.76	3.98	1.2997	$\bar{4}$	0	2
72.90	13.67	1.2976	$\bar{2}$	6	2
73.16	6.81	1.2936	4	0	0
73.17	7.73	1.2935	$\bar{3}$	3	9
73.21	1.83	1.2928	1	3	13
73.56	6.04	1.2875	0	6	8
74.11	8.31	1.2794	3	3	7
74.49	7.45	1.2738	$\bar{2}$	6	4
74.86	3.12	1.2684	0	4	13
74.98	4.35	1.2667	4	0	2
75.93	2.15	1.2532	$\bar{2}$	2	14
76.15	5.32	1.2501	$\bar{1}$	7	1
76.24	1.41	1.2489	$\bar{4}$	2	2
76.24	3.06	1.2488	$\bar{4}$	0	6
76.27	1.32	1.2484	$\bar{3}$	5	2
76.42	3.97	1.2463	0	0	16
76.58	1.63	1.2442	0	6	9
76.63	1.29	1.2435	4	2	0
76.68	5.50	1.2427	2	6	4
76.74	3.66	1.2419	$\bar{3}$	5	3
77.28	2.08	1.2346	$\bar{1}$	7	3
77.91	3.38	1.2263	3	5	2
78.10	3.49	1.2237	1	7	3
78.29	5.27	1.2212	$\bar{4}$	2	5
78.55	3.73	1.2178	2	2	13
78.73	1.53	1.2155	$\bar{3}$	5	5
79.40	3.09	1.2069	$\bar{3}$	3	11
79.46	2.46	1.2061	1	7	4
79.67	1.16	1.2035	$\bar{4}$	2	6
79.83	1.18	1.2015	4	2	3
79.90	1.85	1.2006	0	6	10

(continued)

(continued)

80.24	2.09	1.1964	$\bar{3}$	5	6
80.54	2.21	1.1926	3	3	9
80.73	1.80	1.1903	$\bar{2}$	2	15
80.81	4.66	1.1894	3	5	4
81.16	2.31	1.1851	1	7	5
81.38	2.17	1.1824	$\bar{4}$	2	7
81.51	2.87	1.1809	3	1	11
81.58	1.88	1.1800	4	2	4
81.72	1.07	1.1783	$\bar{2}$	4	13
83.29	1.09	1.1601	1	3	15
83.31	1.12	1.1600	$\bar{1}$	1	17
83.67	1.29	1.1558	4	2	5
83.69	3.04	1.1555	$\bar{1}$	7	7
85.88	1.10	1.1317	3	1	12
86.08	1.26	1.1295	2	6	8
86.11	3.75	1.1292	$\bar{2}$	4	14
86.78	1.30	1.1222	4	4	0
87.40	1.22	1.1159	$\bar{4}$	4	4
87.47	10.24	1.1152	$\bar{2}$	6	10
87.87	1.81	1.1111	$\bar{1}$	5	14
88.61	5.50	1.1037	4	0	8

X-ray diffraction pattern
$(K_{0.91}Na_{0.09}Al_{1.70}\ Mg_{0.08}Ti_{0.02}\ Fe^{3+}_{0.14}Fe^{2+}_{0.06}\ (Si_{3.12}Al_{0.88})O_{10}(OH)_2)$

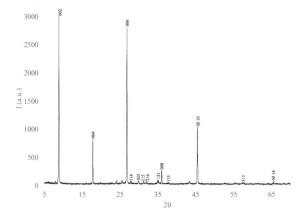

ORTOCHLASE (K(AlSi₃O₈)), monoclinic (C2/m).

Structure
Tseng, H.Y., Heaney, P.J., Onstott, T.C. Physics and Chemistry of Minerals 22 (1995) 399–405.
Characterization of lattice strain induced by neutron irradiation
Cell parameters: 8.6000 Å, 12.9978 Å, 7.2004 Å, 90°, 116.049°, 90°.

ATOM	X	Y	Z	OCCUPANCY	ISO(B)
K	0.28320	0.00000	0.13580	1.000	0.474
Al	0.00900	0.18420	0.22260	0.500	0.158
Si	0.00900	0.18420	0.22260	0.500	0.158
Si	0.70780	0.11848	0.33950	1.000	0.158
O	0.00000	0.14870	0.00000	1.000	0.947
O	0.63280	0.00000	0.27820	1.000	0.947
O	0.83070	0.14660	0.22180	1.000	0.947
O	0.03640	0.31260	0.25460	1.000	0.947
O	0.18210	0.12540	0.40630	1.000	0.947

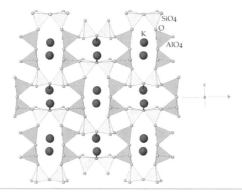

X-ray diffraction
Diffraction data computed using the structure data from the above paper listed, along with the
cell parameters refined from the powder pattern of R040055 (RRUFF database).
X-ray wavelength: 1.541838 Å.

2θ	INTENSITY	D-SPACING	H	K	L
13.33	4.63	6.6416	1	1	0
13.63	6.10	6.4989	0	2	0
15.09	9.69	5.8727	$\bar{1}$	1	1
19.36	1.65	4.5848	0	2	1
20.98	60.88	4.2350	$\bar{2}$	0	1
22.52	19.34	3.9486	1	1	1
23.02	4.86	3.8632	2	0	0
23.54	75.78	3.7790	1	3	0
24.60	16.57	3.6191	$\bar{1}$	3	1
25.10	12.44	3.5481	$\bar{2}$	2	1
25.68	49.72	3.4687	$\bar{1}$	1	2
26.85	100.00	3.3208	2	2	0

(continued)

(continued)

27.09	56.96	3.2915	$\bar{2}$	0	2
27.45	29.79	3.2494	0	4	0
27.58	80.77	3.2345	0	0	2
29.83	63.31	2.9949	1	3	1
30.44	7.45	2.9364	$\bar{2}$	2	2
30.79	25.20	2.9037	0	4	1
30.88	10.05	2.8957	0	2	2
32.34	22.21	2.7686	$\bar{1}$	3	2
34.36	21.60	2.6097	$\bar{3}$	1	2
34.71	4.22	2.5844	2	2	1
34.80	37.93	2.5780	$\bar{2}$	4	1
35.15	8.55	2.5532	1	1	2
35.53	9.69	2.5264	3	1	0
36.12	3.30	2.4867	2	4	0
37.19	9.60	2.4179	$\bar{1}$	5	1
37.65	10.42	2.3890	$\bar{3}$	3	1
38.73	7.58	2.3248	$\bar{1}$	1	3
39.72	2.44	2.2693	$\bar{3}$	3	2
40.37	1.29	2.2343	$\bar{2}$	2	3
40.98	4.46	2.2022	1	5	1
41.69	26.89	2.1663	0	6	0
42.47	10.47	2.1284	2	4	1
42.56	3.61	2.1242	$\bar{4}$	0	1
42.70	6.08	2.1175	$\bar{4}$	0	2
43.63	1.02	2.0745	$\bar{1}$	3	3
43.68	2.48	2.0722	2	0	2
43.72	3.26	2.0704	3	1	1
44.08	6.27	2.0542	0	6	1
45.03	12.86	2.0133	$\bar{4}$	2	2
45.97	11.24	1.9743	2	2	2
46.38	4.64	1.9576	$\bar{3}$	3	3
47.04	11.19	1.9316	4	0	0
47.22	3.29	1.9248	$\bar{3}$	5	1
47.43	5.54	1.9166	$\bar{4}$	0	3
48.16	2.45	1.8895	2	6	0
48.21	2.03	1.8876	3	3	1
49.10	8.56	1.8555	1	1	3
49.21	2.09	1.8515	4	2	0
49.55	2.11	1.8397	1	5	2
49.59	1.60	1.8384	$\bar{4}$	2	3
49.84	3.76	1.8296	3	5	0
50.43	4.76	1.8096	$\bar{2}$	6	2
50.72	9.82	1.7999	0	6	2

(continued)

(continued)

50.73	24.04	1.7996	$\bar{2}$	0	4
50.82	10.18	1.7967	0	4	3
51.39	7.70	1.7780	$\bar{4}$	4	1
51.51	2.62	1.7741	$\bar{4}$	4	2
52.37	4.57	1.7472	2	4	2
52.98	1.07	1.7285	$\bar{1}$	1	4
53.24	1.80	1.7205	1	3	3
53.81	2.53	1.7036	$\bar{5}$	1	2
54.74	5.86	1.6768	$\bar{3}$	5	3
55.70	3.78	1.6502	$\bar{1}$	7	2
56.37	8.20	1.6322	3	5	1
56.38	1.47	1.6319	$\bar{3}$	3	4
56.44	1.17	1.6304	5	1	3
56.65	2.27	1.6247	0	8	0
57.79	6.33	1.5954	$\bar{4}$	2	4
58.84	6.63	1.5694	0	2	4
59.25	2.02	1.5595	2	2	3
60.23	8.81	1.5366	$\bar{5}$	3	3
60.59	2.81	1.5283	0	6	3
61.10	9.06	1.5167	$\bar{4}$	6	1
61.96	24.42	1.4977	2	8	0
62.71	3.14	1.4815	$\bar{5}$	1	4
63.75	3.11	1.4600	$\bar{1}$	7	3
63.90	1.04	1.4569	$\bar{2}$	8	2
63.97	3.58	1.4555	5	3	0
64.14	1.11	1.4519	0	8	2
64.21	3.94	1.4505	1	1	4
64.32	1.19	1.4482	$\bar{1}$	5	4
64.65	2.54	1.4417	4	6	0
64.73	9.74	1.4401	2	4	3
64.97	2.27	1.4355	$\bar{4}$	6	3
65.07	2.53	1.4335	$\bar{5}$	5	2
65.15	7.48	1.4318	$\bar{6}$	0	2
65.22	1.03	1.4305	$\bar{2}$	0	5
65.78	1.99	1.4196	1	9	0
66.38	7.58	1.4083	4	0	2
67.10	2.33	1.3948	$\bar{4}$	0	5
67.42	2.86	1.3891	$\bar{5}$	5	3
67.66	1.02	1.3847	$\bar{6}$	0	1
67.68	4.47	1.3843	$\bar{2}$	6	4
67.74	1.59	1.3832	1	3	4
67.95	1.26	1.3795	$\bar{6}$	2	3
68.45	1.22	1.3707	$\bar{1}$	1	5

(continued)

(continued)

68.84	3.03	1.3638	$\bar{4}$	2	5
69.39	3.36	1.3543	$\bar{6}$	2	1
70.22	6.17	1.3403	$\bar{1}$	9	2
70.73	2.77	1.3320	$\bar{6}$	0	4
70.95	1.11	1.3283	5	5	0
71.87	7.09	1.3136	$\bar{1}$	3	5
72.11	2.16	1.3099	$\bar{5}$	1	5
72.15	1.87	1.3092	$\bar{2}$	4	5
72.18	2.53	1.3088	5	3	1
72.61	2.68	1.3020	2	0	4
72.76	1.62	1.2998	0	10	0
72.90	1.69	1.2976	0	8	3
73.01	1.38	1.2959	0	6	4
73.25	1.29	1.2922	4	4	2
73.36	1.46	1.2905	$\bar{4}$	8	1
73.37	1.06	1.2904	2	6	3
73.46	11.39	1.2890	$\bar{4}$	8	2
73.55	2.41	1.2877	6	0	0
73.95	2.57	1.2817	$\bar{4}$	4	5
74.16	12.08	1.2786	2	8	2
74.48	4.04	1.2739	$\bar{6}$	4	1
74.77	3.00	1.2696	$\bar{3}$	9	2
74.83	1.56	1.2689	0	2	5
75.22	1.35	1.2632	6	2	0
75.36	1.35	1.2612	$\bar{5}$	7	2
75.46	3.72	1.2597	$\bar{5}$	3	5
75.63	4.09	1.2573	$\bar{3}$	5	5
76.28	1.11	1.2482	3	7	2
78.39	1.72	1.2199	3	5	3
78.48	1.47	1.2187	$\bar{6}$	0	5
78.84	1.29	1.2140	5	5	1
79.24	1.80	1.2089	$\bar{2}$	10	2
79.47	1.32	1.2060	0	10	2
80.18	1.17	1.1972	6	4	0
80.39	1.40	1.1945	$\bar{6}$	6	2
80.45	1.83	1.1937	$\bar{2}$	6	5
81.72	1.20	1.1784	$\bar{7}$	$\bar{3}$	3
83.63	5.44	1.1562	$\bar{4}$	8	4
84.53	5.07	1.1462	0	8	4
85.04	1.20	1.1407	$\bar{2}$	10	3
86.41	4.05	1.1260	4	4	3

(continued)

(continued)

87.39	1.08	1.1159	2	6	4
87.75	2.05	1.1123	$\bar{1}$	9	4
89.01	2.61	1.0998	7	1	0
89.03	1.96	1.0996	6	4	1
89.78	1.14	1.0923	5	5	2

X-ray diffraction pattern
$K_{0.96}$ $(Al_{0.96}Si_{0.03}Fe_{0.02})Si_{3.00}O_8$

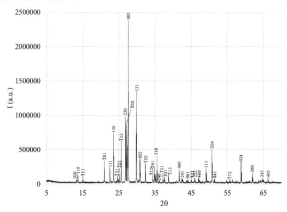

Quartz low (SiO_2), trigonal $(P3_121, P3_221)$.

Structure
Speziale, S., Duffy, T.S. (2002). Single-crystal elastic constants of fluorite (CaF_2) to 9.3 GPa.
Physics and Chemistry of Minerals, 29, 465–472
Cell parameters: 4.9134 Å, 4.9134, Å 5.4042 Å, 90°, 90°, 120°.

(continued)

(continued)

ATOM	X	Y	Z	OCCUPANCY	ISO(B)
Si	0.46970	0.00000	0.00000	1.000	0.616
O	0.41350	0.26690	0.11910	1.000	1.057

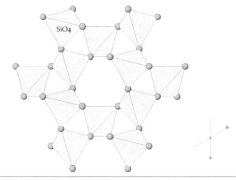

X-ray diffraction

Diffraction data computed using the structure data from the above paper listed, along with the cell parameters refined from the powder pattern of R040031 (RRUFF database).

X-ray wavelength: 1.541838 Å.

2θ	INTENSITY	D-SPACING	H	K	L
20.88	19.92	4.2551	1	0	0
26.66	69.90	3.3432	0	1	1
26.66	30.10	3.3432	1	0	1
36.58	7.13	2.4567	1	1	0
39.51	1.01	2.2810	0	1	2
39.51	6.18	2.2810	1	0	2
40.33	3.19	2.2365	1	1	1
42.49	5.24	2.1276	2	0	0
45.84	1.00	1.9797	0	2	1
45.84	2.09	1.9797	2	0	1
50.19	12.58	1.8177	1	1	2
54.93	3.09	1.6716	0	2	2
55.38	1.61	1.6589	0	1	3
60.02	4.93	1.5415	1	2	1
60.02	4.11	1.5415	2	1	1
64.10	1.72	1.4527	1	1	3
67.81	1.38	1.3820	2	1	2
67.81	4.18	1.3820	1	2	2
68.22	1.36	1.3748	0	2	3
68.22	5.31	1.3748	2	0	3
68.38	4.13	1.3719	0	3	1
73.55	1.73	1.2877	1	0	4
75.74	1.91	1.2559	3	0	2
79.97	2.30	1.1997	2	1	3
81.26	2.41	1.1838	1	1	4

(continued)

(continued)

X-ray diffraction pattern

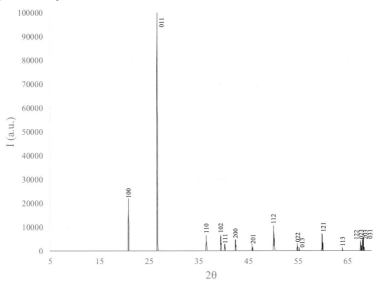

Sillimanite (Al_2SiO_5), orthorhombic (P*nma*).

Structure
Winter, J.K., Ghose, S. (1979). Thermal expansion and high-temperature crystal chemistry of the Al_2SiO_5 polymorphs. American Mineralogist, 64, 573–586.
Cell parameters: 7.486484, 7.674338 Å, 5.772069 Å, 90°, 90°, 90°.

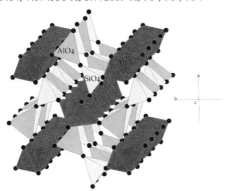

X-ray diffraction
Diffraction data computed using the structure data from the above paper listed, along with the cell parameters refined from the powder pattern of R050034 (RRUFF database).
X-ray wavelength: 1.541838 Å.

(continued)

(continued)

2θ	INTENSITY	D-SPACING	H	K	L
16.57	18.42	5.3589	1	1	0
19.45	3.24	4.5712	1	0	1
23.21	1.36	3.8372	0	2	0
23.80	7.29	3.7432	2	0	0
26.12	80.51	3.4148	1	2	0
26.52	100.00	3.3644	2	1	0
30.43	1.05	2.9390	1	2	1
31.01	17.11	2.8860	0	0	2
33.46	26.03	2.6795	2	2	0
35.34	48.45	2.5410	1	1	2
37.16	12.07	2.4207	1	3	0
37.93	1.31	2.3732	3	1	0
39.07	1.72	2.3065	0	2	2
39.44	9.37	2.2856	2	0	2
40.95	65.27	2.2042	1	2	2
42.83	13.79	2.1120	2	3	0
43.26	3.05	2.0920	3	2	0
45.75	1.35	1.9834	2	3	1
46.24	1.14	1.9636	2	2	2
48.65	8.38	1.8716	4	0	0
49.74	7.30	1.8330	3	1	2
50.17	1.33	1.8183	4	1	0
51.14	2.96	1.7863	3	3	0
53.68	8.14	1.7074	2	4	0
54.15	6.99	1.6938	3	2	2
54.55	13.21	1.6822	4	2	0
57.69	16.52	1.5977	0	4	2
58.80	8.31	1.5703	4	0	2
59.12	2.67	1.5626	1	4	2
60.14	1.56	1.5384	4	1	2
60.99	38.86	1.5189	3	3	2
64.06	4.23	1.4533	4	2	2
64.57	16.52	1.4430	0	0	4
65.74	3.72	1.4201	2	5	0
67.09	5.81	1.3949	5	2	0
70.24	3.97	1.3397	4	4	0
70.62	10.99	1.3335	1	5	2
70.88	6.14	1.3292	1	2	4
71.06	7.19	1.3262	2	1	4
72.10	2.72	1.3096	5	1	2
72.24	1.09	1.3074	3	5	0
74.11	1.58	1.2791	0	6	0
74.43	13.37	1.2742	2	5	2
74.69	3.22	1.2705	2	2	4
75.71	10.28	1.2559	5	2	2

(continued)

(continued)

76.29	3.55	1.2477	6	0	0
76.89	3.52	1.2395	1	3	4
77.37	1.17	1.2330	3	1	4
77.48	1.70	1.2316	6	1	0
78.72	2.55	1.2152	4	4	2
80.60	1.43	1.1915	2	3	4
81.60	3.21	1.1794	5	3	2
84.80	1.84	1.1428	4	0	4
87.32	1.01	1.1162	2	6	2
88.73	2.71	1.1021	2	4	4
89.18	1.35	1.0976	4	5	2
89.20	1.05	1.0975	6	2	2
89.43	4.31	1.0953	4	2	4
89.71	1.03	1.0925	5	4	2
89.71	1.03	1.0925	5	4	2

X-ray diffraction pattern
$Al_{0.99}Fe^{3+}_{0.01}SiO_5$

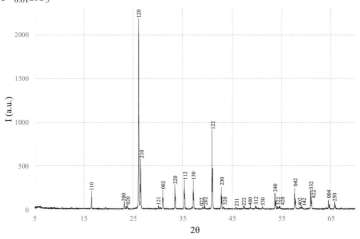

Staurolite $((Fe^{++}Mg)_2Al_9(Si,Al)_4O_{20}(O,OH)_4)$, monoclinic $(C2/m)$.

Structure
Smith, J.V. (1968). The crystal structure of staurolite. American Mineralogist, 53, 1139–1155
Cell parameters: 7.871300 Å, 16.62040 Å, 5.656000 Å, 90°, 90°, 90°.

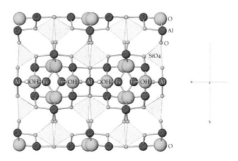

X-ray diffraction
Diffraction data computed using the structure data from the above paper listed, along with the
cell parameters refined from the powder pattern of R050079 (RRUFF database).
X-ray wavelength: 1.541838 Å.

2θ	INTENSITY	D-SPACING	H	K	L
10.68	9.63	8.3094	0	2	0
12.48	20.13	7.1113	1	1	0
21.41	11.78	4.1547	0	4	0
22.62	1.73	3.9340	2	0	0
25.06	10.89	3.5556	2	2	0
25.21	3.89	3.5356	$\bar{1}$	3	1
25.21	4.48	3.5356	1	3	1
27.64	2.98	3.2297	$\bar{2}$	0	1
27.64	1.29	3.2297	2	0	1
29.18	9.99	3.0618	1	5	0
29.69	31.29	3.0103	$\bar{2}$	2	1
29.69	38.80	3.0103	2	2	1
31.33	7.35	2.8566	2	4	0
31.65	17.57	2.8282	0	0	2
32.33	30.82	2.7698	0	6	0
33.29	37.54	2.6926	$\bar{1}$	5	1
33.29	36.76	2.6926	1	5	1

(continued)

(continued)

33.48	6.97	2.6773	0	2	2
34.64	3.50	2.5906	3	1	0
35.21	13.32	2.5499	$\bar{2}$	4	1
35.21	16.37	2.5499	2	4	1
37.50	48.14	2.3990	$\bar{1}$	3	2
37.50	48.32	2.3990	1	3	2
37.97	35.93	2.3704	3	3	0
38.22	20.04	2.3553	3	1	1
38.22	21.57	2.3553	$\bar{3}$	1	1
38.51	1.39	2.3379	0	4	2
39.24	5.24	2.2964	$\bar{2}$	0	2
39.24	5.17	2.2964	2	0	2
39.66	1.83	2.2729	1	7	0
39.81	8.10	2.2648	2	6	0
42.88	13.30	2.1090	$\bar{1}$	7	1
42.88	12.60	2.1090	1	7	1
43.98	1.27	2.0589	3	5	0
45.86	53.67	1.9789	0	6	2
46.15	27.38	1.9670	4	0	0
51.39	3.88	1.7778	4	4	0
51.58	1.01	1.7717	1	7	2
52.36	2.65	1.7471	$\bar{2}$	8	1
52.36	3.52	1.7471	2	8	1
53.48	2.23	1.7132	$\bar{1}$	9	1
53.48	2.36	1.7132	1	9	1
54.83	2.07	1.6742	0	8	2
55.13	4.80	1.6657	$\bar{2}$	2	3
55.13	4.10	1.6657	2	2	3
55.17	1.53	1.6645	3	5	2
55.17	1.22	1.6645	$\bar{3}$	5	2
55.27	2.40	1.6619	0	10	0
57.02	4.94	1.6148	4	0	2
57.02	4.91	1.6148	$\bar{4}$	0	2
57.38	8.65	1.6055	$\bar{1}$	5	3
57.38	9.61	1.6055	1	5	3
57.45	2.93	1.6037	4	6	0
58.66	2.95	1.5736	$\bar{2}$	4	3
58.66	2.07	1.5736	2	4	3
58.94	1.44	1.5666	5	1	0
59.94	3.62	1.5429	4	6	1
59.94	3.21	1.5429	$\bar{4}$	6	1
60.04	4.95	1.5405	$\bar{2}$	8	2
60.04	4.65	1.5405	2	8	2
60.74	3.77	1.5244	3	1	3
60.74	3.81	1.5244	$\bar{3}$	1	3

(continued)

(continued)

61.06	11.84	1.5171	$\bar{1}$	9	2
61.06	11.79	1.5171	1	9	2
61.22	13.33	1.5137	5	3	0
61.40	5.30	1.5098	5	1	1
61.40	4.81	1.5098	$\bar{5}$	1	1
61.39	5.21	1.5098	3	9	0
62.87	2.66	1.4777	$\bar{2}$	10	1
62.87	3.44	1.4777	2	10	1
63.62	1.09	1.4623	$\bar{5}$	3	1
64.16	3.86	1.4511	$\bar{1}$	7	3
64.16	4.46	1.4511	1	7	3
65.08	2.42	1.4328	0	10	2
66.05	29.60	1.4141	0	0	4
67.07	50.05	1.3951	4	6	2
67.07	49.95	1.3951	$\bar{4}$	6	2
67.63	23.25	1.3849	0	12	0
68.44	1.41	1.3704	5	1	2
68.44	1.44	1.3704	$\bar{5}$	1	2
71.71	1.08	1.3157	$\bar{2}$	8	3
72.13	1.16	1.3091	3	11	0
73.02	1.63	1.2953	6	2	0
74.35	3.00	1.2754	$\bar{3}$	11	1
74.35	2.84	1.2754	3	11	1
75.45	3.18	1.2594	0	6	4
76.57	1.55	1.2438	0	12	2
78.22	1.98	1.2216	$\bar{4}$	6	3
78.22	1.47	1.2216	4	6	3
78.27	1.27	1.2210	6	4	1
78.27	1.52	1.2210	$\bar{6}$	4	1
78.77	3.94	1.2144	$\bar{3}$	3	4
78.77	4.08	1.2144	3	3	4
79.52	1.82	1.2049	5	1	3
79.52	1.80	1.2049	$\bar{5}$	1	3
80.09	5.23	1.1977	5	9	0
80.84	1.29	1.1885	$\bar{2}$	10	3
80.84	1.15	1.1885	2	10	3
81.51	1.51	1.1804	$\bar{5}$	3	3
81.51	1.51	1.1804	5	3	3
84.31	2.17	1.1482	4	0	4
84.31	2.27	1.1482	$\bar{4}$	0	4
85.38	1.36	1.1364	2	14	0
85.76	2.98	1.1324	4	12	0

(continued)

(continued)

86.36	1.50	1.1261	$\bar{3}$	13	1
86.36	1.35	1.1261	3	13	1
87.51	1.26	1.1142	2	14	1
87.51	1.03	1.1142	$\bar{2}$	14	1
87.51	1.03	1.1142	$\bar{2}$	14	1

X-ray diffraction pattern
$(Fe^{++}_{0.77}Mg_{0.23})_2Al_9Ti_{0.01}(Si_{0.99},Al_{0.01})_4O_{20}(O_{0.76},OH_{0.24})_4$
The (00 l) peaks of muscovite are also present.

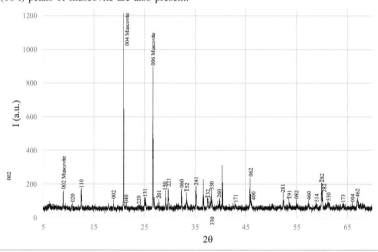

Tremolite $(Ca_2Mg_5Si_8O_{22}(OH)_2)$, monoclinic (C2/m).

Structure
Sueno, S., Cameron, M., Papike, J.J., Prewitt, C.T. (1973). The high temperature crystal chemistry of tremolite. American Mineralogist, 58, 649–664
Cell parameters: 9.860000 Å, 18.11800 Å, 5.285000 Å, 90°, 104.57°, 90°.

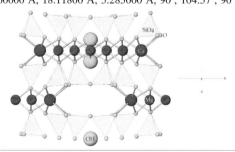

X-ray diffraction
Diffraction data computed using the structure data from the above paper listed, along with the cell parameters refined from the powder pattern of R040045 (RRUFF database).
X-ray wavelength: 1.541838 Å.

(continued)

(continued)

2θ	INTENSITY	D-SPACING	H	K	L
9.83	48.33	9.0288	0	2	0
10.54	67.99	8.4201	1	1	0
17.41	5.70	5.1031	0	0	1
17.46	13.38	5.0873	1	3	0
18.21	35.15	4.8780	$\bar{1}$	1	1
18.67	6.45	4.7591	2	0	0
19.69	27.35	4.5144	0	4	0
20.01	1.37	4.4426	0	2	1
21.13	9.26	4.2101	2	2	0
22.08	2.86	4.0293	$\bar{2}$	0	1
22.33	3.70	3.9846	1	1	1
22.97	28.29	3.8761	$\bar{1}$	3	1
26.39	56.05	3.3803	1	3	1
27.25	39.81	3.2753	2	4	0
28.58	64.21	3.1249	3	1	0
28.74	2.49	3.1082	2	0	1
29.56	2.00	3.0238	$\bar{3}$	1	1
30.43	38.12	2.9389	2	2	1
30.41	6.67	2.9408	$\bar{1}$	5	1
31.90	8.96	2.8067	3	3	0
32.79	22.87	2.7328	$\bar{3}$	3	1
33.12	100.00	2.7058	1	5	1
34.61	29.22	2.5923	0	6	1
35.18	8.12	2.5516	0	0	2
35.45	56.17	2.5332	$\bar{2}$	0	2
36.61	1.79	2.4554	0	2	2
36.86	2.83	2.4390	$\bar{2}$	2	2
37.33	1.25	2.4095	3	1	1
37.30	3.37	2.4112	$\bar{2}$	6	1
37.43	1.27	2.4036	$\bar{4}$	0	1
37.75	6.77	2.3836	3	5	0
38.52	31.64	2.3378	$\bar{3}$	5	1
38.78	14.70	2.3227	$\bar{4}$	2	1
39.20	16.72	2.2988	$\bar{1}$	7	1
39.62	18.08	2.2755	$\bar{3}$	1	2
40.86	6.39	2.2092	$\bar{2}$	4	2
41.40	5.07	2.1812	1	7	1
41.71	1.56	2.1658	1	3	2
41.78	26.06	2.1621	2	6	1
42.17	3.06	2.1434	$\bar{3}$	3	2
42.44	2.63	2.1303	$\bar{1}$	5	2
44.35	14.78	2.0427	2	0	2
44.95	12.41	2.0168	3	5	1

(continued)

(continued)

45.00	7.50	2.0147	$\bar{4}$	0	2
45.31	6.28	2.0015	3	7	0
45.97	1.41	1.9742	$\bar{3}$	7	1
46.24	5.14	1.9632	1	9	0
46.17	1.37	1.9663	$\bar{4}$	2	2
46.93	3.70	1.9362	$\bar{3}$	5	2
47.15	1.63	1.9274	4	2	1
48.06	8.06	1.8932	5	1	0
48.47	1.77	1.8781	$\bar{4}$	6	1
48.82	11.13	1.8654	$\bar{1}$	9	1
48.94	2.59	1.8610	2	4	2
49.41	3.58	1.8444	$\bar{1}$	7	2
49.55	3.30	1.8398	$\bar{4}$	4	2
50.27	4.39	1.8150	5	3	0
50.54	1.85	1.8058	0	10	0
52.38	4.88	1.7465	$\bar{5}$	1	2
53.30	1.29	1.7185	$\bar{5}$	5	1
53.45	1.44	1.7141	$\bar{3}$	7	2
53.85	1.78	1.7023	0	10	1
54.25	1.43	1.6906	0	8	2
54.44	6.83	1.6852	$\bar{2}$	8	2
54.50	6.94	1.6836	$\bar{1}$	3	3
54.92	5.69	1.6716	0	2	3
55.70	23.44	1.6500	4	6	1
56.16	6.71	1.6376	4	8	0
56.91	11.33	1.6177	1	11	0
58.14	3.44	1.5864	6	0	0
58.50	20.20	1.5775	$\bar{1}$	5	3
59.16	1.03	1.5614	2	10	1
59.32	1.19	1.5575	$\bar{5}$	7	1
59.47	4.17	1.5541	4	0	2
60.35	9.28	1.5335	$\bar{6}$	0	2
61.03	3.85	1.5179	1	9	2
61.29	17.77	1.5122	$\bar{2}$	6	3
61.62	9.23	1.5048	0	12	0
61.76	4.45	1.5017	5	5	1
61.70	2.29	1.5030	$\bar{4}$	8	2
63.27	3.67	1.4695	4	4	2
63.22	1.37	1.4704	$\bar{2}$	10	2
63.41	1.44	1.4666	2	2	3
63.50	1.20	1.4648	3	7	2
63.83	3.50	1.4580	3	11	0
64.12	3.01	1.4520	$\bar{6}$	4	2
64.21	2.87	1.4502	$\bar{1}$	7	3
64.35	1.03	1.4473	$\bar{3}$	11	1

(continued)

(continued)

64.80	25.58	1.4384	$\bar{6}$	6	1
65.45	1.50	1.4257	$\bar{5}$	3	3
66.86	1.26	1.3990	6	2	1
68.80	10.32	1.3641	5	1	2
69.07	4.09	1.3595	$\bar{5}$	5	3
69.28	2.43	1.3559	7	1	0
70.20	6.21	1.3403	1	11	2
70.24	1.03	1.3396	$\bar{1}$	13	1
70.58	2.38	1.3340	5	3	2
70.52	3.43	1.3349	$\bar{3}$	11	2
70.66	4.50	1.3327	2	6	3
71.39	1.98	1.3209	$\bar{5}$	9	2
72.06	7.05	1.3102	$\bar{7}$	5	1
72.35	5.36	1.3057	$\bar{1}$	1	4
72.85	1.15	1.2979	6	8	0
72.96	6.00	1.2962	0	12	2
73.12	12.51	1.2937	$\bar{2}$	12	2
73.24	1.38	1.2920	$\bar{3}$	9	3
73.95	2.61	1.2814	6	6	1
74.09	1.46	1.2792	$\bar{1}$	3	4
74.32	3.03	1.2758	0	0	4
74.41	1.13	1.2745	$\bar{3}$	3	4
74.95	4.44	1.2666	$\bar{4}$	0	4
77.54	1.16	1.2307	4	2	3
77.71	1.43	1.2284	$\bar{2}$	14	1
80.01	7.43	1.1988	$\bar{5}$	11	2
80.74	1.78	1.1898	8	0	0
83.21	3.70	1.1606	2	0	4
83.14	1.32	1.1614	$\bar{8}$	4	2
83.72	1.07	1.1547	1	15	1
84.30	1.43	1.1483	4	6	3
84.36	1.19	1.1476	2	10	3
86.41	2.49	1.1256	7	9	0
87.77	1.51	1.1116	$\bar{6}$	4	4
88.27	1.08	1.1066	$\bar{7}$	7	3
90.03	1.39	1.0894	3	1	4
90.03	1.39	1.0894	3	1	4

(continued)

(continued)

X-ray diffraction pattern
$Ca_{2.01}(Fe_{0.01}Mg_{0.99})_5(Si_{0.99}Fe_{0.01})_8O_{22}(OH)_2$

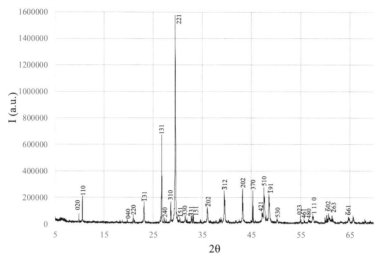

Zircon ($ZrSiO_4$), tetragonal ($I4_1/amd$).

Structure
Hazen, R.M., Finger, L.W. (1979). Crystal structure and compressibility of zircon at high pressure. American Mineralogist, 64, 196–201
Cell parameters: 6.6077 Å, 6.6077 Å, 5.9957 Å, 90°, 90°, 90°.

ATOM	X	Y	Z	OCCUPANCY	ISO(B)
Zr	0.00000	0.75000	0.12500	1.000	0.276
Si	0.00000	0.25000	0.37500	1.000	0.311
O	0.00000	0.06600	0.19510	1.000	0.503

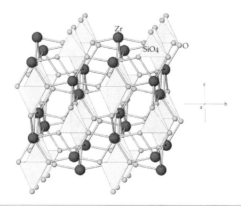

X-ray diffraction
Diffraction data computed using the structure data from the above paper listed, along with the cell parameters refined from the powder pattern of R050034 (RRUFF database).
X-ray wavelength: 1.541838 Å.

(continued)

(continued)

2θ	INTENSITY	D-SPACING	H	K	L
20.00	42.61	4.4402	1	0	1
26.99	100.00	3.3038	2	0	0
33.82	7.57	2.6506	2	1	1
35.58	58.08	2.5232	1	1	2
38.54	12.84	2.3362	2	2	0
40.64	6.07	2.2201	2	0	2
43.79	20.35	2.0675	3	0	1
47.53	13.63	1.9130	1	0	3
52.19	14.22	1.7526	3	2	1
53.45	54.07	1.7142	3	1	2
55.51	3.55	1.6555	2	1	3
55.64	14.85	1.6519	4	0	0
59.73	2.73	1.5482	4	1	1
61.90	3.08	1.4989	0	0	4
62.78	2.91	1.4801	3	0	3
62.90	9.75	1.4775	4	2	0
67.81	16.91	1.3821	3	3	2
68.77	10.22	1.3650	2	0	4
73.36	5.10	1.2906	4	3	1
75.33	8.79	1.2616	2	2	4
76.14	2.51	1.2503	4	1	3
80.80	9.59	1.1895	5	1	2
82.60	2.93	1.1681	4	4	0
87.86	2.56	1.1111	2	1	5
87.97	5.10	1.1101	4	0	4
88.75	1.96	1.1023	5	0	3
88.86	4.57	1.1013	6	0	0

X-ray diffraction pattern

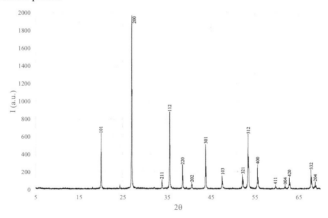

Appendix VII

N values for the crystal systems

N	hkl Cubic $N = h^2 + k^2 + l^2$	hk0 Tetragonal $N = h^2 + k^2$	Hexagonal $N = h^2 + hk + k^2$	N	hkl Cubic $h^2 + k^2 + l^2$	hk0 Tetragonal $N = h^2 + k^2$	Hexagonal $N = h^2 + hk + k^2$
1	100	100	100	81	900; 841; 744; 663	900	900
2	110	110		82	910; 833	910	
3	111		110	83	911; 753		
4	200	200	200	84	842		820
5	210	210		85	920; 760	920; 760	
6	211			86	921; 761; 655		
7			210	87			
8	220	220		88	664		
9	300; 221	300	300	89	922; 850; 843; 762	850	
10	310	310		90	930; 851; 754	930	
11	311			91	931		650; 930
12	222		220	92			
13	320	320	310	93	852		740
14	321			94	932; 763		
15				95			
16	400	400	400	96	844		
17	410; 322	410		97	940; 665	940	830
18	411; 330	330		98	941; 853; 770	770	
19	331			99	933; 771; 755		
20	420	420		100	10,00; 860	10,00; 860	10,00
21	421		210	101	10,10; 942; 861; 764	10,10	
22	332			102	10,11; 772		
23				103			920
24	422			104	10,20; 862	10,20	

(continued)

C. Marcos, *Crystallography*, Springer Textbooks in Earth Sciences, Geography and Environment, https://doi.org/10.1007/978-3-030-96783-3

(continued)

N	hkl Cubic $N = h^2 + k^2 + l^2$	hk0 Tetragonal $N = h^2 + k^2$	Hexagonal $N = h^2 + hk + k^2$		hkl Cubic $h^2 + k^2 + l^2$	hk0 Tetragonal $N = h^2 + k^2$	Hexagonal $N = h^2 + hk + k^2$
25	500; 430	500	500	105	10,21; 854		
26	510; 431	510		106	950; 943	9,50	
27	511; 333		330	107	951; 773		
28			420	108	10,22; 666		660
29	520; 432	520		109	10,30; 863	10,30	750
30	521			110	10,31; 952; 765		
31			510	111			10,10
32	440	440		112			840
33	522; 441			113	10,32; 944; 870	870	
34	530; 433	530		114	871; 855; 774		
35	531			115	953		
36	600; 442	600	600	116	10,40	10,40	
37	610	610	430	117	10,41; 960;872	960	930
38	611; 532			118	10,33; 961		
39			520	119			
40	620	620		120	10,42		
41	621; 540; 443	540		121	11,00; 962; 766	11,00	11,00
42	541			122	11,10; 954; 873	11,10	10,20
43	533			123	11,11; 775		
44	622		610	124			
45	630; 542	630		125	11,20; 10,50; 10,43; 865	11,20; 10,50	
46	631			126	11,21; 10,51; 963		
47				127			760
48	444		440	128	880	880	
49	700; 632	700	700; 530	129	11,22; 10,52; 881; 874		850
50	710; 550; 543	710;550		130	11,30; 970	11,30; 970	
51	711;551			131	11,31; 971; 955		
52	640	640	620	132	10,44; 882		
53	720; 641	720		133	964		11,10; 940
54	721; 633; 552			134	11,32; 10,53; 972; 776		
55				135			
56	642			136	10,60; 866	10,60	
57	722; 544		710	137	11,40; 10,61; 883	11,40	
58	730	730		138	11,41; 875		
59	731; 553			139	11,33; 973		10,30
60				140	10,62		
61	650; 643	650	540	141	11,42; 10,54		
62	732; 651			142	965		
63			630	143			

(continued)

(continued)

N	hkl	hk0			hkl	hk0	
	Cubic $N = h^2 + k^2 + l^2$	Tetragonal $N = h^2 + k^2$	Hexagonal $N = h^2 + hk + k^2$		Cubic $h^2 + k^2 + l^2$	Tetragonal $N = h^2 + k^2$	Hexagonal $N = h^2 + hk + k^2$
64	800	800	800	144	12,00; 884	12,00	12,00
65	810; 740; 652	810; 740		145	12,10; 10,63	12,10; 980	
66	811; 741; 554			146	12,11; 11,50; 11,43; 981; 974	11,50	
67	733		720	147	11,51; 777		11,20; 770
68	820; 644	820		148	12,20	12,20	860
69	821; 742			149	12,21; 10,70; 982; 876	10,70	
70	653			150	11,52; 10,71; 10,55		
71				151			950
72	822; 660	660		152	1222; 1144; 1072; 966; 885	12,30	
73	830; 661	830	810	153	1230; 1144; 1072; 966; 885	12, 30	
74	831; 750; 743	750		154	1231; 983		
75	751; 55 5		550	155	1153; 975		
76	662		64	156			10,40
77	832; 654			157	12,32; 11,60	11,60	12,10
78	752			158	11,61; 10,73	11,60	
79			730	159			
80	840	840					

Printed in the United States
by Baker & Taylor Publisher Services